Dynamics of Lattice Materials

Dynamics of Lattice Materials

Edited by
A. Srikantha Phani
University of British Columbia, Canada

Mahmoud I. Hussein
University of Colorado Boulder, USA

Registered Office(s)
John Wiley & Sons, Inc., 111 River Street, Hoboken, NJ 07030, USA
John Wiley & Sons Ltd, The Atrium, Southern Gate, Chichester, West Sussex, PO19 8SQ, UK

Editorial Office
The Atrium, Southern Gate, Chichester, West Sussex, PO19 8SQ, UK

For details of our global editorial offices, customer services, and more information about Wiley products visit us at www.wiley.com.

Wiley also publishes its books in a variety of electronic formats and by print-on-demand. Some content that appears in standard print versions of this book may not be available in other formats.

Library of Congress Cataloging-in-Publication Data

Names: Phani, A. Srikantha, editor. | Hussein, Mahmoud I., editor.
Title: Dynamics of lattice materials / [edited by] A. Srikantha Phani,
    Mahmoud I. Hussein.
Description: Chichester, West Sussex, United Kingdom : John Wiley & Sons,
    Inc., 2017. | Includes bibliographical references and index.
Identifiers: LCCN 2016042860 | ISBN 9781118729595 (cloth) | ISBN 9781118729571
    (epub) | ISBN 9781118729564 (Adobe PDF)
Subjects: LCSH: Lattice dynamics.
Classification: LCC QC176.8.L3 D85 2017 | DDC 530.4/11–dc23 LC record
available at https://lccn.loc.gov/2016042860

Cover design by Wiley
Cover image: Courtesy of the author

Set in 10/12pt Warnock by SPi Global, Chennai, India

10 9 8 7 6 5 4 3 2 1

To Ananya and Krishna
To Alaa and Ismail

Contents

List of Contributors

M. Arya
University of Toronto
Ontario
Canada

S. Arabnejad
McGill University
Montreal
Quebec
Canada

K. Bertoldi
Harvard University
Cambridge
Massachusetts
USA

O.R. Bilal
University of Colorado Boulder
Colorado
USA

W.J. Cantwell
Khalifa University of Science
Technology and Research
Abu Dhabi
UAE

F. Casadei
Harvard University
Cambridge
Massachusetts
USA

Z.W. Guan
University of Liverpool
UK

G. Hibbard
University of Toronto
Ontario
Canada

G.M. Hulbert
University of Michigan
Ann Arbor
USA

M.I. Hussein
University of Colorado Boulder
Colorado
USA

D. Krattiger
University of Colorado Boulder
Colorado
USA

A.T. Lausic
University of Toronto
Ontario
Canada

M.J. Leamy
Georgia Institute of Technology
Atlanta
USA

K. Manktelow
Georgia Institute of Technology
Atlanta
USA

A.N. Norris
Rutgers University
Piscataway
New Jersey
USA

D. Pasini
McGill University
Montreal
Quebec
Canada

M. Ruzzene
Georgia Institute of Technology
Atlanta
USA

M. Smith
University of Sheffield
Rotherham
UK

A.S. Phani
University of British Columbia
Vancouver
Canada

C.A. Steeves
University of Toronto
Ontario
Canada

C. Yilmaz
Bogazici University
Istanbul
Turkey

P. Wang
Harvard University
Cambridge
Massachusetts
USA

Foreword

When Srikantha Phani and Mahmoud Hussein asked me if I would write a foreword to their book, I was at first a bit hesitant as I was very busy working on a new book (*Extending the Theory of Composites to Other Areas of Science*, now submitted for publication with chapters coauthored by Maxence Cassier, Ornella Mattei, Mordehai Milgrom, Aaron Welters and myself). But then, when I saw the high quality of the chapters submitted by various people, I was happy to agree.

In 1928 the doctoral thesis of Felix Bloch established the quantum theory of solids, using Bloch waves to describe the electrons. Following this, in 1931–1932, Alan Herries Wilson explained how energy bands of electrons can make a material a conductor, a semiconductor or an insulator. Subsequently there was a tremendous effort directed towards calculating the electronic properties of crystals by calculating their band structure; that is, through solving Schrödinger's equation in a periodic system. So it is rather surprising that it took until the late 1980s for similar calculations to be done for wave equations in man-made periodic structures (with the exception of the layered materials that Lord Rayleigh in 1887 had shown exhibited a band gap). Subsequently there was exponentially growing interest in the subject, as illustrated by the graph in the extensive "Photonic and sonic band gap and metamaterial bibliography" of Jonathan Dowling [1], which he maintained until 2008. Now there seems to be a similar migration of ideas from people who have studied topological insulators in the context of the quantum 2D Hall effect to the study of similar effects in man-made periodic structures where there is some time-symmetry breaking. In the context of elasticity this time-symmetry breaking can be achieved with gyroscopic metamaterials [2] and, most significantly, waves can only travel in one direction around the boundary.

This book sheds light on the dynamics of lattice materials from different perspectives. As I read through it, connections with other work (sometimes mine) came to mind. I suspect this is probably a reflection of my background, as the writer (or writers) may have been exposed to different schools of thought than myself, but I believe the cross-pollination of ideas is always beneficial to the advancement of science. Therefore I hope the collection of remarks I have made here will lead the reader (if they have the time to explore the references I have given) on some excursions of the mind that complement those provided by the authors of the individual chapters.

Chapter 1 provides the setting for the book, giving a brief but excellent introduction to lattice materials. Maxwell's rule for determining the stiffness of a structure is discussed and, as the authors mention, Maxwell realized this is only a necessary condition for a structure to be stiff. The exact condition is nontrivial to determine, but in a 2D structure one can play the "pebble game" to resolve the question [3]. Maxwell's counting rule has been generalized for periodic lattices [4]. Perhaps there is a generalization (maybe an obvious one) of the "pebble game" to periodic 2D lattices, but I have not fully explored the literature. The possible motions of kinematically indeterminant periodic arrays of rigid rods with flexible joints are of considerable interest to me, and the case in which the macroscopic motions are affine is described in the literature [5, 6], and references therein.

Pasini and Arabnejad's Chapter 2 provides an excellent survey of homogenization methods for the elastostatics of lattice materials. This is still very much an active area of research. It is an important one, because not only do the homogenized equations govern the macroscopic response but also, as emphasized by Pasini and Arabnejad, the solution of the so-called cell-problem (that is needed to calculate the effective moduli) can provide useful estimates of the maximum fields in the material, which are helpful in knowing if plastic yielding or cracking might occur. While Pasini and Arabnejad's review concentrates on periodic lattice materials, it is worth mentioning that, curiously, for random composites the justification of successive terms in the asymptotic expansions, such as their Eq. (2.12), requires successively higher dimensions of space [7, and references therein]; while 2D and 3D composites are those of practical interest, one may of course think of composites in higher dimensions too. Also, it is important to remember that with high-contrast linear elastic materials one can theoretically achieve almost any homogenized response compatible with the natural constraint of positivity of the elastic energy [8]: non-local interactions in the homogenized equations can be achieved with dumbbell shaped inclusions where the diameter of the bar is so small that it does not couple with the surrounding medium except in the near vicinity of the bar. These results are only in the framework of linear elasticity, because such bars can easily buckle when the dumbbell is under compression. Some beautiful examples of exotic elastic behavior, which go beyond that of Cosserat theory, are given by Seppecher, Alibert, and Dell Isola [9].

Chapter 3, by Phani, gives a great introduction to the elastodynamics of lattice materials. I especially like their use of simple mass-spring models. My coauthors and I find mass-spring models, with the addition of rigid elements, to be very helpful in explaining concepts such as negative effective mass, anisotropic mass density, and (when the springs have some viscous damping) complex effective mass density [10, 11]. In fact it is possible (with the framework of linear elasticity) to give a complete characterization of the possible dynamic responses of multiterminal mass-spring networks [12]. The presentation by Phani of the deformation modes associated with the branches in the dispersion diagram in Figures 3.13–3.17 is beautiful, and sheds a lot more light on the behavior than is contained in dispersion curves, which frequently is all most scientists present. Also, I would mention that a dramatic illustration of the directionality of wave propagation is in phonon focussing [13]. If at low temperatures one heats a crystal from below by directing a laser at a point on the surface, then the distribution of heat on the top surface (as seen by the height of liquid helium on the surface that, due to the fountain effect, flows towards the heat) has amazing patterns, due to caustics in the "slowness" surface associated with the direction of elastic wave propagation in crystals that is governed simply by the elasticity tensor of the crystal. The elastic waves carry the heat (phonons). It is worth remarking that, subsequent to pioneering work by Bensoussan, Lions, and Papanicolaou in Chapter 4 of their book [14], there has been a resurgence of interest in high-frequency homogenization at stationary points in the dispersion diagram, which may be local minima or maxima, or even saddle points [15–21]. The wave is a modulated Bloch wave and modulation satisfies appropriate effective equations. The most interesting effects occur when one has a saddle point: then the effective equation is hyperbolic and there are associated characteristic directions. One may also employ homogenization techniques for travelling waves at other points in the dispersion diagram [22–26].

Chapter 4, by Krattiger, Phani, and Hussein examines wave propagation in damped lattice materials, both for passive waves and driven waves. One rarely sees dispersion diagrams with damping, but of course for many materials damping is a significant factor. Their dispersion diagrams with driven waves (Figures 4.2 and 4.4) have an interesting and complex structure. It is interesting that some periodic materials with damping can have trivial dispersion relations, with a dispersion diagram equivalent to that of a homogeneous damped material [27, 28]: this happens when the moduli are analytic functions, not of the frequency, but of the complex variable $x_1 + ix_2$, where $i = \sqrt{-1}$ and for a 2D material x_1 and x_2 are the Cartesian spatial coordinates. Closely related materials were discovered by Horsley, Artoni, and La Rocca, who realized they would not reflect radiation incident from one side, whatever the angle of incidence [29].

In Chapter 5 by Manktelow, Ruzzene, and Leamy we encounter the exciting topic of wave propagation in nonlinear lattice materials. The study of nonlinear effect in composites is largely a wide-open area of research: there are so many interesting and novel directions that could be explored, and it is a certainty that surprises await. One surprise we found is as follows [30]. When one mixes linear conducting composites in fixed proportions, if one wants to maximize the current in the direction of the electric field then it is best to layer the materials with the layer boundaries parallel to the applied field; by contrast, in some nonlinear materials we found that the maximum current sometimes occurs when the layer interfaces are normal to the applied field. Manktelow, Ruzzene, and Leamy talk about higher harmonic generation in nonlinear materials. Anyone who has used an inexpensive green laser may be interested to know that the green light comes from frequency doubling the infrared light from a neodymium-ion oscillator as it passes through a nonlinear crystal, and this can pose a danger if the conversion is faulty because the infrared light can easily damage eyes [31].

Chapter 6 by Casadei, Wang and Bertoldi also deals with nonlinearity, but in the context of buckling creating a pattern transformation that can be used to tune the propagation of elastic waves. This is fantastic work, and in an entirely new direction. Buckling instabilities are well known in Bertoldi's group: they created the Buckliball a structured sphere that remains approximately spherical, but much reduced in size, as it buckles [32]. Much remains to be explored in this area: one especially significant result that I have found is that materials that combine a stable phase with an unstable one could have a stiffness greater than diamond in dynamic bending experiments [33]. It had been hoped that one could get stiffnesses dramatically higher than that of the components in stable static materials too [34], but this was ruled out when it was realized that the well-known elastic variational principles still hold even when some of the components are in isolation unstable (that is, they have negative elastic moduli) [35] .

I found interesting the work in Chapter 7 of Smith, Cantwell, and Guan on the impact and blast response of lattice materials. A feature of their experiments is that the stress has a plateau as the lattice structure is crumpled. This is exactly what one needs if the aim is to minimize the maximum force felt by an object colliding with the structure, subject to the constraint that the object should decelerate over a fixed distance. We recently encountered similar questions when trying to determine the optimal non-linear rope for a falling climber [36]. The answer turned out to be a "rope" with a stress plateau, like a shape memory wire (and with a big hysterisis loop to absorb the energy). It is pretty amazing to see the progress that has been made recently with impact-resistant composites: a good example is the composite metal foam of Afsaneh Rabiei, which literally obliterates bullets [37].

Pentamode materials, as discussed by Norris in Chapter 8, are a class of materials close to my heart. When we invented them, back in 1995 [38], we never dreamed they would actually be made, but that is exactly what the group of Martin Wegener did, in an amazing feat of 3D lithography [39]. Their lattice structure is similar to diamond, with a stiff double-cone structure replacing each carbon bond. This structure ensures that the tips of four double-cone structures meet at each vertex. This is the essential feature: treating the double-cone structures as struts, the tension in one determines uniquely the tension in the other three. This is simply balance of forces. Thus the structure as a whole can essentially only support one stress, but that stress can be any desired symmetric matrix if the pentamode lattice structure is appropriately tailored. Water is a bit like a pentamode, but unlike water, which can only support a hydrostatic stress, pentamodes can support any desired stress matrix, in other words, a desired mixture of shear and compression. They are the building blocks for constructing any desired elasticity matrix C_* that is positive definite. Elasticity tensors of 3D materials are actually fourth-order tensors, specifically linear maps on the space of symmetric matrices, but using a basis on the 6D space of symmetric matrices, they can be represented by a 6-by-6 matrix as is common in engineering notation. Expressing C_* in terms of its eigenvectors and eigenvalues,

$$C_* = \sum_{i=1}^{6} \lambda_i v_i \otimes v_i. \tag{1}$$

The idea, roughly speaking, is to find six pentamode structures, each supporting a stress represented by the vector v_i, $i = 1, 2, \ldots, 6$. The stiffness of the material and the necks of the junction regions at the vertices need to be adjusted so each pentamode structure has an effective elasticity tensor close to

$$C_*^{(i)} = \lambda_i v_i \otimes v_i. \tag{2}$$

Then one successively superimposes all these six pentamode structures, with their lattice structures being offset to avoid collisions. Additionally, one may need to deform the structures appropriately to avoid these collisions [38], and when one does this it is necessary to readjust the stiffness of the material in the structure to maintain the value of λ_i. Then the remaining void in the structure is replaced by an extremely compliant material. Its presence is just needed for technical reasons, to ensure that the assumptions of homogenization theory are valid so that the elastic properties can be described by an effective tensor. But it is so compliant that essentially the effective elasticity tensor is just a sum of the effective elasticity tensors of the six pentamodes; in other words, the elastic interaction between the six pentamodes is neglible. In this way we arrive at a material with (approximately) the desired elasticity tensor C_*. Now, Andrew Norris and the group of Martin Wegener have become the leading experts on pentamodes and their 2D equivalents, which strictly speaking should be called bimodes. One important observation that Norris makes (see his Eq. (8.5)) is that if a pentamode is macroscopically inhomogeneous then the stress field it supports should be divergence-free in the absence of body forces such as gravitational forces. The new and important ingredient in the chapter of Norris is the analytic inclusion of bending effects, to better analyse the elements of the effective elasticity tensor.

Chapter 9, by Krattinger and Hussein, uses a reduced number of modes in a Bloch mode expansion to treat the vibration of plates within a frequency range of interest. Expanding on the ideas of structural mechanics, where one splits a structure into sub-structures, conducts a modal analysis on each of these, and then links the modes through interface boundary conditions, they develop a similar procedure at the unit-cell level for very efficiently calculating the band structure, which they call "Bloch mode synthesis." I very much like the word "platonic crystal" [40] – crafted after the terms photonic crystals, phononic crystals, and plasmonic crystals – which Ross McPhedran coined for such studies of the propagation of flexural waves through plates with periodic structure. The term has caught on in Australia, France, New Zealand and the UK (where Ross is a frequent visitor) but not yet in the U.S.

Chapter 10 by Bilal and Hussein deals with topology optimization of lattice materials. Their pixel-based designs remind me very much of the digital metamaterials of my colleague Rajesh Menon (also produced by topology optimization, but in the context of electromagnetism rather than elasticity), which have been incredibly successful, for example resulting in the world's smallest polarization beam-splitter [41]. The field of topology optimization has seen some amazing achievements, producing stuctures with fascinating and sometimes unexpected geometries that optimize performance in some respect. In particular, the group of Ole Sigmund in Denmark is well known for mastering this art, and recently they have used it for acoustic design [42]; the next wave of symphony halls will probably use the technique in their designs.

Chapter 11 presents work by Yilmaz and Hulbert on the dynamics of locally resonant and inertially amplified lattice materials. Nano-sized silver and gold metal spheres, that are resonant to light account for the beautiful colors of the Roman Lycurgus cup [43], and many stained glass windows gain their colors from such local resonances [44]. Resonant arrays of metallic split rings may lead to artifical magnetism [45], with the effective magnetic permeability taking negative values in appropriate frequency ranges [46]. Low-frequency spectral gaps were noticed by Zhikov [47, 48]. Negative effective mass densities, due to local resonances, were discovered in 2000 [49], although it was not until later that the experiments were correctly interpreted [50]. In periodic arrays of split cylinders, negative magnetic permeability can be related to the negative effective mass density in antiplane vibrations, due to the fact that both are governed by the Helmholtz equation [51]. The generation of band gaps through inertial amplification is nicely explained through essentially 1D models by Yilmaz and Hulbert in Section 11.3.1: the key aspect is that small macroscopic movements cause large amplitude movements of the internal masses. They then explore both 2D and 3D lattices. One would suspect that nonlinear effects could be very important in these models, even for quite small amplitudes of vibrations, although I do not know whether this has been explored.

Chapter 12 by Steeves, Hibbard, Arya, and Lausic provides an absolutely superb introduction to 3D printing, with a step-by-step explanation of the processes involved, highlighting the advantages of metal-coated polymer structures. In Figure 12.2 the improvement of adding a metal coating does not look particularly dramatic, until you realize there is a different scale (on the right-hand side of the graph), so in fact the improvement is about an order of magnitude in the tensile stress of the structure can support. Estimates for the elastic properties are obtained and the problem of optimizing the band gap to be as wide as possible, and at the desired frequencies, is discussed. There has been a lot of numerical work on optimizing band gaps. What I find most interesting is that it is possible to derive upper bounds on the width of band gaps that are sharp when the contrast between phases is low [52].

That ends my foreword, and now I hope the reader will go on and thoroughly enjoy the book.

Graeme. W. Milton
Salt Lake City, Utah

References

1 J. Dowling, "Photonic and sonic band-gap and metamaterial bibliography," nd. URL: http://www.phys.lsu.edu/~jdowling/pbgbib.html.

2 L. M. Nash, D. Kleckner, A. Read, V. Vitelli, A. M. Turner, and W. T. M. Irvine, "Topological mechanics of gyroscopic metamaterials," *Proceedings of the National Academy of Sciences*, vol. 112, no. 47, pp. 14495–14500, 2015.

3 D. J. Jacobs and B. Hendrickson, "An algorithm for two-dimensional rigidity percolation: The pebble game," *Journal of Computational Physics*, vol. 137, no. 2, pp. 346–365, 1997.

4 S. D. Guest and P. W. Fowler, "Symmetry-extended counting rules for periodic frameworks," *Philosophical Transactions of the Royal Society of London A: Mathematical, Physical and Engineering Sciences*, vol. 372, no. 2008, p. 20120029, 2013.

5 G. W. Milton, "Complete characterization of the macroscopic deformations of periodic unimode metamaterials of rigid bars and pivots," *Journal of the Mechanics and Physics of Solids*, vol. 61, no. 7, pp. 1543–1560, 2013.

6 G. W. Milton, "Adaptable nonlinear bimode metamaterials using rigid bars, pivots, and actuators," *Journal of the Mechanics and Physics of Solids*, vol. 61, no. 7, pp. 1561–1568, 2013.

7 Y. Gu, "High order correctors and two-scale expansions in stochastic homogenization," *arxiv.org*, pp. 1–28, 2016.

8 M. Camar-Eddine and P. Seppecher, "Determination of the closure of the set of elasticity functionals," *Archive for Rational Mechanics and Analysis*, vol. 170, no. 3, pp. 211–245, 2003.

9 P. Seppecher, J.-J. Alibert, and F. D. Isola, "Linear elastic trusses leading to continua with exotic mechanical interactions," *Journal of Physics: Conference Series*, vol. 319, no. 1, p. 012018, 2011.

10 G. W. Milton, M. Briane, and J. R. Willis, "On cloaking for elasticity and physical equations with a transformation invariant form," *New Journal of Physics*, vol. 8, no. 10, p. 248, 2006.

11 G. W. Milton and J. R. Willis, "On modifications of Newton's second law and linear continuum elastodynamics," *Proceedings of the Royal Society A: Mathematical, Physical, & Engineering Sciences*, vol. 463, no. 2079, pp. 855–880, 2007.

12 F. Guevara Vasquez, G. W. Milton, and D. Onofrei, "Complete characterization and synthesis of the response function of elastodynamic networks," *Journal of Elasticity*, vol. 102, no. 1, pp. 31–54, 2011.

13 B. Taylor, H. J. Maris, and C. Elbaum, "Phono. focusing in solids," *Physical Review Letters*, vol. 23, no. 8, pp. 416–419, 1969.

14 A. Bensoussan, J.-L. Lions, and G. C. Papanicolaou, *Asymptotic Analysis for Periodic Structures*, vol. 5 of Studies in Mathematics and its Applications. Amsterdam: North-Holland Publishing Co., 1978.

15 M. S. Birman and T. A. Suslina, "Homogenization of a multidimensional periodic elliptic operator in a neighborhood of the edge of an internal gap," *Journal of Mathematical Sciences (New York, NY)*, vol. 136, no. 2, pp. 3682–3690, 2006.

16 R. V. Craster, J. Kaplunov, and A. V. Pichugin, "High frequency homogenization for periodic media," *Proceedings of the Royal Society A: Mathematical, Physical, & Engineering Sciences*, vol. 466, no. 2120, pp. 2341–2362, 2010.

17 M. A. Hoefer and M. I. Weinstein, "Defect modes and homogenization of periodic Schrödinger operators," *SIAM Journal on Mathematical Analysis*, vol. 43, no. 2, pp. 971–996, 2011.

18 T. Antonakakis and R. V. Craster, "High frequency asymptotics for microstructured thin elastic plates and platonics," *Proceedings of the Royal Society A: Mathematical, Physical, & Engineering Sciences*, vol. 468, no. 2141, pp. 1408–1427, 2012.

19 T. Antonakakis, R. V. Craster, and S. Guenneau, "Asymptotics for metamaterials and photonic crystals," *Proceedings of the Royal Society of London. Series A*, vol. 469, no. 2152, p. 20120533, 2013.

20 T. Antonakakis, R. V. Craster, and S. Guenneau, "Homogenization for elastic photonic crystals and metamaterials," *Journal of the Mechanics and Physics of Solids*, vol. 71, pp. 84–96, 2014.

21 L. Ceresoli, R. Abdeddaim, T. Antonakakis, B. Maling, M. Chmiaa, P. Sabouroux, G. Tayeb, S. Enoch, R. V. Craster, and S. Guenneau, "Dynamic effective anisotropy: Asymptotics, simulations and microwave experiments with dielectric fibres," *Physical Review B: Condensed Matter and Materials Physics*, vol. 92, no. 17, p. 174307, 2015.

22 G. Allaire, M. Palombaro, and J. Rauch, "Diffractive behaviour of the wave equation in periodic media: weak convergence analysis," *Annali di Matematica Pura ed Applicata. Series IV*, vol. 188, no. 4, pp. 561–590, 2009.

23 M. Brassart and M. Lenczner, "A two-scale model for the periodic homogenization of the wave equation," *Journal de Mathématiques Pures et Appliquées*, vol. 93, no. 5, pp. 474–517, 2010.

24 G. Allaire, M. Palombaro, and J. Rauch, "Diffractive geometric optics for Bloch waves," *Archive for Rational Mechanics and Analysis*, vol. 202, no. 2, pp. 373–426, 2011.

25 G. Allaire, M. Palombaro, and J. Rauch, "Diffraction of Bloch wave packets for Maxwell's equations," *Communications in Contemporary Mathematics*, vol. 15, no. 06, p. 1350040, 2013.

26 D. Harutyunyan, R. V. Craster, and G. W. Milton, "High frequency homogenization for travelling waves in periodic media," *Proceedings of the Royal Society A: Mathematical, Physical, & Engineering Sciences*, vol. 472, no. 2191, p. 20160066, 2016.

27 G. W. Milton, "Exact band structure for the scalar wave equation with periodic complex moduli," *Physica. B, Condensed Matter*, vol. 338, no. 1–4, pp. 186–189, 2003.

28 G. W. Milton, "The exact photonic band structure for a class of media with periodic complex moduli," *Methods and Applications of Analysis*, vol. 11, no. 3, pp. 413–422, 2004.

29 S. A. R. Horsley, M. Artoni, and G. C. La Rocca, "Spatial Kramers–Kronig relations and the reflection of waves," *Nature Photonics*, vol. 9, pp. 436–439, 2015.

30 G. W. Milton and S. K. Serkov, "Bounding the current in nonlinear conducting composites," *Journal of the Mechanics and Physics of Solids*, vol. 48, no. 6/7, pp. 1295–1324, 2000.

31 C. W. C. Jemellie Galang, Alessandro Restelli and E. Hagley, "The dangerous dark companion of bright green lasers," *SPIE Newsroom, 10 January*, 2011. URL: http://spie.org/newsroom/3328-the-dangerous-dark-companion-of-bright-green-lasers.

32 Bertoldi Group webpage. URL: http://bertoldi.seas.harvard.edu/pages/buckliball-buckling-induced-encapsulation.

33 T. Jaglinski, D. Kochmann, D. Stone, and R. S. Lakes, "Composite materials with viscoelastic stiffness greater than diamond," *Science*, vol. 315, no. 5812, pp. 620–622, 2007.

34 R. S. Lakes and W. J. Drugan, "Dramatically stiffer elastic composite materials due to a negative stiffness phase?," *Journal of the Mechanics and Physics of Solids*, vol. 50, no. 5, pp. 979–1009, 2002.

35 D. M. Kochmann and G. W. Milton, "Rigorous bounds on the effective moduli of composites and inhomogeneous bodies with negative-stiffness phases," *Journal of the Mechanics and Physics of Solids*, vol. 71, pp. 46–63, 2014.

36 D. Harutyunyan, G. W. Milton, T. J. Dick, and J. Boyer, "On ideal dynamic climbing ropes," *Journal of Sports Engineering and Technology*, 2016.

37 M. Shipman, "Metal foam obliterates bullets – and that's just the beginning," *NC State News, 5 April*, 2016. URL: https://news.ncsu.edu/2016/04/metal-foam-tough-2016/.

38 G. W. Milton and A. V. Cherkaev, "Which elasticity tensors are realizable?," *ASME Journal of Engineering Materials and Technology*, vol. 117, no. 4, pp. 483–493, 1995.

39 M. Kadic, T. Bückmann, N. Stenger, M. Thiel, and M. Wegener, "On the practicability of pentamode mechanical metamaterials," *Applied Physics Letters*, vol. 100, no. 19, p. 191901, 2012.

40 Wikipedia article, "Platonic crystal." URL: https://en.wikipedia.org/wiki/Platonic_crystal.

41 B. Shen, P. Wang, R. Polson, and R. Menon, "An integrated-nanophotonics polarization beamsplitter with $2.4 \times 2.4\,\mu m^2$ footprint," *Nature Photonics*, vol. 9, no. 2, pp. 378–382, 2015.

42 R. E. Christiansen, O. Sigmund, and E. Fernandez-Grande, "Experimental validation of a topology optimized acoustic cavity," *The Journal of the Acoustical Society of America*, vol. 138, no. 6, pp. 3470–3474, 2015.

43 Wikipedia article, "Lycurgus cup." URL: https://en.wikipedia.org/wiki/Lycurgus_Cup.

44 J. C. Maxwell Garnett, "Colours in metal glasses and in metallic films," *Philosophical Transactions of the Royal Society A: Mathematical, Physical, and Engineering Sciences*, vol. 203, no. 359–371, pp. 385–420, 1904.

45 S. A. Schelkunoff and H. T. Friis, *Antennas: The Theory and Practice*, pp. 584–585–. New York / London / Sydney, Australia: John Wiley and Sons, 1952.

46 J. B. Pendry, A. J. Holden, D. J. Robbins, and W. J. Stewart, "Magnetism from conductors and enhanced nonlinear phenomena," *IEEE transactions on microwave theory and techniques*, vol. 47, no. 11, pp. 2075–2084, 1999.

47 V. V. Zhikov, "On an extension and an application of the two-scale convergence method," *Matematicheskii Sbornik*, vol. 191, no. 7, pp. 31–72, 2000.

48 V. V. Zhikov, "On spectrum gaps of some divergent elliptic operators with periodic coefficients," *Algebra i Analiz*, vol. 16, no. 5, pp. 34–58, 2004.

49 Z. Liu, X. Zhang, Y. Mao, Y. Y. Zhu, Z. Yang, C. T. Chan, and P. Sheng, "Locally resonant sonic materials," *Science*, vol. 289, no. 5485, pp. 1734–1736, 2000.

50 Z. Liu, C. T. Chan, and P. Sheng, "Analytic model of phononic crystals with local resonances," *Physical Review B: Condensed Matter and Materials Physics*, vol. 71, no. 1, p. 014103, 2005.

51 A. B. Movchan and S. Guenneau, "Split-ring resonators and localized modes," *Physical Review B: Condensed Matter and Materials Physics*, vol. 70, no. 12, p. 125116, 2004.

52 M. C. Rechtsman and S. Torquato, "Method for obtaining upper bounds on photonic band gaps," *Physical Review B: Condensed Matter and Materials Physics*, vol. 80, no. 15, p. 155126, 2009.

Preface

A lattice material may be viewed as an enlarged and carefully tuned crystal, artificially constructed to function precisely as desired in engineering applications. It is formed from a spatially periodic network of interconnected rods, beams, plates or other slender structures. The ability to tailor the unit-cell architecture of a lattice material makes it possible to attain superior mechanical, elastodynamic and acoustic properties for numerous industrial applications – properties that may not be achievable using conventional materials. Naturally inspired by concepts from crystal physics, the methods and analysis techniques used in the study of lattice materials directly apply to periodic materials in general, including phononic crystals and elastic metamaterials that exhibit local resonances and/or other unique features.

In this book, we have sought to provide a comprehensive coverage of the emerging field of the dynamics of lattice materials. Co-written by a selection of leading researchers in the field, spanning three continents, the book gently introduces key concepts and fundamental theories in the discipline, while also boldly considering, often in considerable depth, the state of the art.

The topics covered include elastostatics (Chapter 2) and elastodynamics (Chapter 3), the effects of damping (Chapter 4), nonlinearity (Chapter 5), instabilities (Chapter 6) and impact loads (Chapter 7); exotic dynamics such as pentamodes (Chapter 8); model reduction (Chapter 9) and optimization (Chapter 10); metamaterial concepts including local resonance and inertial amplification (Chapter 11); and nano lattices (Chapter 12). Guided by an introductory chapter (Chapter 1), a systematic and unified synthesis of these topics pertaining to lattice materials is provided to help the reader consolidate concepts across the chapters.

The book is suitable for and accessible to graduate students and research scientists with backgrounds in dynamics, vibrations, and acoustics; mechanics and strength of materials; and condensed matter physics and materials science. It serves as a useful reference to researchers based in academia and practitioners in industrial research laboratories and design centers. It may also be used as a textbook for graduate courses on the mechanics of lattice materials, or a more focused course on wave propagation in periodic materials.

Many people have contributed to this book, directly or indirectly. First and foremost, colleagues and contributors to each chapter are acknowledged for their insightful presentations and diligent responses to requests from the editors. While credit for success goes to the contributing authors and their tireless efforts, the editors are responsible for any lingering typos or unintended omissions. We would like to extend our special thanks to Prof. Graeme Milton for his scholarly and insightful foreword. Credit also goes to the Wiley publishing team, especially to Paul Petralia for his initiative and sustained leadership, and to Nandhini Thandavamoorthy for her hard work and patience during the various stages of the evolution of this project. SP would like to acknowledge the funding for his research from the Natural Sciences and Engineering Research Council (NSERC) of Canada through its various programs, the assistance from his graduate students Behrooz Yousefzadeh, Lalitha Raghavan, Prateek Chopra and Ehsan Moosavimehr, and the support from his family members, particularly Ananya and Krishna. MIH, on his part, acknowledges funding for his research from several United States federal agencies, particularly the National Science Foundation, numerous seed grants from the University of Colorado Boulder, and the generous support provided through his H. Joseph Smead Faculty Fellowship. In addition to his student co-authors, Dimitri Krattiger and Osama Bilal, MIH is also grateful to current or former doctoral students Bruce Davis, Michael Frazier, Romik Khajehtourian, Clémence Bacquet, Hossein Honarvar, Alec Kucala, and Mary Bastawrous, and former postdoctoral fellow Lina Yang, for their assistance at the CU-Boulder Phononics Laboratory. Dimitri's efforts in creating the lattice image used on the front cover is much appreciated. Most of all, MIH is grateful to family members Alaa and Ismail (Jr.) in Boulder and Heba, Nahla, Iziz and Ismail (Sr.) in Cairo, Egypt.

The overarching goal of this book is to spur fundamental and applied research in the design, manufacturing, and utilization of lattice materials and structures across not only numerous existing applications, but also applications that are yet to be conceived.

February 2017

> A. Srikantha Phani
> *Vancouver, British Columbia*
>
> Mahmoud I. Hussein
> *Boulder, Colorado*

1

Introduction to Lattice Materials

A. Srikantha Phani[1] and Mahmoud I. Hussein[2]

[1] Department of Mechanical Engineering, University of British Columbia, Vancouver, Canada
[2] Department of Aerospace Engineering Sciences, University of Colorado Boulder, USA

1.1 Introduction

The word "lattice" implies a certain ordered pattern characterized by spatial periodicity, and hence symmetry. In crystalline solids, for example, atoms are arranged in a spatially periodic pattern or a lattice. Such a crystal lattice is specified by a unit cell and the associated basis vectors defining the directions of tessellation [1, 2]. Spatially repetitive patterns are not unique to atomic length scales. They appear over a wide range of length scales, spanning several disciplines and areas of application; see Figure 1.1 for a representative list. Carbon nanotubes [3] and single-layer graphene sheets [4] are periodic materials with nanoscale features. Microelectromechanical systems (MEMS) for radio frequency applications use microscale periodic architectures to form mechanical filters [5]. Biomedical implants such as cardiovascular stents are periodic cylindrical mesh structures [6, 7]. At macro and mega scales, periodic structural construction is widely used in composites in materials engineering [8, 9], turbomachinery in aerospace engineering [10, 11], and bridge and tower structures in civil engineering [12]. Aircraft surfaces typically use a skin-stinger configuration in the form of a uniform shell, reinforced at regular spatial intervals by identical stiffener/stingers. Similarly, rib-skin aircraft structural components, used in tails and fins, comprise two skins (plates) interconnected by ribs [13]. Interested readers are referred to the book by Gibson and Ashby [14] for further studies on lattice materials and the reviews by Mead and by Hussein et al. [15, 16] on the dynamics of periodic materials in general.

In a closely related research discipline, periodic materials are referred to as *phononic crystals* [17, 18], where strong analogies are drawn with their electromagnetic counterpart, photonic crystals. While there is a significant overlap between lattice materials and phononic crystals [19, 20], the former category is mostly associated with low-density construction and utilization in structural mechanics applications, whereas the latter is mostly connected to applications in applied physics, including filtering [21], waveguiding [22], sensing [23], imaging [24], and, more recently, vibrational energy harvesting [25], thermal transport management at the nanoscale [26], and control of wall-bounded flows [27]. Another class of artificial materials that possess unique

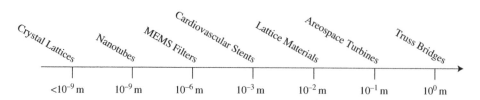

Figure 1.1 Periodic materials and structures across different length scales and disciplines. (MEMS: microelectromechanical systems.)

wave-propagation properties is referred to as *acoustic/elastic metamaterials* [28]. These are similar to phononic crystals, with the added feature of local resonators – small oscillating substructures integrally embedded within, or attached to the medium of the host material [29, 30]. However, unlike lattice materials and phononic crystals, periodicity is not a necessity for metamaterials. In addition to controlling sound and vibration, locally resonant "nanophononic metamaterials" have been shown to reduce thermal conductivity [31]. A recent book [32] and review article [16] provide historical background, the state of the art in the analysis and design of phononic crystals and metamaterials, together with their applications. In recent years, a new research community has formed around this discipline, now more broadly termed *phononics*, which incorporates the study and manipulation of "sound" waves in general and across the various spatial and temporal scales [33, 34].

The dynamic response of lattice materials, and structures, and by association phononic crystals and metamaterials, is the overarching theme of the book. We begin with a brief overview of periodic materials and structures, with emphasis on lattice materials, which are considered a new class of periodic materials. A formal classification is presented, followed by a discussion of manufacturing techniques and applications. A link to phononic crystals and acoustic/elastic metamaterials – also a new development in periodic materials – is presented when appropriate. We conclude this introductory chapter with an overview of the book.

1.2 Lattice Materials and Structures

A lattice material is defined as a spatially periodic network of structural elements, such as rods, beams, plates, or shells, whose constituent length scales are generally larger than the load-deformation length scales[1]; see Figure 1.2 for example. It possesses a spatially ordered pattern specified by a unit cell and associated tessellation directions (lattice basis vectors). The unit cell itself is an interconnected network of structural elements. Let us consider a network of flexural beams as an example. The material constituent of each beam can be a single homogeneous isotropic material (such as steel or aluminum) or a hierarchical anisotropic composite. Thus lattice materials, in the form of an interconnected spatially periodic network of composite beams, can be viewed as discrete multiscale materials with hierarchy. The ability to fabricate a spatially periodic network of beams using advanced manufacturing methods has spurred interest in lattice materials; see Fleck et al. [35] for a recent review. When viewed as a porous

---

1 This condition does not necessarily hold for lattice metamaterials where the size of unit-cell may be smaller than the deformation length scales.

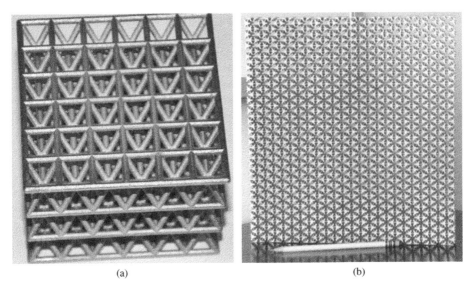

Figure 1.2 Lattice materials formed from a periodic network of beams: (a) ultralight nanometal truss hybrid lattice; (b) pentamode lattice.

solid, or a hybrid material (of fluid and metal) [36], the high-porosity limit yields a network of beams while the low-porosity limit leads to a continuum with pores. Most of the discussion and examples covered in this book are focused on material configurations at the high-porosity end of this range, although the ideas are usually relevant to low-porosity configurations as well.

1.2.1 Material versus Structure

A spatially periodic network of structural elements, such as beams, can be viewed both as a material and a structure for the following reasons. In engineering applications, employing a truss lattice of beams as a core in a sandwich panel, the length of each lattice beam is of the order of the thickness of the panel, and the thickness of each beam is typically an order of magnitude less. When the deformation processes of interest are at a length scale much larger than the individual beam length, a spatially periodic network of beams is termed a "lattice material" and has its own effective properties. At length scales of the order of the individual beam length, a spatially periodic network of beams behaves as a structure, such as a frame in a building or a truss in a bridge. Thus principles of structural mechanics can be applied to the design of lattice materials [37]. Another avenue for distinguishing between material and structure is in terms of the number unit cells, as well as the internal unit-cell symmetry. It is generally recognized that a for a finite system to exhibit material characteristics, at least a handful of unit cells are needed [38, 39]. In addition, a finite structure based on a repetition of a unit cell with symmetrical internal features is more likely to respond to dynamic loading in a manner consistent with the dispersion band structure of a material theoretically consisting of an infinite number of this unit cell [40, 41].

1.2.2 Motivation

The development of lattice materials is motivated by a desire to design multifunctional materials and structures that are not only light and stiff but also possess a desirable vibroacoustic response and thermal-transport properties, among other features. The need to overcome the limitations of metal foams [42, 43] has propelled the development of lattice materials, a process that has benefited from insights already acquired through studies of cellular solids [37, 44–48]. Similarly, accumulated research on the dynamics of periodic materials and structures (such as aircraft components and conventional composite materials) has provided a valuable knowledge base to build on for the study of wave-propagation characteristics in lattice materials. The following list provides an incomplete but indicative summary of efforts and motivations for current research in lattice materials.

1. Design lightweight and stiff/strong structures with optimal lattice core for multifunctional applications [49–52]. In this line of research, ongoing efforts aim to tailor the effective stiffness and strength of the truss lattice core to achieve high performance with the lowest possible density. The discovery of new unit-cell geometries using topology optimization and other computational methods is a promising avenue for further improvements [53, 54].
2. Advance mathematical modeling and analysis of complex lattice structures. This involves developing homogenization techniques for lattices [55, 56] and in-depth studies on the influence of damping [57–59] and nonlinearities [60–62] on the dispersive behavior of lattices.
3. Develop lattice unit-cell structures with tunable elastodynamic [63–65] and stability [66] properties.
4. Develop lattice-styled metamaterials based on periodic micro-architectures with extraordinary dynamic (acoustic and/or elastic) effective properties, not achievable using conventional materials [67, 68].
5. Create innovative nanostructured lattice materials based on periodic architectures for mechanical [50, 69, 70] and thermal [26, 31, 71] applications.

1.2.3 Classification of Lattices and Maxwell's Rule

Lattices can be classified based on their geometric or their mechanical deformation properties. Geometry-based classification is universally accepted in mathematics and solid-state physics. In 2D, planar lattices are classified into two categories: regular and semi-regular [72]. Regular lattices are obtained by tessellating a single, regular, polygonal unit cell to fill a plane. Here, a regular polygon is defined to be equiangular (all angles are equal) and equilateral (all lengths are equal). Square, triangle, and hexagon are the only plane-filling regular polygons, so there are only three regular planar lattices: square lattice, triangular lattice, and hexagonal lattice. In contrast to regular lattices, semi-regular lattices are obtained by tessellating a unit cell, containing more than one regular polygon, to fill a plane. There are only eight such semi-regular lattices; see Cundy and Rollett [72] for more detail. Kagome or triangular-hexagon lattice is a semi-regular lattice that is widely used in weaving baskets and in architectural construction. A detailed classification of 3D lattices and polyhedra can be found in the literature [72, 73].

Lattices can also be classified into bending- or stretching-dominated categories [37, 73] on the basis of their rigidity. A bending-dominated lattice responds to external loads by cell-wall bending, whereas a stretching-dominated lattice deforms predominantly by stretching. Bending-dominated lattices are less stiff and strong than stretching-dominated lattices, for the same porosity or relative density. Here, relative density, $\bar{\rho}$, is defined as the non-dimensional ratio of the density of the lattice material to the density of the solid. A low value of $\bar{\rho}$ indicates high porosity and $\bar{\rho} = 1$ indicates zero porosity. Thus it is important to identify whether a given lattice is bending- or stretching-dominated.

Maxwell's rule for simply stiff frames [74] provides a rigorous mathematical framework to decide if a given lattice is simply stiff or not. According to Maxwell's rule [74], a *finite* freely-supported pin-jointed lattice with b bars and j frictionless joints is simply stiff provided $b = 2j - 3$ in 2D and $b = 3j - 6$ in 3D. Here, a simply stiff lattice is defined to be both statically and kinematically determinate. Static determinacy implies that all bar tensions due to external forces can be computed from the available number of independent equilibrium equations. States of self-stress are present in statically indeterminate lattices. Similarly, kinematic determinacy implies that joint locations are uniquely determined by the individual lengths of the bars. Mechanisms are present in kinematically indeterminate lattices, such as in a pin-jointed square lattice. Thus it follows from Maxwell's rule that a lattice having j joints requires $3j - 6$ bars in 3D to render it simply stiff. If the bars are fewer, mechanisms exist rendering the lattice kinematically indeterminate. Likewise, if the number of bars exceeds $3j - 6$, states of self-stress exist rendering the lattice statically indeterminate.

Maxwell's rule is a necessary condition. Exceptions to these necessary conditions were recognized by Maxwell in his original work [74, 75] and are put on a firm footing by the generalized Maxwell's rule derived by Pellegrino and Calladine using matrix methods [76]. For example, certain tensegrity structures have fewer bars than are necessary to satisfy Maxwell's rule, and yet are not mechanisms. Their stiffness has been anticipated to be of lower order by Maxwell. Such exceptional cases permit at least one state of self-stress that offers first-order stiffness to one or more infinitesimal mechanisms [75]. This fact has been exploited by Buckminster Fuller in some of his tensegrity structures. Not surprisingly, exceptions to Maxwell's rule occur in biological fibrous structures as well [77, 78].

The generalized Maxwell's rule, derived using matrix algebra [75, 76], is $b - 2j + 3 = s - m$ for 2D lattices, and $b - 3j + 6 = s - m$ for 3D lattices, where s and m are the number of states of self-stress and mechanisms. We note that for a special class of lattices with similarly situated nodes (the lattice appears the same when viewed from any node), necessary and sufficient conditions for rigidity have been shown to be $Z = 6$ for planar lattices and $Z = 12$ for 3D lattices, where Z is the nodal connectivity, defined to be the number of bars emanating from a joint [73]. Finally, it has been shown that an *infinite* lattice cannot be simultaneously statically and kinematically determinate [79].

To conclude the discussion on Maxwell's rule, consider the planar Kagome or triangular-hexagonal lattice. It achieves the Hashin–Shtrikman upper bound for effective elastic properties. It has four bars at every joint and hence $Z = 4$. It still exhibits stretching-dominated deformation behaviour under macroscopic loading, and its effective in-plane moduli are linearly proportional to the relative density. This is due to the presence of periodic collapse mechanisms that produce no macroscopic strain [80]. However, in the presence of imperfections, a Kagome lattice exhibits large regions of bending-dominated boundary layers emanating from the edges and interfaces [35, 81]. Such boundary layers, with bending-dominated deformation, have been shown to enhance fracture toughness [82] and hence flaw-tolerance.

1.2.4 Manufacturing Methods

Advances in manufacturing techniques are central to the development of lattice materials, particularly those with complex features of the unit cell. A truss is a natural choice of construction that uses minimum material to fill maximum space, without sacrificing the stiffness and strength requirements. However, manufacturing trusses at mesoscales, with strut lengths of the order of millimetres and diameters of the order of micrometers, is a non-trivial challenge. Several advanced manufacturing techniques have emerged from the microelectronics industry [83] and traditional metal and composite manufacturing industries [84]. Methods for manufacturing metallic lattices include sheet forming, wire assembly, perforated sheet folding/drawing, investment casting, and wire assembly; a comprehensive overview of different techniques can be found in the literature [84–86]. While planar lattices are achievable using microfabrication techniques such as LIGA (LIthographie, Galvano, and Abformung or lithography, electrodeposition, and molding), achieving three-dimensionality is a challenge. 3D lattices can be made either through layer-by-layer addition or by using serial techniques, such as laser micromachining, that carve microstructures from solid objects or "write" 3D microstructures. A small, lightweight, space-filling truss structure is fabricated using soft lithography (rapid prototyping and micro contact printing) in combination with electrodeposition [87]. Three-dimensionality is achieved by folding a 2D silver grid at a specific angle using brass dies, and then assembling to create a 3D silver template. Electrodeposition of nickel on the silver template joins the three layers, two planar lattices, and the truss lattice core grids, and strengthens the overall 3D truss system.

More recently, ultralight ($<10\,\text{mg/cm}^3$) hierarchical metallic microlattices with hollow struts have been fabricated, starting with a template formed by self-propagating photopolymer waveguide prototyping, coating the template using electroless nickel plating, and subsequently etching away the template [50]. Three levels of hierarchy and associated length scales are present: unit cell (mm to cm), hollow tube lattice member (mm), and hollow tube wall (nm to mm). Exceptional control over the design and properties of the resulting microlattice is possible due to independent control of each architectural element. Ceramic scaffolds that mimic the length scales and hierarchy of biological materials have been demonstrated [88], leading to the design of biologically inspired hierarchical damage-tolerant engineering materials. Hollow-tube alumina nanolattices have been fabricated using two-photon lithography, atomic layer deposition, and oxygen plasma etching [89]. Nanocrystalline hybrid lattices are shown to have exceptional stiffness and strength properties [69]. Furthermore, advances in 3D printing have been utilized in fabricating hierarchical lattices with fractral microstructures [90] and lattice-based materials with continuously variable mechanical properties [91, 92]. These ongoing developments have prompted studies that consider integration of 3D printing into the design process [93]. This represents a promising research direction for realizing lattice materials and structures with both good manufacturability and performance attributes. In summary, rapid advances in micro and nano manufacturing technologies are enabling the fabrication and exploitation of truss lattice materials at multiple length scales and will continue to open new application areas for lattice materials.

1.2.5 Applications

Applications of lattice materials and structures span several areas of technological interest. This section provides a brief sketch of the rapidly expanding applications base of lattice materials and structures.

Lightweight structures are in high demand in the aerospace, automotive, marine, and other industries. The employment of lattice materials and structures in such multifunctional structural applications [35, 94, 95] is possible due to the open-core architecture that truss lattices provide when used as a core in sandwich structures. Effective thermoelastic stiffness and strength properties of lattices are topology-dependent, stretching-dominated lattices being the stiffest for a given relative density. For example, in Figure 1.3, the in-plane Young's modulus and shear modulus are compared for different planar lattices for a fixed relative density. It can be seen that the symmetry of the lattice topology governs whether a given lattice is isotropic or anisotropic. Further, it is possible to tailor effective properties such as the Poisson ratio with both extreme positive and negative values [46, 96–98]. Lattice morphologies with low coefficients of thermal expansion (CTE) have been studied for aerospace applications [99–101].

Lattices naturally act as filters for mechanical waves in both the frequency (temporal) and wavenumber (spatial) domains due to their spatial periodicity and the induced scattering at unit-cell interfaces [15, 18, 102–104]. Other notable engineering applications employing lattices include reconfigurable lattices [105, 106] and morphing/shape changing structures achieved with inclusion of actuators within a lattice [107–110]. Micro-truss lattice architectures have been used in nano-engineered composites to simultaneously obtain high stiffness and damping [70]. A judicious combination of

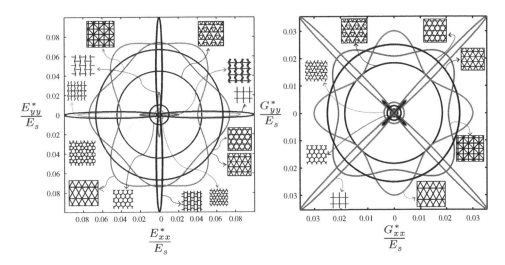

Figure 1.3 Unit-cell geometry-dependent effective in-plane elastic moduli of planar lattices for the same mass: Young's modulus tensor component E_{ii}^* (left) and shear modulus tensor component G_{ii}^* (right) along the Cartesian x (positive to the right) and y (positive upwards) axes. The Young's modulus of the isotropic parent solid is denoted E_s.

carbon nanotube engineered trusses held in a dissipative polymer has been used to design a composite material that simultaneously exhibits both high stiffness and damping, quite high in any single monolithic material. Nano-engineered lattice materials hold significant promise for future structural applications.

Truss lattice materials are also being used in energy-absorption applications, notably the design of impact- and blast-tolerant structures [111–117]. It has been shown that sandwich structures with lattice or foam cores perform better than monolithic plates of the same mass. Fluid–structure interactions in water blast have been shown to significantly enhance the relative performance, due to a reduction in the momentum acquired by the sandwich plate. This enhancement is more pronounced for structures subject to explosive loading in water than in air.

More recently, periodic-architectured biomedical implants have emerged [118]. Various geometrical arrays of cellular, reticulated mesh, and open-cell foams with interconnected porosities have been manufactured in complex, functional, monolithic structures. These complex arrays have the potential for unique bone compatibility as well as the accommodation of more natural bone tissue ingrowth, including vascular system development. Cardiovascular stents also utilize lattice structures. Shape-optimization methods have been developed to design a stress concentration-free lattice for self-expandable Nitinol stent grafts [119]. A systematic study of the influence of cell geometry on expansion characteristics has been reported [120, 121]. Regular and auxetic lattice geometries have been combined to obtain a stent with zero net foreshortening when subjected to large radial plastic deformation during its expansion inside a human artery. These selected examples illustrate the potential for lattice structures in the field of biomedical implant design, including tissue-engineering scaffolds.

Phononic crystals and acoustic/elastic metamaterials [15, 16, 32] add a whole range of applications, as listed in Section 1.1. The close connection between lattice materials/structures and phononic crystals/metamaterials stems mostly from the utilization of periodicity, which is a common theme in both groups and itself draws inspiration (and some analysis tools) from the vast literature on crystalline materials [1, 2].

1.3 Overview of Chapters

The dynamic response, both linear and nonlinear, of lattice materials is the central theme of this book, and where appropriate a structural perspective is also provided. The book is divided into twelve chapters, each dealing with a specific aspect of lattice materials. We begin with the elastostatic response of lattice materials in Chapter 2, with emphasis on homogenization methods. Homogenization methods provide an effective continuum description of lattices, with consequent analytical and computational advantages. Chapter 3 deals with the elastodynamic response of 1D and 2D lattice materials. Of particular interest are elastic wave propagation and related phenomena that are unique to periodic structures. Progressing from 1D lattice configurations, the chapter ends with 2D lattice materials and establishes relevant connections between the solid-state physics and structural dynamics literature along the way. Dissipation-driven phenomena are described in Chapter 4 using state-space transformations, quadratic eigenvalue analysis, and the Bloch–Rayleigh perturbation method. Attention is given to the effects of damping on dispersion relations, considering free waves and frequency-driven waves. Of particular interest is how dissipation alters the wave motion anisotropies stemming from Bragg scattering. Chapter 5 explores weakly nonlinear lattices and their effective exploitation and tailoring for novel functionalities. Analysis tools are described and results are presented from example studies that highlight the potential for exploiting nonlinearity in the design of materials, systems, and devices. It emerges that nonlinear lattices promise to provide a rich platform for developing tunable and controllable phononic devices.

Mechanical instabilities in periodic porous elastic structures may lead to the formation of homogeneous patterns, opening avenues for a wide range of applications that are related to the geometry of the system. Chapter 6 shows how instabilities via buckling can be used to tune the propagation of elastic waves through a square lattice of elastic beams under equibiaxial compression, enhancing the tunability of its dynamic response. Impact and blast responses of body-centered-cubic lattice structures and sandwich structures is assessed in Chapter 7. Tests over a range of strain rates have shown that the lattice structures are rate-sensitive, with the plateau stress increasing by over 20% over roughly six orders of strain rate. A post-test analysis of the samples indicates that the failure mechanisms did not change with strain rate, for a given lattice architecture. Dynamic tests on sandwich panels with carbon-fibre-reinforced polymer skins have shown that the outer composite layers serve to constrain deformation, thereby improving the properties of the lattice structure.

Static and dynamic properties of continuous pentamode materials are reviewed in Chapter 8. Pentamode materials (PM) possess only one stiffness mode. This is analogous to a liquid, which can easily shear but resists compression, but the stiffness of the junctions in pentamode lattice structures ensures small but finite rigidity and hence structural stability. The effective moduli of PMs are determined for periodic lattice structures using simple beam theory for unit cells with coordination number $d + 1$, where $d = 2, 3$ is the spatial dimension. The main result is an explicit relation for the quasi-static stiffness C. Example applications to specific lattice microstructures illustrate both isotropic and anisotropic PMs.

Models of lattice unit cells can be rather sizable, especially for 3D lattices. Large computational resources are needed to obtain band structures and other dynamical properties. This problem is even more significant when numerous repetitions of the calculations are needed, such as in parametric sweeps and unit-cell topology optimization. Chapter 9 introduces two reduced-order modeling techniques for band structure calculations and discusses the trade-offs between computational savings and the accuracy of the predictions for each technique. Chapter 10 extends the theme of rapid lattice-material band structure calculations and examines the topic of unit-cell design and optimization. A genetic algorithm uniquely tailored for lattice-material band structures is provided and an optimized 2D lattice topology is reported.

Chapter 11 investigates lattice materials involving antiresonances in the form of local resonances or inertial amplification. A detailed mathematical treatment for obtaining dispersion curves for locally resonant and inertially amplified lattices is provided. The dynamical response of these systems, and how they contrast with each other, is described and discussed using 1D, 2D, and 3D lattice examples considering both infinite and finite systems. Chapter 12 focuses on the nanotruss lattices. These nanostructures are fabricated using polymer 3D printing – convenient for generation of complex geometries – enhanced with electrodeposition of nanocrystalline metals for high strength. Being hybrid polymer–nanometal structures, they have excellent mechanical properties for their mass when optimally designed. Because of the nearly limitless geometric flexibility made available through 3D printing, such lattices can be designed to achieve additional functional goals. This chapter examines the use of polymer–nanometal hybrids in conditions where wave propagation is significant. Concepts for linking Floquet–Bloch analysis to the design of nanolattices with desirable properties are outlined.

Acknowledgment

The editors would like to thank each of the the contributors for their tireless efforts and cooperation. Assistance provided by Mr. Prateek Chopra with Figure 1.3 is acknowledged.

References

1 L. Brillouin, *Wave Propagation in Periodic Structures*, 2nd edn. Dover Publications, 1953.

2 C. Kittel, *Introduction to Solid State Physics*, 8th edn. Wiley Publishers, 2004.

3 S. Iijima, "Helical microtubules of graphitic carbon," *Nature*, vol. 354, pp. 56–58, 1991.

4 K. S. Novoselov, A. K. Geim, S. V. Morozov, D. Jiang, Y. Zhang, S. V. Dubonos, I. V. Grigorieva, and A. A. Firsov, "Electric field effect in atomically thin carbon films," *Science*, vol. 306, pp. 666–669, 2004.

5 I. El-Kady, R. H. Olsson, and J. G. Fleming, "Phononic band-gap crystals for radio frequency communications," *Applied Physics Letters*, vol. 92, p. 233504, 2008.

6 J. Vincent, *Structural Biomaterials*, 3rd edn. Princeton University Press, 2012.

7 L. Gibson, M. Ashby, and B. Harley, *Cellular Materials in Nature and Medicine*. Cambridge University Press, 2010.

8 C. T. Sun, J. D. Achenbach, and G. Herrmann, "Time-harmonic waves in a stratified medium propagating in the direction of the layering," *Journal of Applied Mechanics-Transactions of the ASME*, vol. 35, pp. 408–411, 1968.

9 R. Esquivel-Sirvent and G. Cocoletzi, "Band-structure for the propagation of elastic-waves in superlattices," *Journal of the Acoustical Society of America*, vol. 95, pp. 86–90, 1994.

10 D. J. Ewins, "Vibration characteristics of bladed disk assemblies," *Journal of Mechanical Engineering Science*, vol. 15, pp. 165–186, 1973.

11 M. P. Castanier, G. Ottarsson, and C. Pierre, "A reduced order modeling technique for mistuned bladed disks," *Journal of Vibration and Acoustics-Transactions of the ASME*, vol. 119, pp. 439–447, 1997.

12 M. Brun, A. B. Movchan, and I. S. Jones, "Phononic band gap systems in structural mechanics: Finite slender elastic structures and infinite periodic waveguides," *Journal of Vibration and Acoustics-Transactions of the ASME*, vol. 135, p. 041013, 2013.

13 A. L. Abrahamson, *The response of periodic structures to aero-acoustic pressures with particular reference to aircraft skin-rib-spar structures*. PhD Thesis, University of Southampton, 1973.

14 L. Gibson and M. Ashby, *Cellular Solids: Structure and Properties*, 2nd edn. Cambridge Solid State Science Series, Cambridge University Press, 1999.

15 D. Mead, "Wave propagation in continuous periodic structures: Research contributions from Southampton 1964–1995," *Journal of Sound and Vibration*, vol. 190, no. 3, pp. 495–524, 1996.

16 M. I. Hussein, M. J. Leamy, and M. Ruzzene, "Dynamics of phononic materials and structures: Historical origins, recent progress, and future outlook," *Applied Mechanics Reviews*, vol. 66, no. 4, pp. 040802–040802, 2014.

17 M. M. Sigalas and E. N. Economou, "Elastic and acoustic wave band structure," *Journal of Sound and Vibration*, vol. 158, no. 2, pp. 377–382, 1992.

18 M. S. Kushwaha, P. Halevi, L. Dobrzynski, and B. Djafari-Rouhani, "Acoustic band structure of periodic elastic composites," *Physical Review Letters*, vol. 71, pp. 2022–2025, 1993.

19 E. Yablonovitch, "Inhibited spontaneous emission in solid-state physics and electronic," *Physical Review Letters*, vol. 58, pp. 2059–2062, 1987.

20 S. John, "Strong localization of photons in certain disordered dielectric superlattices," *Physical Review Letters*, vol. 58, pp. 2486–2489, 1987.

21 M. I. Hussein, G. M. Hamza, Hulbert, R. A. Scott, and K. Saitou, "Multiobjetive evolutionary optimization of periodic layered materials for desired wave dispersion characteristics," *Structural and Multidisciplinary Optimization*, vol. 31, pp. 60–75, 2006.

22 M. M. Sigalas, "Defect states of acoustic waves in a two-dimensional lattice of solid cylinders," *Journal of Applied Physics*, vol. 84, pp. 2026–3030, 1998.

23 R. Lucklum and J. Li, "Phononic crystals for liquid sensor applications," *Measurement Science and Technology*, vol. 20, p. 124014, 2009.

24 L. S. Chen, C. H. Kuo, and Z. Ye, "Acoustic imaging and collimating by slabs of sonic crystals made from arrays of rigid cylinders in air," *Applied Physics Letters*, vol. 85, pp. 1072–1074, 2004.

25 M. Carrara, M. R. Cacan, M. J. Leamy, M. Ruzzene, and A. Erturk, "Dramatic enhancement of structure-borne wave energy harvesting using an elliptical acoustic mirror," *Applied Physics Letters*, vol. 100, no. 20, 2012.

26 J. K. Yu, S. Mitrovic, D. Tham, J. Varghese, and J. R. Heath, "Reduction of thermal conductivity in phononic nanomesh structures," *Nature Nanotechnology*, vol. 5, pp. 718–721, 2010.

27 M. I. Hussein, S. Biringen, O. R. Bilal, and A. Kucala, "Flow stabilization by subsurface phonons," *Proceedings of the Royal Society A*, vol. 471, p. 20140928, 2015.

28 Z. Y. Liu, X. X. Zhang, Y. W. Mao, Y. Y. Zhu, Z. Y. Yang, C. T. Chan, and P. Sheng, "Locally resonant sonic materials," *Science*, vol. 289, no. 5485, pp. 1734–1736, 2000.

29 Y. Pennec, B. Djafari-Rouhani, H. Larabi, J. O. Vasseur, and A.-C. Ladky-Hennion, "Low-frequency gaps in a phononic crystal constituted of cylindrical dots deposited on a thin homogeneous plate," *Physical Review B*, vol. 78, p. 104105, 2008.

30 T. T. Wu, Z. G. Huang, T. C. Tsai, and T. C. Wu, "Evidence of complete band gap and resonances in a plate with periodic stubbed surface," *Applied Physics Letters*, vol. 93, p. 111902, 2008.

31 B. L. Davis and M. I. Hussein, "Nanophononic metamaterial: Thermal conductivity reduction by local resonance," *Physical Review Letters*, vol. 112, p. 055505, 2014.

32 P. A. Deymier, *Acoustic Metamaterials and Phononic Crystals*. Springer, 2013.

33 M. I. Hussein and I. El-Kady, "Preface to special topic: Selected articles from Phononics 2011: The First International Conference on Phononic Crystals, Metamaterials and Optomechanics, 29 May–2 June, 2011, Santa Fe, NM," *AIP Advances*, vol. 1, p. 041301, 2011.

34 M. I. Hussein, I. El-Kady, B. Li, and J. Sánchez-Dehesa, "Preface to special topic: Selected articles from Phononics 2013: The Second International Conference on Phononic Crystals/Metamaterials, Phonon Transport and Optomechanics, 2–7 June, 2013, Sharm El-Sheikh, Egypt," *AIP Advances*, vol. 4, p. 124101, 2014.

35 N. A. Fleck, V. S. Deshpande, and M. F. Ashby, "Micro-architectured materials: past, present and future," *Proceedings of the Royal Society A*, vol. 466, no. 2121, pp. 2495–2516, 2010.

36 M. F. Ashby, "Hybrids to fill holes in material property space," *Philosophical Magazine*, vol. 85, no. 26-27, pp. 3235–3257, 2005.

37 M. F. Ashby, "The properties of foams and lattices," *Philosophical Transactions of the Royal Society: Mathematical, Physical and Engineering Sciences*, vol. 364, no. 1838, pp. 15–30, 2006.

38 D. Mead, "Wave propagation and natural modes in periodic systems: I. Mono-coupled systems," *Journal of Sound and Vibration*, vol. 40, pp. 1–18, 1975.

39 M. I. Hussein, G. M. Hulbert, and R. A. Scott, "Dispersive elastodynamics of 1D banded materials and structures: Analysis," *Journal of Sound and Vibration*, vol. 289, pp. 779–806, 2006.

40 D. Mead, "Wave propagation and natural modes in periodic systems: II. Multi-coupled systems, with and without damping," *Journal of Sound and Vibration*, vol. 40, pp. 19–39, 1975.

41 L. Raghavan and A. S. Phani, "Local resonance bandgaps in periodic media: Theory and experiment," *Journal of the Acoustical Society of America*, vol. 134, pp. 1950–1959, 2013.

42 J. Banhart, "Light-metal foams – history of innovation and technological challenges," *Advanced Engineering Materials*, vol. 15, no. 3, pp. 82–111, 2013.

43 M. F. Ashby, T. Evans, N. A. Fleck, L. J. Gibson, J. W. Hutchinson, and H. N. G. Wadley, *Metal Foams: A Design Guide*. Butterworth-Heinemann, 2000.

44 R. Lakes, "Materials with structural hierarchy," *Nature*, vol. 361, no. 6412, pp. 511–515, 1993.

45 L. J. Gibson, M. F. Ashby, G. S. Schajer, and C. I. Robertson, "The mechanics of two-dimensional cellular materials," *Proceedings of the Royal Society of London A*, vol. 382, no. 1782, pp. 25–42, 1982.

46 L. J. Gibson and M. F. Ashby, *Cellular Solids: Structure and Properties*, 2nd edn. Cambridge Solid State Science Series, Cambridge University Press, 1997.

47 S. Torquato, L. Gibiansky, M. Silva, and L. Gibson, "Effective mechanical and transport properties of cellular solids," *International Journal of Mechanical Sciences*, vol. 40, no. 1, pp. 71–82, 1998.

48 R. Christensen, "Mechanics of cellular and other low-density materials," *International Journal of Solids and Structures*, vol. 37, no. 1–2, pp. 93–104, 2000.

49 V. S. Deshpande, N. A. Fleck, and M. F. Ashby, "Effective properties of the octet-truss lattice material," *Journal of the Mechanics and Physics of Solids*, vol. 49, pp. 1747–1769, 2001.

50 T. A. Schaedler, A. J. Jacobsen, A. Torrents, A. E. Sorensen, J. Lian, J. R. Greer, L. Valdevit, and W. B. Carter, "Ultralight metallic microlattices," *Science*, vol. 334, pp. 962–965, 2011.

51 X. Zheng, H. Lee, T. H. Weisgraber, M. Shusteff, J. DeOtte, E. B. Duoss, J. D. Kuntz, M. M. Biener, Q. Ge, J. A. Jackson, S. O. Kucheyev, N. X. Fang, and C. M. Spadaccini1, "Ultralight, ultrastiff mechanical metamaterials," *Science*, vol. 344, pp. 1373–1377, 2014.

52 L. R. Meza, A. J. Zelhofer, N. Clarke, A. J. Mateos, D. M. Kochmann, and J. R. Greer, "Resilient 3D hierarchical architected metamaterials," *Proceedings of the National Academy of Sciences*, vol. 112, pp. 11502–11507, 2015.

53 O. Sigmund, "Tailoring materials with prescribed elastic properties," *Mechanics of Materials*, vol. 20, no. 4, pp. 351–368, 1995.

54 A. Evans, J. Hutchinson, N. Fleck, M. Ashby, and H. Wadley, "The topological design of multifunctional cellular metals," *Progress in Materials Science*, vol. 46, no. 3–4, pp. 309–327, 2001.

55 S. Gonella and M. Ruzzene, "Homogenization of vibrating periodic lattice structures," *Applied Mathematical Modelling*, vol. 32, pp. 459–482, 2008.

56 S. Nemat-Nasser, J. R. Willis, A. Srivastava, and A. V. Amirkhizi, "Homogenization of periodic elastic composites and locally resonant sonic materials," *Physical Review B*, vol. 83, p. 104103, 2011.

57 R. P. Moiseyenko and V. Laude, "Material loss influence on the complex band structure and group velocity in phononic crystals," *Physical Review B*, vol. 83, p. 064301, 2011.

58 A. S. Phani and M. I. Hussein, "Analysis of damped Bloch waves by the Rayleigh perturbation method," *Journal of Vibration and Acoustics-Transactions of the ASME*, vol. 135, p. 041014, 2013.

59 M. J. Frazier and M. I. Hussein, "Metadamping: An emergent phenomenon in dissipative metamaterials," *Journal of Sound and Vibration*, vol. 332, pp. 4767–4774, 2013.

60 R. K. Narisetti, M. J. Leamy, and M. Ruzzene, "A perturbation approach for predicting wave propagation in one-dimensional nonlinear periodic structures," *Journal of Vibration and Acoustics-Transactions of the ASME*, vol. 132, p. 031001, 2010.

61 R. Khajehtourian and M. I. Hussein, "Dispersion characteristics of a nonlinear elastic metamaterial," *AIP Advances*, vol. 4, p. 124308, 2014.

62 B. Yousefzade and A. S. Phani, "Energy transmission in finite dissipative nonlinear periodic structures from excitation within a stop band," *Journal of Sound and Vibration*, vol. 354, pp. 180–195, 2015.

63 M. Ruzzene and F. Scarpa, "Directional and band-gap behavior of periodic auxetic lattices," *Physica Status Solidi (B)*, vol. 242, pp. 665–680, 2005.

64 O. Sigmund and J. Søndergaard Jensen, "Systematic design of phononic band gap materials and structures by topology optimization," *Philosophical Transactions of the Royal Society of London A*, vol. 361, no. 1806, pp. 1001–1019, 2003.

65 O. R. Bilal and M. I. Hussein, "Ultrawide phononic band gap for combined in-plane and out-of-plane waves," *Physical Review E*, vol. 84, p. 065701, 2011.

66 K. Bertoldi and M. C. Boyce, "Mechanically triggered transformations of phononic band gaps in periodic elastomeric structures," *Physical Review B*, vol. 77, p. 052105, 2008.

67 D. Torrent and J. Sánchez-Dehesa, "Acoustic cloaking in two dimensions: a feasible approach," *New Journal of Physics*, vol. 10, p. 063015, 2008.

68 A. N. Norris, "Acoustic cloaking theory," *Proceedings of the Royal Society of London A*, vol. 464, pp. 2411–2434, 2008.

69 B. Bouwhuis, J. McCrea, G. Palumbo, and G. Hibbard, "Mechanical properties of hybrid nanocrystalline metal foams," *Acta Materialia*, vol. 57, no. 14, pp. 4046–4053, 2009.

70 J. Meaud, T. Sain, B. Yeom, S. J. Park, A. B. Shoultz, G. Hulbert, Z.-D. Ma, N. A. Kotov, A. J. Hart, E. M. Arruda, and A. M. Waas, "Simultaneously high stiffness and damping in nanoengineered microtruss composites," *ACS Nano*, vol. 8, no. 4, pp. 3468–3475, 2014. PMID: 24620996.

71 C. A. Steeves and A. G. Evans, "Optimization of thermal protection systems utilizing sandwich structures with low coefficient of thermal expansion lattice hot faces," *Journal of the American Ceramic Society*, vol. 94, pp. s55–s61, 2011.

72 H. Cundy and A. Rollett, *Mathematical Models*. Tarquin Publications, 1981.

73 V. Deshpande, M. Ashby, and N. Fleck, "Foam topology: bending versus stretching dominated architectures," *Acta Materialia*, vol. 49, no. 6, pp. 1035–1040, 2001.

74 J. C. Maxwell, "On the calculation of the equilibrium and stiffness of frames," *Philosophical Magazine Series 6*, vol. 27, pp. 294–299, 1864.

75 C. Calladine, "Buckminster Fuller's 'tensegrity' structures and Clerk Maxwell's rules for the construction of stiff frames," *International Journal of Solids and Structures*, vol. 14, no. 2, pp. 161–172, 1978.

76 S. Pellegrino and C. Calladine, "Matrix analysis of statically and kinematically indeterminate frameworks," *International Journal of Solids and Structures*, vol. 22, no. 4, pp. 409–428, 1986.

77 C. P. Broedersz, X. Mao, T. C. Lubensky, and F. C. MacKintosh, "Criticality and isostaticity in fibre networks," *Nature Physics*, vol. 7, no. 12, pp. 983–988, 2011.

78 E. van der Giessen, "Materials physics: Bending Maxwell's rule," *Nature Physics*, vol. 7, no. 12, pp. 923–924, 2011.

79 S. Guest and J. Hutchinson, "On the determinacy of repetitive structures," *Journal of the Mechanics and Physics of Solids*, vol. 51, no. 3, pp. 383–391, 2003.

80 R. Hutchinson and N. Fleck, "The structural performance of the periodic truss," *Journal of the Mechanics and Physics of Solids*, vol. 54, no. 4, pp. 756–782, 2006.

81 A. S. Phani and N. A. Fleck, "Elastic boundary layers in two-dimensional isotropic lattices," *Journal of Applied Mechanics*, vol. 75, p. 021020, 2008.

82 N. A. Fleck and X. Qiu, "The damage tolerance of elastic-brittle, two-dimensional isotropic lattices," *Journal of the Mechanics and Physics of Solids*, vol. 55, no. 3, pp. 562–588, 2007.

83 M. Madou, *Fundamentals of Microfabrication: The Science of Miniaturization*, 2nd edn. Taylor & Francis, 2002.

84 H. Wadley, "Cellular metals manufacturing," *Advanced Engineering Materials*, vol. 4, no. 10, pp. 726–733, 2002.

85 H. N. Wadley, "Multifunctional periodic cellular metals," *Philosophical Transactions of the Royal Society of London A*, vol. 364, no. 1838, pp. 31–68, 2006.

86 K.-J. Kang, "Wire-woven cellular metals: The present and future," *Progress in Materials Science*, vol. 69, no. 0, pp. 213–307, 2015.

87 S. T. Brittain, Y. Sugimura, O. J. Schueller, A. Evans, and G. M. Whitesides, "Fabrication and mechanical performance of a mesoscale space-filling truss system," *Journal of Microelectromechanical Systems*, vol. 10, no. 1, pp. 113–120, 2001.

88 D. Jang, L. R. Meza, F. Greer, and J. R. Greer, "Fabrication and deformation of three-dimensional hollow ceramic nanostructures," *Nature Materials*, vol. 12, no. 10, pp. 893–898, 2013.

89 L. R. Meza, S. Das, and J. R. Greer, "Strong, lightweight, and recoverable three-dimensional ceramic nanolattices," *Science*, vol. 345, no. 6202, pp. 1322–1326, 2014.

90 R. Oftadeh, B. Haghpanah, D. Vella, A. Boudaoud, and A. Vaziri, "Optimal fractal-like hierarchical honeycombs," *Physical Review Letters*, vol. 113, p. 104301, 2014.

91 P. S. Chang and D. W. Rosen, "The size matching and scaling method: A synthesis method for the design of mesoscale cellular structures," *International Journal of Computer Integrated Manufacturing*, vol. 26, pp. 907–927, 2013.

92 T. Stankovic, J. Mueller, P. Egan, and K. Shea, "A generalized optimality criteria method for optimization of additively manufactured multimaterial lattice structures," *Journal of Mechanical Design-Transactions of the ASME*, vol. 137, p. 111405, 2015.

93 J. Mueller, K. Shea, and C. Daraio, "Mechanical properties of parts fabricated with inkjet 3D printing through efficient experimental design," *Materials and Design*, vol. 86, pp. 902–912, 2015.

94 N. Wicks and J. W. Hutchinson, "Optimal truss plates," *International Journal of Solids and Structures*, vol. 38, pp. 5165–5183, 2001.

95 A. Evans, J. Hutchinson, and M. Ashby, "Multifunctionality of cellular metal systems," *Progress in Materials Science*, vol. 43, no. 3, pp. 171–221, 1998.

96 R. Lakes, "Foam structures with a negative Poisson's ratio," *Science*, vol. 235, no. 4792, pp. 1038–1040, 1987.

97 K. E. Evans and A. Alderson, "Auxetic materials: Functional materials and structures from lateral thinking!," *Advanced Materials*, vol. 12, no. 9, pp. 617–628, 2000.

98 G. W. Milton, "Composite materials with Poisson's ratios close to −1," *Journal of the Mechanics and Physics of Solids*, vol. 40, no. 5, pp. 1105–1137, 1992.

99 R. Lakes, "Cellular solid structures with unbounded thermal expansion," *Journal of Materials Science Letters*, vol. 15, no. 6, pp. 475–477, 1996.

100 C. A. Steeves, S. L. dos Santos e Lucato, M. He, E. Antinucci, J. W. Hutchinson, and A. G. Evans, "Concepts for structurally robust materials that combine low thermal expansion with high stiffness," *Journal of the Mechanics and Physics of Solids*, vol. 55, no. 9, pp. 1803–1822, 2007.

101 C. Steeves, C. Mercer, E. Antinucci, M. He, and A. Evans, "Experimental investigation of the thermal properties of tailored expansion lattices," *International Journal of Mechanics and Materials in Design*, vol. 5, no. 2, pp. 195–202, 2009.

102 R. Langley, N. Bardell, and H. M. Ruivo, "The response of two-dimensional periodic structures to harmonic point loading: A theoretical and experimental study of a beam grillage," *Journal of Sound and Vibration*, vol. 207, no. 4, pp. 521–535, 1997.

103 A. S. Phani, J. Woodhouse, and N. A. Fleck, "Wave propagation in two-dimensional periodic lattices," *The Journal of the Acoustical Society of America*, vol. 119, no. 4, pp. 1995–2005, 2006.

104 M. Ruzzene, F. Scarpa, and F. Soranna, "Wave beaming effects in two-dimensional cellular structures," *Smart Materials and Structures*, vol. 12, no. 3, p. 363, 2003.

105 J. Shim, S. Shan, A. Kosmrlj, S. Kang, E. Chen, J. Weaver, and K. Bertoldi, "Harnessing instabilities for design of soft reconfigurable auxetic/chiral materials," *Soft Matter*, vol. 9, pp. 8198–8202, 2013.

106 A. Q. Liu, W. M. Zhu, D. P. Tsai, and N. I. Zheludev, "Micromachined tunable metamaterials: a review," *Journal of Optics*, vol. 14, no. 11, p. 114009, 2012.

107 S. dos Santos e Lucato, J. Wang, P. Maxwell, R. McMeeking, and A. Evans, "Design and demonstration of a high authority shape morphing structure," *International Journal of Solids and Structures*, vol. 41, no. 13, pp. 3521–3543, 2004.

108 S. Mai and N. Fleck, "Reticulated tubes: effective elastic properties and actuation response," *Proceedings of the Royal Society of London A*, vol. 465, no. 2103, pp. 685–708, 2009.

109 N. Wicks and S. Guest, "Single member actuation in large repetitive truss structures," *International Journal of Solids and Structures*, vol. 41, no. 3–4, pp. 965–978, 2004.

110 R. Hutchinson, N. Wicks, A. Evans, N. Fleck, and J. Hutchinson, "Kagome plate structures for actuation," *International Journal of Solids and Structures*, vol. 40, no. 25, pp. 6969–6980, 2003.

111 N. A. Fleck and V. S. Deshpande, "The resistance of clamped sandwich beams to shock loading," *Journal of Applied Mechanics*, vol. 71, no. 3, pp. 386–401, 2004.

112 J. W. Hutchinson and Z. Xue, "Metal sandwich plates optimized for pressure impulses," *International Journal of Mechanical Sciences*, vol. 47, no. 4–5, pp. 545–569, 2005.

113 K. P. Dharmasena, H. N. Wadley, Z. Xue, and J. W. Hutchinson, "Mechanical response of metallic honeycomb sandwich panel structures to high-intensity dynamic loading," *International Journal of Impact Engineering*, vol. 35, no. 9, pp. 1063–1074, 2008.

114 Z. Xue and J. W. Hutchinson, "A comparative study of impulse-resistant metal sandwich plates," *International Journal of Impact Engineering*, vol. 30, no. 10, pp. 1283–1305, 2004.

115 N. Kambouchev, R. Radovitzky, and L. Noels, "Fluid–structure interaction effects in the dynamic response of free-standing plates to uniform shock loading," *Journal of Applied Mechanics*, vol. 74, no. 5, pp. 1042–1045, 2006.

116 J. Harrigan, S. Reid, and A. S. Yaghoubi, "The correct analysis of shocks in a cellular material," *International Journal of Impact Engineering*, vol. 37, no. 8, pp. 918–927, 2010.

117 A. Evans, M. He, V. Deshpande, J. Hutchinson, A. Jacobsen, and W. Carter, "Concepts for enhanced energy absorption using hollow micro-lattices," *International Journal of Impact Engineering*, vol. 37, no. 9, pp. 947–959, 2010.

118 L. E. Murr, S. M. Gaytan, F. Medina, H. Lopez, E. Martinez, B. I. Machado, D. H. Hernandez, L. Martinez, M. I. Lopez, R. B. Wicker, and J. Bracke, "Next-generation biomedical implants using additive manufacturing of complex, cellular and functional mesh arrays," *Philosophical Transactions of the Royal Society of London A*, vol. 368, no. 1917, pp. 1999–2032, 2010.

119 E. M. K. Abad, D. Pasini, and R. Cecere, "Shape optimization of stress concentration-free lattice for self-expandable nitinol stent-grafts," *Journal of Biomechanics*, vol. 45, no. 6, pp. 1028–1035, 2012.

120 G. R. Douglas, A. S. Phani, and J. Gagnon, "Analyses and design of expansion mechanisms of balloon expandable vascular stents," *Journal of Biomechanics*, vol. 47, no. 6, pp. 1438–1446, 2014.

121 T. W. Tan, G. R. Douglas, T. Bond, and A. S. Phani, "Compliance and longitudinal strain of cardiovascular stents: Influence of cell geometry," *Journal of Medical Devices*, vol. 5, no. 4, pp. 041002–041002, 2011.

2

Elastostatics of Lattice Materials

D. Pasini and S. Arabnejad

Department of Mechanical Engineering, McGill University, Montreal, Quebec, Canada

2.1 Introduction

Cellular materials can be loosely categorized according to their cell arrangements. In foams, cells generally cluster in a disorderly fashion, whereas in a lattice the filling of cells is normally ordered [1]. A lattice can be described as a periodic reticulated framework, generated by tessellating a primitive unit cell along periodic directions. To be considered as a material, its structure should have cells with characteristic length at least one or two orders of magnitude below its global length scale. If this occurs, the properties of the whole lattice can be determined from a limited portion of it: the representative volume element (RVE). Once the physical response of the lattice and its RVE is identical, the properties of the RVE can simply replace those of a homogeneous continuum with *effective* properties equivalent to those of the lattice. This is the underlying principle of the process of homogenization, which enables a direct approach, where each element of the lattice is discretized, to be avoided, thus saving time and computational resources.

In general for a lattice, the RVE can be defined by a volume (or an area in 2D) that can contain the primitives of either one distinct unit cell, as in a regular lattice, or dissimilar cells, as in a semi-regular lattice. However, if the non-linear behavior of the lattice is the subject of interest, then the size of the RVE might be not limited to one single cell, but rather multiple cells might be needed to make up the RVE [2].

A number of factors control the global mechanical response of a lattice. A major one is the unit cell topology, which describes the element arrangement, much equivalent to the member layout in a periodic truss. Besides cell topology, however, nodal connectivity and volume fraction, along with material properties, play a role too; together they govern the deformation modes developed in the cell elements, which can either stretch or bend or withstand a combination thereof. This is not irrelevant, since deformation mode impacts structural efficiency: elements that stretch, in fact, yield higher structural efficiency than those that bend. To distinguish these, we can resort to fundamental concepts of structural mechanics, such as the determinacy analysis of a finite truss, and properly apply them to an unbounded truss [3–7]. Such analysis allows identification of states of self-stress and internal mechanisms via the calculation of the rank k of the equilibrium matrix of the unit cell. For instance, $k = 3$ (or 6) is the necessary and sufficient condition for rigidity of 2D (or 3D) pin-jointed frameworks [3]. Examples of rigid lattices

Dynamics of Lattice Materials, First Edition. Edited by A. Srikantha Phani and Mahmoud I. Hussein.
© 2017 John Wiley & Sons Ltd. Published 2017 by John Wiley & Sons Ltd.

that satisfy this condition are the fully triangulated (in 2D) and the Octet truss (in 3D). If k exceeds these values, the lattice becomes redundant and statically indeterminate. On the other hand in a 3D lattice, $k < 6$ means that there are $(6 - k)$ independent mechanisms, an example being the cubic pin-jointed lattice, which has $k = 3$ and thus three mechanisms [3].

To assess the mechanical properties of a lattice, a wealth of approaches exists, ranging from theoretical and computational methods to experimental investigations [8–26]. Noteworthy theoretical contributions are those of Gibson and Ashby [27], Masters and Evans [9], Christensen [10], and Wang and McDowell [11, 28], which provide closed-form expressions for the effective mechanical properties obtained from those of the RVE. These methods often rely on certain premises, which will be briefly reported here. A uniform traction is generally applied to the RVE boundaries, and the cell walls within it are assumed to behave like Euler–Bernoulli beams. Internal forces and moments in the individual cell elements are then determined by solving deformation and equilibrium problems, from which normal and shear strains are obtained and then used to calculate the effective elastic properties and yield strength. Named here a "force-based" approach, this homogenization technique works well for topologies that are statically and kinematically determinate; they require, however, additional efforts for unit cells containing one or more states of self-stresses or internal mechanisms [11], as with overdetermined structural frames. Among other theoretical methods, we recall also the use of micropolar theory [13–17, 29], which requires the introduction of independent microscopic rotations, besides translational deformations [30, 31]. This scheme allows the micropolar elastic constants of the stiffness matrix to be obtained through either an explicit structural analysis of the representative unit cell [16, 17] or an energy-based approach [12–15].

Among the computational approaches, asymptotic homogenization (AH) has been successfully applied to predict the effective mechanical properties of materials with a periodic composite layout [19, 32–34]. AH has been widely used not only to characterize porous materials, such as tissue scaffolds [20, 35–37], but also composite materials, as well as being used in the topology optimization of structures [38–41]. AH assumes that any field quantity, such as the displacement, can be described as an asymptotic expansion, which, when placed in the governing equations of equilibrium, allows the effective properties to be derived [32, 33]. Validation of AH results in experiments has shown that AH can provide reliable and accurate predictions for the properties of heterogeneous periodic materials [42–47].

Other approaches combining Bloch's theorem and the Cauchy–Born hypothesis have been successfully applied to study the elastostatic responses of a set of planar lattices [48, 49]. Hutchinson and Fleck [48] first formulated the microscopic nodal deformations of a lattice in terms of a macroscopic strain field, from which the material macroscopic stiffness properties can be derived. The methodology proposed to characterize cell topologies with a certain level of symmetry, such as the Kagome lattice and the triangular–triangular lattice, was later extended to deal with planar topologies with arbitrary cell geometry [49].

Another method to list here involves the analysis of the partial differential equations of motion associated with the homogenized continuum [50]. The equations of the discrete lattice are formulated with a finite element (FE) discretization of the unit cell, from which the governing equations of a periodic medium can be obtained. The homogenized properties of a unit cell can then be assessed by a direct comparison between the coefficients of the homogenized equations and the elasticity equations of the continuum. More recently, a more general matrix-based procedure has been introduced for the analysis of arbitrary bidimensional and tridimensional cell topologies for bending- and stretching-dominated lattices [51, 52]. This scheme is multiscale and has been used for both linear and nonlinear analysis of lattice materials to simultaneously determine their effective material stiffness and yield strength.

As briefly summarized by the not-inclusive list above, a wide range of methods exists to assess the elastostatics of lattice materials. Some of them were specifically developed for cellular materials; others stem from a wide range of disciplines dealing with heterogeneous media, material spatial randomness, solid-state physics, and applied mechanics. Aware that an attempt to classify all of them poses challenges, we select in this chapter a set of methods that to the authors knowledge, seem to be among the most relevant. In the next section, we start by providing a description of the RVE, along with the main types of boundary condition that can be applied to it. Then we give a brief review of each homogenization method with an emphasis on the underlying assumptions, advantages, and limitations. A comparative study follows, in which a set of selected methods is tested for the analysis of a hexagonal lattice. Their accuracy in predicting effective elastic constants and mechanical strength is presented for given ranges of relative density.

2.2 The RVE

The concept of the RVE is fundamental to determining the equivalent properties of a lattice. The notion stems from the self-consistent scheme developed by Eshelby [53], who studied the mechanics of an ellipsoidal inclusion in an infinite matrix under homogeneous boundary conditions. Formally introduced by Hill [54], followed by Hashin [55, 56], Nemat-Nasser [57], and Willis [58], the concept applies to heterogenous materials with inhomogeneities at least a couple of orders of magnitude below the characteristic length of the macroscopic medium. The hypothesis is that the properties of a heterogeneous medium can be obtained from the analysis of a limited region of it. This portion, the RVE, can be conveniently used to obtain the effective properties of an *equivalent homogeneous medium* via a process of homogenization. To do so, the RVE should contain the main microstructural features of the heterogonous material and respond as the infinite medium when uniform strain or stress is applied to its boundaries [26, 59, 60]. Homogenization is advantageous because it permits lengthy full-scale simulations to be avoided; these would involve the explicit description of each microheterogeneity.

Figure 2.1 illustrates schematically the homogenization process applied to a lattice with square unit cells. A body Ω with a periodic microarchitecture subjected to a traction t at the traction boundary Γ_t, a displacement d at the displacement boundary Γ_d, and a body force f is replaced by a homogenized body $\overline{\Omega}$. The external and traction boundaries are prescribed and applied to both the discrete and homogenized medium, the latter being a homogeneous continuum with no voids. The effective mechanical properties of the RVE should be determined in such a way that the macroscopic behavior of Ω and $\overline{\Omega}$ are equivalent.

The simplest homogenization scheme that can be applied to a lattice, as to any composite material, is the rule of mixture. This method assumes that the equivalent

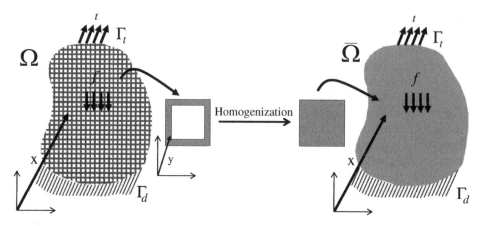

Figure 2.1 A schematic of the homogenization process of a square lattice material (left) into an equivalent homogeneous medium (right) with effective properties obtained from the RVE analysis.

properties can be simply obtained from the weighted mean of the heterogeneities only, with no need to account for boundary conditions, cell topology, and other (important) factors. On the other hand, the RVE definition for a periodic medium Ω has more strings attached to it. One requirement is that the constitutive equations and material properties of the homogenized counterpart $\overline{\Omega}$ are position independent. Another factor deals with the equivalence between the mechanical behavior of the RVE and that of the effective medium. This condition should hold for a given volume with prescribed shape and size, and it should be satisfied if the average of a given mechanical field, such as stress and strain, is equal for both the volume elements (Figure 2.1). The calculation to obtain the effective properties of these volumes can be accomplished with respect to either the surface or the volume of the RVE [61–65]. With the former, stress and strain averages are computed on the surface of the RVE, whereas with the latter the average strain energy is obtained over the volume. Surface and volume average are two homogenization approaches, examined below in Sections 2.3 and 2.4 with reference to a lattice. The descriptions of other methods follow.

2.3 Surface Average Approach

The surface average approach assumes the application of either stress or strain distributions to the surface of the RVE. These distributions should generate either an average stress (if tractions are applied) or an average strain (for applied displacements) in a homogeneous material with the size of the RVE. The relationship between the average stress and the boundary tractions can be written as:

$$\overline{\sigma} = \frac{1}{V_{RVE}} \int_{V_{RVE}} \sigma dV_{RVE} = \frac{1}{V_{RVE}} \int_{\Gamma_{RVE}} \frac{1}{2}(t_i y_j + t_j y_i)d\Gamma_{RVE} \tag{2.1}$$

where $\overline{\sigma}$ is the macroscopic average stress tensor, σ is the local stress tensor in the RVE, t_i is the traction imposed on the RVE boundary, y_i are local coordinates of the RVE boundary, and Γ_{RVE} is the RVE boundary. Equation (2.1) describes the distribution of the stress components along the surfaces of the RVE. Note that the subscripts i and j follow the Einstein notation.

The second requirement is the equivalence of the strain tensor generated in the RVE, ε, and in the equivalent medium, $\bar{\varepsilon}$. The strain average in the RVE is given by the difference of the displacements at the opposite surfaces of the volume element. If this difference is not constant along the surface, the surface integral of this quantity can be taken. Using the divergence theorem, the relationship between the average strain tensor and the displacement boundary conditions can be written as

$$\bar{\varepsilon} = \frac{1}{V_{RVE}} \int_{V_{RVE}} \varepsilon \, dV_{RVE} = \frac{1}{V_{RVE}} \int_{\Gamma_{RVE}} \frac{1}{2}(u_i n_j + u_j n_i) d\Gamma_{RVE} \tag{2.2}$$

where u_i is the displacement vector imposed on the RVE boundary, and n_i are the components of the vector normal to the RVE boundary. It is important to note that the relationship between the average stress (or strain), and the boundary tractions (or displacements) is not unique. Dissimilar boundary displacements integrated over the boundary can provide identical average strain. Hence when in-situ boundary conditions are unknown, the displacements in Eq. (2.1), or the tractions in Eq. (2.2), are generally chosen to be uniform.

In general, to predict the RVE properties, alternative boundary conditions can be applied. The main ones are the Dirichlet (DBC), Neumann (NBC), mixed (MBC), and periodic (PBC) boundary conditions. MBCs can be deduced from DBC and NBC, as it includes both of them. For these boundary conditions, we can write the governing equations of the elastic problem as:

$$\sigma(x)_{ij,j} = 0, \quad x \in \Omega \quad \begin{cases} u_i|_\Gamma = \bar{\varepsilon}_{ij}^{kl} x_j|_{\Gamma_{RVE}} & DBC \\ T_i|_\Gamma = \bar{\sigma}_{ij}^{kl} n_j|_{\Gamma_{RVE}} & NBC \\ u_i^{A+} - u_i^{A-} = \bar{\varepsilon}_{ij}^{kl}(x_j^{A+} - x_j^{A-}) & PBC \end{cases} \tag{2.3}$$

where σ_{ij} are the components of the stress tensor in the RVE domain, $(\cdot)_{,i}$ is the gradient of a field quantity with respect to the global coordinate system, and x_i are the nodal coordinates. $\bar{\varepsilon}_{ij}^{kl}$ and $\bar{\sigma}_{ij}^{kl}$ are the macroscopic unit strain and stress for the kl-th traction (or strain) over the RVE boundaries. Indices "$A+$" and "$A-$" represent the Ath pair of two opposite parallel surfaces for the RVE boundary. With the DBC and PBC, a unit strain is applied on the nodes of the boundary defining the RVE, whereas with the NBC uniform tractions are applied. In an FE analysis, the PBC can be easily applied as nodal displacement constraints that automatically guarantee traction continuity on the RVE boundaries.

With the DBC and PBC, only three uniform strain states (six for 3D) need to be applied to the RVE. We can then introduce a local structural strain tensor $\mathbf{M}$ to express the average strain $\bar{\varepsilon}^{kl}$ as a function of the local (microstructural) strain ε^{kl} as

$$\varepsilon^{kl} = \mathbf{M}\bar{\varepsilon}^{kl} \tag{2.4}$$

Using $\mathbf{M}$, the local strain at any point within the RVE may be calculated from an arbitrary average strain as:

$$\varepsilon = \mathbf{M}\bar{\varepsilon} \tag{2.5}$$

Hooke's law is then used to obtain the microscopic stress distribution over the RVE.

$$\boldsymbol{\sigma} = \mathbf{E}\boldsymbol{\varepsilon} \tag{2.6}$$

where $\mathbf{E}$ is the stiffness tensor of the base material of the lattice. Substituting Eqs. (2.5) and (2.6) into Eq. (2.1), the average of stress over the RVE can be computed from:

$$\overline{\boldsymbol{\sigma}} = \frac{1}{V_{RVE}} \int_{V_{RVE}} \boldsymbol{\sigma} dV_{RVE} = \frac{1}{V_{RVE}} \int_{V_{RVE}} \mathbf{E}\mathbf{M} dV_{RVE} \overline{\boldsymbol{\varepsilon}} \tag{2.7}$$

from which the effective stiffness tensor may be defined as

$$\overline{\mathbf{E}} = \frac{1}{V_{RVE}} \int_{V_{RVE}} \mathbf{E}\mathbf{M} dV_{RVE} \tag{2.8}$$

where $\overline{\mathbf{E}}$ is the stiffness tensor describing the effective elastic properties of the RVE.

With the NBC, three uniform stress states (six for 3D) are applied to the RVE, and the local problem is then solved to obtain microscopic stress distribution. Assuming a linear elastic medium, a local structural stress tensor $\mathbf{N}$ can be defined to relate the average stress $\overline{\boldsymbol{\sigma}}^{kl}$ and the local or microstructural stress $\boldsymbol{\sigma}^{kl}$ for the kl-th traction over the domain:

$$\boldsymbol{\sigma}^{kl} = \mathbf{N}\overline{\boldsymbol{\sigma}}^{kl} \tag{2.9}$$

The local structural stress tensor can be also used to obtain the local stress at any point within the RVE from an arbitrary macroscopic stress. To obtain the effective material properties, the strain average over the RVE should be computed from Eq. (2.2). The macroscopic strain and stress can then be related through the effective compliance matrix $\overline{\mathbf{S}}$:

$$\overline{\boldsymbol{\varepsilon}} = \overline{\mathbf{S}}\overline{\boldsymbol{\sigma}} \tag{2.10}$$

Since uniform macroscopic stresses are applied to the RVE, each column of the effective compliance matrix is the macroscopic strain computed from Eq. (2.2).

It is important to note the role of boundary conditions here, essential to any homogenization method. As in any boundary-value problem, the type of boundary applied to the RVE influences the accuracy of the solution. The DBC and NBC have no rigorous physical foundation. They do not guarantee the continuity of the displacement field throughout the lattice. In addition, the results obtained by applying them generally depend on the RVE size and shape, an undesirable outcome. It has been shown that the DBC produces results close to the Voigt (upper) bound [66], whereas the NBC result is close to the Reuss (lower) bound [67]. A more accurate type of condition applied to the RVE boundary is the MBC [26, 68], which produces results very close to those obtained using the PBC. The MBC, however, has limitations, as it can be applied only to lattice topologies having at least orthotropic symmetry. The PBC, on the other hand, has no limitation on the type of cell topology because it assures the continuity of displacement and traction on the RVE boundaries, thereby guaranteeing congruent deformation among unit cells after deformation. Further comments on the comparison of alternative boundary conditions namely the DBC, NBC, and PBC, are given in the case study at the end of the chapter.

2.4   Volume Average Approach

For certain cell topologies, such as those that are nonorthotropic, the surface average approach might yield errors in the prediction of the effective strain energy. For example, if stress couples act at the element joints and on the RVE surfaces, the stress average on the surface of the RVE, computed from Eq. (2.1), would be zero. This means that the effective properties calculated via the surface average scheme cannot account for the contribution of stress couples. Although for some basic cellular geometries, this limitation can be avoided by choosing the RVE in such a way that no stress couples act on the external surface Γ_{RVE} [69–71], for more sophisticated cell geometries, the surface average method cannot be used. In such a case, a volume average based approach can be applied [64, 65].

The volume average based approach relies on the following assumption. The mechanical responses of the RVE consisting of the given microstructure, and the effective medium can be considered equivalent if the RVE strain energy of the former is equal to that of the latter. This can be expressed as:

$$W_{RVE} = \frac{1}{2} \int_V \sigma^{kl} \varepsilon^{kl} dV = \frac{V}{2} \bar{\varepsilon}^{kl} \overline{E} \bar{\varepsilon}^{kl} = \frac{V}{2} \bar{\sigma}^{kl} \overline{S} \bar{\sigma}^{kl} \tag{2.11}$$

where W_{RVE} and V are the strain energy and volume of the RVE.

For 2D problems with orthotropic cell topologies, the effective stiffness and compliance matrix is defined by four independent components. To obtain the effective stiffness tensor, four uniform strain states need to be applied on the RVE with either the NBC or PBC: two uniaxial, one biaxial, and one shear stress. For example, if $\bar{\varepsilon}^{11} = \begin{bmatrix} 1 & 0 & 0 \end{bmatrix}$ is applied to the unit cell, the first component of the stiffness tensor is given by $\overline{E}_{1111} = (2W_{RVE})/V$. On the other hand, by applying a uniform stress of $\bar{\sigma}^{11} = \begin{bmatrix} 1 & 0 & 0 \end{bmatrix}$ to the RVE boundaries, we can obtain the first component of the compliance tensor as $\overline{S}_{1111} = (2W_{RVE})/V$. Similar to the DBC, the number of local RVE problems to solve corresponds to the number of independent coefficients of the compliance tensor. Therefore, for a general 2D and 3D anisotropic lattice, 6 and 21 uniform macroscopic strains/stresses should be applied to the RVE to obtain the effective stiffness/compliance tensor.

We remark here that the volume average approach has often been used to predict the effective properties of cellular materials [72–74]. It has a solid foundation since the method uses the basic law of continuum mechanics: the conservation of energy. It has also no limit of application with respect to the cell topology and its geometric symmetry.

2.5   Force-based Approach

A classical force-based approach has been extensively applied to predict the linear elastic behaviour of cellular materials [9, 27, 75–81]. While in early studies the linear elastic deformation was calculated from the axial extension of the cell walls only [75, 82], later models rectified this deficiency to account also for bending and shear deformation [9, 77, 79–81]. This method has often been applied with the assumption that the cell walls are slender [9, 11, 27]. For a bending-dominated lattice, Euler–Bernoulli beam and Timoshenko beam elements have been applied to model cell-wall deformation

[9, 11, 27]. For stretching-dominated unit cells, cell walls have been treated as 1D rod elements and characterized by a dominant extensional deformation mode.

The force-based approach applied to a lattice generally assumes that field quantities are uniform over the RVE. A uniform traction is thus commonly applied to its boundaries. If the lattice topology is stretching-dominated and statically determinate – such as triangle, Kagome, and diamond for 2D lattices – the forces in each cell member can be simply obtained from the equilibrium equations and periodic boundary conditions, as no internal mechanisms and state of self-stress exist. On the other hand, for statically indeterminate lattices, such as mixed square/triangular topologies and other more complex cell topologies, one or more states of self-stresses can emerge, an instance that has been often neglected. In this case, since the static equilibrium equations are insufficient to determine the internal forces and reactions in the cell walls, the compatibility equations at the lattice joints should be used to compute the normal strain in the unit cell as a function of the internal forces. This system of linear equations can thus be solved to compute microscopic stresses and strains in the lattice, a step necessary to find the effective elastic properties and yield strength.

The main advantage of a force-based approach is that the effective properties can be conveniently described with closed-form expressions, which become handy to generate material property charts. The effective elastic modulus and yield strength can be readily obtained as a function of relative density and used for the analysis and design of lattice materials. It has been shown that the use of these expressions is generally restricted to lower relative densities ($\rho < 0.3$) [11, 27]. For higher relative densities, the predictions of rod/beam theory become less accurate due to significant axial and shear deformation at the cell edges and vertices [83].

2.6 Asymptotic Homogenization Method

Asymptotic homogenization (AH) is a well-developed theory, with a sound mathematical foundation. Developed in 1970s, this theory has been used to obtain the effective properties of periodic materials [84–86] in several areas of physics and engineering, such as heat transfer, wave propagation, and electromagnetism [34, 87–89]. The results obtained with AH have been proved to be consistently aligned with those obtained from experiments [42–47]. The underlying assumption of AH is that each field quantity depends on two different scales: one on the macroscopic level x, and the other on the microscopic level, $y = x/\varepsilon$. ε is a magnification factor that scales the dimensions of the unit cell to the dimensions of the material at the macroscale. AH also assumes that field quantities, such as displacement, vary smoothly at the macroscopic level, and are periodic at the microscale [32, 90]. Based on AH, each physical field, such as the displacement field u, in a porous elastic body, can be expanded into a power series with respect to ε:

$$u_i^\varepsilon(x) = u_{0i}(x, y) + \varepsilon u_{1i}(x, y) + \varepsilon^2 u_{2i}(x, y) + \dots \tag{2.12}$$

where the functions $u_{0i}, u_{1i}, u_{2i} \dots$ are y-periodic with respect to the local coordinate y, which means they yield identical values on the opposing sides of the unit cell. u_{1i} and u_{2i} are perturbations in the displacement field due to the microstructure. u_{0i} can be shown to depend only on the macroscopic scale and to be the average value of the displacement field [32]. Taking the derivative of the asymptotic expansion of the displacement field with respect to the global coordinate system and using the chain rule allows the small deformation strain tensor to be written as:

$$\varepsilon = \bar{\varepsilon} + \varepsilon^*, \quad \bar{\varepsilon}_{ij} = \frac{1}{2}\left(\frac{\partial u_{0i}}{\partial x_j} + \frac{\partial u_{0j}}{\partial x_i}\right), \quad \varepsilon^*_{ij} = \frac{1}{2}\left(\frac{\partial u_{1i}}{\partial y_j} + \frac{\partial u_{1j}}{\partial y_i}\right) \tag{2.13}$$

where $\bar{\varepsilon}$ is the average or macroscopic strain and ε^* is the fluctuating strain fluctuating periodically at the microscale. Substituting the strain tensor into the standard weak form of the equilibrium equations for a cellular body Ω, yields [90]:

$$\int_\Omega E(\varepsilon^0(v) + \varepsilon^1(v))(\bar{\varepsilon} + \varepsilon^*)\, d\Omega = \int_{\Gamma_t} \mathbf{t}v\, d\Gamma \tag{2.14}$$

where E is the local elasticity tensor that depends on the position within the RVE, $\varepsilon^0(v)$ and $\varepsilon^1(v)$ are the virtual macroscopic and microscopic strains, respectively, and $\mathbf{t}$ is the traction at the traction boundary Γ_t. Being the virtual displacement, $\mathbf{v}$ may be chosen to vary only on the microscopic level and to be constant on the macroscopic level. Based on this assumption, the microscopic equilibrium equation can be obtained as:

$$\int_\Omega E\varepsilon^1(v)(\bar{\varepsilon} + \varepsilon^*)\, d\Omega = 0 \tag{2.15}$$

Taking the integral over the RVE volume (V_{RVE}), Eq. (2.15) may be rewritten as:

$$\int_{V_{RVE}} E\varepsilon^1(v)\varepsilon^*\, d\Omega = -\int_{V_{RVE}} E\varepsilon^1(v)\bar{\varepsilon}\, d\Omega \tag{2.16}$$

Equation (2.16) represents a local problem defined on the RVE. For a given applied macroscopic strain, the material can be characterized if the fluctuating strain ε^* is known. The periodicity of the strain field is ensured by imposing periodic boundary conditions on the RVE edges; the nodal displacements on the opposite edges are set to be equal [90, 91]. Equation (2.16) can be discretized and solved via FE analysis, as described elsewhere [19, 33, 90, 92]. For this purpose, Eq. (2.16) can be simplified to obtain a relation between the microscopic displacement field $\mathbf{d}$ and the force vector $\mathbf{f}$ as:

$$\mathbf{Kd} = \mathbf{f} \tag{2.17}$$

where $\mathbf{K}$ is the global stiffness matrix, defined as:

$$\mathbf{K} = \sum_{e=1}^{m} \mathbf{k}^e, \quad \mathbf{k}^e = \int_{Y^e} \mathbf{B}^T\, \mathbf{EB}dY^e \tag{2.18}$$

with $\sum_{e=1}^{m}(\cdot)$ the finite element assembly operator, m the number of elements, $\mathbf{B}$ the strain-displacement matrix, and Y^e the element volume. The force vector $\mathbf{f}$ in Eq. (2.19) is expressed as:

$$\mathbf{f} = \sum_{e=1}^{m} \mathbf{f}^e, \quad \mathbf{f}^e = \int_{Y^e} \mathbf{BE}\bar{\varepsilon}(u)\, dY^e \tag{2.19}$$

Equation (2.19) can be used either for the linear elastic analysis of the microstructure or to model the effect of material nonlinearity as a result of elastoplasticity deformation of the unit cell. The material yield strength and the effective elastic modulus can be characterized by the linear analysis of the microstructure, while the ultimate strength of the material can be obtained through elastoplasticity analysis. The elastoplasticity behavior of a lattice using AH has been recently addressed for 2D topologies [93].

Considering the assumption of small deformation and elastic material behavior, the solution of Eq. (2.17) leads to a linear relation between the macroscopic $\bar{\varepsilon}(u)$ and microscopic $\varepsilon(u)$ strain through the local structural strain tensor M:

$$\varepsilon(u) = \mathbf{M}\bar{\varepsilon}(u) \tag{2.20}$$

For a 2D case, three independent unit strains are required to construct the **M** matrix.

$$\bar{\varepsilon}^{11} = \begin{bmatrix} 1 & 0 & 0 \end{bmatrix}^T, \ \bar{\varepsilon}^{22} = \begin{bmatrix} 0 & 1 & 0 \end{bmatrix}^T, \ \bar{\varepsilon}^{12} = \begin{bmatrix} 0 & 0 & 1 \end{bmatrix}^T \tag{2.21}$$

The macroscopic strains are applied to Eq. (2.21) to obtain the force vector for the computation of the microscopic displacements through Eq. (2.17). Using the strain-displacement matrix **B**, the fluctuating strain tensor $\varepsilon^*(u)$ is determined and used to calculate the microscopic strain tensor $\varepsilon(u)$ through Eq. (2.13). The local structural tensor **M** can then be obtained at the element centroid by solving three sets of equations (for 2D) once $\bar{\varepsilon}(u)$ and $\varepsilon(u)$ are known. Here, since three independent unit strains are considered, each column of the matrix **M** represents the microscopic strain tensor $\varepsilon(u)$. The effective stiffness matrix can be simply derived by taking the integral of the microscopic stress over the RVE and dividing by the RVE volume, as shown in Eq. (2.7). The effective stiffness matrix $\bar{\mathbf{E}}$, which can be defined from Eq. (2.8), relates the macroscopic strains to the macroscopic stresses of the homogenized material. AH also allows the macroscopic stresses that lead to the microscopic yield, or the endurance limit, as well as fracture, to be obtained. To calculate the yield strength of the unit cells, the microscopic stress distribution **σ** corresponding to the multiaxial macroscopic stress $\bar{\sigma}$ can be obtained via the following equation:

$$\sigma = \mathbf{E}\mathbf{M}(\bar{\mathbf{E}})^{-1}\bar{\sigma} \tag{2.22}$$

The von Mises stress distribution at the microstructure can then be used to capture the yield surface of the unit cell, expressed as:

$$\bar{\sigma}^y = \frac{\sigma_{ys}}{\max\left\{\sigma_{vM}(\bar{\sigma})\right\}}\bar{\sigma} \tag{2.23}$$

where $\bar{\sigma}^y$ is the yield surface of the unit cell, σ_{ys} is the yield strength of the bulk material, and $\sigma_{vM}(\cdot)$ is the von Mises stress at the microstructure corresponding to the applied macroscopic stress. To calculate the yield strength of the unit cells under uniaxial tension in x and y directions and pure shear, macroscopic unit stresses can be applied to Eq. (2.23). $\bar{\sigma}^y_{xx}$, $\bar{\sigma}^y_{yy}$, and $\bar{\tau}^y_{xy}$ denote, respectively, the macroscopic yield strength of the unit cell under uniaxial tension in x and y directions and under pure shear. To generate the yield surface of the unit cell under multiaxial loadings, the procedure summarized above is repeated for alternative combinations of multiaxial stresses.

Validation of AH results with experiments has shown that AH is a reliable and accurate method to predict the effective mechanical properties of heterogeneous periodic materials [42–47]. Takano et al. [42] applied AH to analyze micro-macro coupled behaviour of knitted fabric composite materials under large deformation conditions. The predicted largely deformed microstructures were compared with the experimental results with very good agreement observed. In another study, to validate the accuracy of AH results, the predicted value of the effective elastic modulus of alumina with 3.1% porosity was compared to the elastic constant measured from experiments [43], and a relative error of 1% was found. Guinovart-Díaz et al. [44, 45] computed the thermoelastic effective coefficients of a two-phase fibrous composite using AH. The results were compared with experimental data, and good agreement was found here too. In a more recent study, AH was applied to predict the failure behavior of 3D woven composites [47], and the predictions of the stress–strain response and failure modes were shown to match experimental results.

Compared to other homogenization schemes, a noteworthy advantage of AH is that the stress distribution in the unit cell can be determined accurately and thus be used for a detailed analysis of the strength of and damage to heterogeneous periodic materials [34, 92, 94]. Furthermore, AH has no limitation on the cell topology or on the range of relative density; essentially, AH can handle any lattice regardless of its relative density, as recently shown in a study where the effective mechanical properties, including effective elastic constants and yield strength, of six 2D cell topologies were obtained for the whole range of relative densities [103]. On the other hand, AH has some limitations to account for. While effective in computing the stress and strain distribution at each scale, care should be given at regions where the underlying assumption, y-periodicity of field quantities, is not satisfied. This can include areas with local heterogeneity in the material, areas with high gradients of field quantities, or zones in the vicinity of borders [95–101]. These boundary effects, however, may be captured by applying either a boundary layer corrector [95, 97] or a spatially decaying stress localization function [102], or by use of multilevel computational methods [96, 98–100].

2.7 Generalized Continuum Theory

In classical continuum theory, the stress state at a given point is assumed to be a function of its strain state only. As a result, since the displacement at a point is the only kinematic condition to account for, the interaction between two neighboring points can be modeled via a force vector only. However, if crack tips, notches, or high strain gradients are present, such an assumption ceases to hold [104, 105]. In these cases, a generalization of classical continuum theory is required, as introduced by E. and F. Cosserat [106] and Eringen [30]. Cosserat theory, also known as micropolar theory, can capture the impact of nonlocality, high strain gradients, and size effects. Its main assumption is that both displacements and rotations of a point are independent kinematic quantities, and a couple vector can be used to represent the interaction between two points in a medium [30, 107].

Applied to a lattice material, micropolar theory assumes that joint displacement and joint rotation contribute to the total joint displacement [16, 17, 108–110]. Figure 2.2 shows joint translation and joint rotation in the initial and deformed geometry of a typical cell element.

In linear micropolar elasticity, the kinematic relations can be expressed as

$$\varepsilon_{ij} = u_{j,i} - e_{kij}\phi_k$$
$$k_{ij} = \phi_{j,i} \tag{2.24}$$

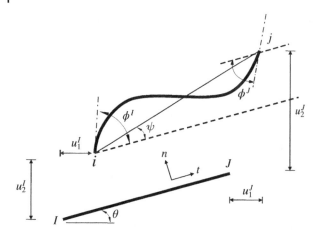

Figure 2.2 Initial (*IJ*) and deformed (*ij*) geometry of a typical cell member of a cellular solid.

where ε_{ij} is the strain tensor, $u_{j,i}$ is the displacement gradient, e_{kij} is the permutation tensor, φ_k is the microrotation, k_{ij} is the curvature strain tensor, and $\phi_{j,i}$ is the microrotation gradient. The generalized strain vector of a micropolar medium is related to the displacement gradients, microrotation, and microrotation gradients and can be written as follows:

$$\boldsymbol{\varepsilon} = \begin{bmatrix} \varepsilon_{11} & \varepsilon_{22} & \varepsilon_{12} & \varepsilon_{21} & k_{13} & k_{23} \end{bmatrix}^T = \begin{bmatrix} u_{1,1} & u_{2,2} & u_{2,1} - \phi & u_{1,2} + \phi & \phi_{,1} & \phi_{,2} \end{bmatrix}^T \quad (2.25)$$

where $\varphi_{,1}$ and $\varphi_{,2}$ are the abbreviations for $\varphi_{3,1}$ and $\varphi_{3,2}$ respectively. The generalized stress vector is given by

$$\boldsymbol{\sigma} = \begin{bmatrix} \sigma_{11} & \sigma_{22} & \sigma_{12} & \sigma_{21} & m_{13} & m_{23} \end{bmatrix}^T \quad (2.26)$$

where m_{13} and m_{23} are the couple stresses in the x and y planes, respectively. Using the stress and strain components, the 2D constitutive relations for anisotropic micropolar solids can be written as:

$$\boldsymbol{\sigma} = \overline{\mathbf{C}} \boldsymbol{\varepsilon} \quad (2.27)$$

where $\overline{\mathbf{C}}$ is the 6×6 matrix of the constitutive law coefficients for a micropolar medium. To represent a cellular material as a micropolar continuum, the coefficients of the constitutive Eq. (2.27) must be obtained. Since the constitutive matrix is symmetric, 13 independent coefficients are required. For a general in-plane orthotropic material, only five micropolar constants are necessary [105].

To obtain the micropolar elastic constants of a lattice, both a force-based approach and an energy method can be used, as shown in previous studies [15–17]. With a force-based approach, the general deformation state of the RVE is obtained by imposing uniform macroscopic stresses on the unit cell. Besides uniaxial tension and shear stress, couple stresses should be also applied to the RVE to compute nodal displacements and microrotations. The deformation is then governed by the displacements and rotations at the nodes where the cell struts (or walls) intersect. The important assumption here is that cell elements are slender so that they can be considered to be Euler–Bernoulli beams. To apply the force-based approach, it is assumed that stresses in the RVE are obtained from the stress resultants acting on the individual struts at the boundary of the RVE.

For a given macroscopic stress, local equilibrium equations on the RVE can be solved, and internal forces and moments for each individual strut obtained. Strains are then calculated via the gradient of the displacements over the unit cell. The effective stresses and strains over the RVE can be calculated, and from them the constitutive equations derived.

Alternatively, if the energy approach is used, the stresses and couple stresses of the unit cell can be found by taking the derivative of the strain energy density, w, with respect to the strain vector components as

$$
\begin{aligned}
\sigma_{11} &= \frac{\partial w}{\partial \varepsilon_{11}} = \frac{\partial w}{\partial u_{1,1}} & \sigma_{22} &= \frac{\partial w}{\partial \varepsilon_{22}} = \frac{\partial w}{\partial u_{2,2}} \\
\sigma_{12} &= \frac{\partial w}{\partial \varepsilon_{12}} = \frac{\partial w}{\partial (u_{2,1} - \phi)} & \sigma_{21} &= \frac{\partial w}{\partial \varepsilon_{21}} = \frac{\partial w}{\partial (u_{1,2} + \phi)} \\
m_{13} &= \frac{\partial w}{\partial k_{13}} = \frac{\partial w}{\partial \phi_{,1}} & m_{23} &= \frac{\partial w}{\partial k_{23}} = \frac{\partial w}{\partial \phi_{,2}}
\end{aligned}
\tag{2.28}
$$

To obtain the constitutive matrix, $\overline{C}$, of a given periodic lattice, the continuum approximation of the strain energy of the unit cell must be computed. This is given by the sum of the strain energy contributions of the individual members in a cell. The strain energy of the cell is then expressed in terms of displacements and rotations of the joints. A Taylor series expansion can be applied to achieve the continuum approximation of the strain energy density in terms of displacement and rotation of the unit cell centre. If only the first-order terms of the Taylor series of the displacements and rotations are considered, the results might not be accurate [13]. It has been shown that to achieve joint equilibrium, both the first-order derivatives as well as those of the second order should be retained [14]. Using the displacements and rotations of each cell joint (Figure 2.2), the total strain energy (W^{IJ}) per unit width of the cell wall in the member I–J can be computed by adding the contributions from the axial and bending deformation as:

$$
W^{IJ} = \frac{E_s}{2} \begin{bmatrix} u_t^I & u_t^J \end{bmatrix} \begin{bmatrix} 1 & -1 \\ -1 & 1 \end{bmatrix} \begin{Bmatrix} u_t^I \\ u_t^J \end{Bmatrix}
$$

$$
+ \frac{E_s h^3}{24L} \begin{bmatrix} u_n^I & \phi^I & u_n^J & \phi^J \end{bmatrix} \begin{bmatrix} 12/L^2 & 6/L & -12/L^2 & 6/L \\ 6/L & 4 & -6/L & 2 \\ -12/L^2 & -6/L & 12/L^2 & -6/L \\ 6/L & 2 & -6/L & 4 \end{bmatrix} \begin{bmatrix} u_n^I \\ \phi^I \\ u_n^J \\ \phi^J \end{bmatrix}
\tag{2.29}
$$

where E_s is the modulus of elasticity of the solid cell-wall material, u_n and u_t are displacement components along the normal (n) and tangential (t) directions, respectively, I and J are indices of the beam joints, and h is the beam thickness. In the above equation, the strain energy due to shear deformation is ignored, and only the strain energy due to axial and bending deformation is stated. Using the summation of the strain energy of each constituent of the RVE, we can calculate the total strain energy of the unit cell. The constitutive equation of the equivalent micropolar medium can thus be obtained from Eq. (2.28), and the equivalent elastic constants of the micropolar medium can be written as:

$$\overline{E}_{xx} = \frac{C_{1111}C_{2222} - C_{1122}C_{1122}}{C_{2222}} \qquad \overline{E}_{yy} = \frac{C_{1111}C_{2222} - C_{1122}C_{1122}}{C_{1111}}$$

$$\overline{v}_{xy} = -\frac{C_{1122}}{C_{2222}} \qquad \overline{v}_{yx} = -\frac{C_{1122}}{C_{1111}}$$

$$\overline{G} = \frac{C_{1212} + C_{1221}}{2} \tag{2.30}$$

Micropolar theory combined with strain energy principles can only be applied to unit cells with a specific structural layout that contains a single joint at the centre of the unit cell. For the analysis of lattice materials with micropolar elasticity, user-defined micropolar elements should be defined. This additional step is required since each node has a microrotation component that acts as an additional degree of freedom.

2.8 Homogenization via Bloch Wave Analysis and the Cauchy–Born Hypothesis

Commonly used to study materials with periodic crystal structure, Bloch wave analysis and the Cauchy–Born hypothesis can be combined with concepts of structural mechanics, in particular equilibrium and kinematic analysis, to calculate the homogenized properties of lattice materials [49, 111, 112]. Bloch's theorem was first introduced to describe the transport of electron particles within the crystal structure of a solid [113]. It is also often used to describe periodic waves propagating in periodic lattices and to find mechanisms corresponding to periodic displacements at the lattice joints. It allows identification of collapse mechanisms and states of self-stress generated in the joints of a lattice material subjected to a uniform macroscopic strain. Details of the background and applications of Bloch's theorem in a wide range of problems in theoretical physics can be found in the literature [113–115]. The Cauchy–Born hypothesis, on the other hand, looks at macroscopic mechanisms induced by an applied strain. The process requires the microscopic nodal deformations of the lattice to be expressed as a function of a macroscopic strain field. In particular, the Cauchy–Born hypothesis [12, 116–118] states that the infinitesimal displacement field of a periodic lattice is the summation of two parts, namely the deformation obtained by a macroscopically-homogeneous strain field $\overline{\varepsilon}$, and the periodic displacement field of each unit cell [49]. This is formally expressed as:

$$\mathbf{d}(\mathbf{j}_l + \overrightarrow{\mathbf{R}}, \overline{\varepsilon}) = \mathbf{d}(\mathbf{j}_l, \overline{\varepsilon} = 0) + \mathbf{R}\overline{\varepsilon} \tag{2.31}$$

where $\mathbf{j}_l$ and $\mathbf{j}_l + \overrightarrow{\mathbf{R}}$ are the position vectors of two periodic joints i and j within a lattice structure, $\mathbf{d}(\mathbf{j}_l, \overline{\varepsilon} = 0)$ is the periodic displacement field of joint $\mathbf{j}_l$, and $\mathbf{R}$ is the matrix that relates the macroscopic strain to the displacement of the lattice joints. Using the transformation matrix $\mathbf{T}_d$ obtained with Bloch's theorem [49], we can reduce, $\mathbf{d}$, the vector of the periodic nodal displacements, to its reduced form, $\tilde{\mathbf{d}}$, as:

$$\mathbf{d} = \mathbf{T}_d \tilde{\mathbf{d}} \tag{2.32}$$

Substituting Eq. (2.32) into Eq. (2.31) yields

$$\mathbf{d} = \mathbf{T}_d \tilde{\mathbf{d}} + \mathbf{R}\overline{\varepsilon} \tag{2.33}$$

We can thus write the kinematic system of a lattice as:

$$\mathbf{B}.\mathbf{d} = \mathbf{e} \tag{2.34}$$

where $\mathbf{B}$ is the kinematic matrix governing the joints' displacement vector, $\mathbf{d}$, as a function of the vector, $\mathbf{e}$, of the bars' deformation. Substituting Eq. (2.33) into the kinematic system of the unit cell, Eq. (2.34), results in the following Cauchy–Born condition:

$$\mathbf{B}\left\{\mathbf{T}_d\tilde{\mathbf{d}} + \mathbf{R}\bar{\varepsilon}\right\} = \mathbf{e} \tag{2.35}$$

Equation (2.35) establishes an explicit relation between the microscopic nodal displacements and a homogeneous macroscopic field of strain, $\bar{\varepsilon}$. Equation (2.35) has been successfully applied to regular lattices: those in which the unit cell is a regular polygon, such as the square and the triangle. Planar lattices, however, with more complex arrangements of cells, with arbitrary cell topology, also exist. These appear, for example, when two or more polygons of dissimilar size can be used to construct a lattice, or when more than one joint connectivity is possible. In this case if the primitive cell used to tessellate the plane – the cell envelope – intersects the unit cell struts, then complete information about its nodal periodicity is required. For this purpose, a "dummy node scheme" has been introduced to enable the analysis of lattice materials with any type of cell topology [49]. From Eq. (2.35), we can write the relation between the macroscopic strain field, $\bar{\varepsilon}$, and the reduced vectors of the element deformations, $\tilde{\mathbf{e}}$, via the matrix $\mathbf{H}$ [49] as:

$$\tilde{\mathbf{e}} = \mathbf{H}\bar{\varepsilon} \tag{2.36}$$

Computing the null space of $\mathbf{H}$ gives the independent modes of the macroscopic strain field compatible with zero member extensions. If the null space of $\mathbf{H}$ is zero, the lattice material can support all macroscopic modes induced by the applied strain field. Through the transformation matrix $\mathbf{e} = \mathbf{T}_e\tilde{\mathbf{e}}$ that relates the full vector of element deformations to its respective reduced periodic vectors, the deformations of each element in the unit cell can be written in terms of macroscopic strain as

$$\mathbf{e} = \mathbf{T}_e\mathbf{H}\bar{\varepsilon} \tag{2.37}$$

Equation (2.37) can be also used to obtain the effective yield strength of the unit cell from applied macroscopic strain. The deformation of each element in the unit cell can be computed, and the local microscopic stresses are obtained from Hooke's law. These microscopic stresses can then be compared with the material yield strength to measure the effective yield strength.

From Eq. (2.37) the macroscopic strain energy density of the lattice can be calculated and used to obtain the homogenized stiffness matrix. Given the cell element deformation of Eq. (2.37), the strain energy density of the unit cell can be obtained as [119]:

$$W = \frac{1}{2}\bar{\sigma}\bar{\varepsilon} = \frac{1}{2|Y|}\sum_{k=1}^{b}t_k e_k \tag{2.38}$$

where $\bar{\sigma}$ is the macroscopic stress field, $|Y|$ is the unit cell area, and is the tension force in a bar element. If the lattice is pin-jointed, the bar elements carry only axial loads, and hence the tension is expressed as:

$$t_k = \left(\frac{E_s A}{L}\right)e_k \tag{2.39}$$

where E_s is the Young's modulus of the solid material, A is the cross-sectional area of the bar element, and L is the bar length. Substituting Eq. (2.39) and then Eq. (2.37) into Eq. (2.38) results in:

$$W = \frac{1}{2}\overline{\sigma} : \overline{\varepsilon} = \frac{E_s A}{2L|Y|} \sum_{k=1}^{b} (\mathbf{H}(k,:)\overline{\varepsilon})^2 \tag{2.40}$$

where $\mathbf{H}(k,:)$ is the k-th row in the matrix $\mathbf{H}$. The effective elasticity tensor can then be obtained by partial differentiation of the strain energy density with respect to the macroscopic strain as:

$$\overline{C}_{iijj} = \frac{\partial^2 W}{\partial \overline{\varepsilon}_{ii} \partial \overline{\varepsilon}_{jj}} \tag{2.41}$$

The above formulation was used to determine the stiffness properties of ten pin-jointed planar lattices [48, 49, 111]. It was shown that the stiffness matrices of stretching-dominated structures are full rank, whereas for bending-dominated structures, such as square and hexagonal lattices, the stiffness matrices are singular. Therefore, pin-jointed lattices are mechanisms under certain loading conditions unless the joints are assumed rigid; in this case, the lattice becomes bending-dominated. An extension to characterize rigid-jointed lattices via the Cauchy–Born hypothesis has been proposed and successfully applied to planar lattices [111, 120]. In this case, cell members are modeled as beams while rotations at the cell joints are considered as independent kinematic variables. The bending moments are then accounted for in the equilibrium equation of the cell with the corresponding stiffness contribution. A modified Cauchy–Born hypothesis should be used to express the microscopic nodal deformations in terms of a homogeneous macroscopic strain field applied to the lattice material. After calculating the nodal forces and deformations, the principle of virtual work can be applied to derive the homogenized stiffness of the material. So far this approach has been developed by assuming cell walls as beam/rod elements. Similar to the force-based approach, these assumptions limit its application to low relative densities ($\rho < 0.3$).

2.9 Multiscale Matrix-based Computational Technique

In the general area of multiscale modelling of multi-phase materials, especially random media, there are several computational schemes [33, 96, 121–135] that can be adopted to model lattice materials. These approaches, often addressed as global–local analyses, are generally applied to heterogeneous media and develop constitutive relationships from the analysis of the RVE. They generally do not develop closed-form expressions of the constitutive equations and work on a two-level process. At the macroscale, an FE model of the component is solved by assuming the material is homogeneous. At the microscale, the boundary problem of the RVE subject to the boundary conditions generated by the macroscopic model is numerically solved to find the stress–strain relationship at every integration point of the macroscale model. The macroscopic stress is then found as the average of the microscopic stress on the RVE.

Based on this local-global analysis, a multiscale scheme has been recently developed and applied to pin-jointed and rigid-jointed lattices [2, 51, 52, 136]. The procedure assumes that the macroscopic stress of a lattice can be determined as the gradient of the strain energy density with respect to the components of the macroscopic displacement gradient. The kinematic assumptions imposed are on the deformation of the lattice; in particular, during the deformation, the periodic directions should be congruent with the macroscopic displacement gradient. In addition, the RVE is subject to the classical periodic equilibrium conditions. This procedure has been used to obtain the effective elastic constants and yield strength of a library of 2D and 3D lattice topologies [51, 52].

Figure 2.3 summarizes the main steps of the multiscale procedure. At the microscopic level, after expressing the nodal degrees of freedom (DoFs) of the unit cell as a function of the components of the macrostrain field, the internal forces in the lattice members are determined to verify whether the solid material of the cells fails. Two boundary value problems are considered: one at the component level, and the other at the microscopic level. The solution is found by defining proper relations between the micro and macroscale models. The steps of the procedure are as follows:

1. At the macroscopic level, the components of the Cauchy strain tensor, ε_M, are obtained from the displacements of the continuous medium, $\mathbf{u}_M$.
2. Under the action of a uniform macroscopic strain field, the lattice deforms. The deformed periodic directions can be related to the components of the macroscopic strain tensor by means of [51, 137]:

$$\mathbf{a}'_i = (\mathbf{I} + \varepsilon_M)\mathbf{a}_i \qquad (2.42)$$

where $\mathbf{a}'_i$ and $\mathbf{a}_i$ are the initial and deformed lattice periodic vectors, respectively, and $\mathbf{I}$ is the unit tensor. It is assumed that although the macroscopic strain distorts the lattice, after deformation the microtruss remains periodic, and the deformed tessellation vectors comply with the macroscopic strain.
3. The microscopic displacement and strain fields of the unit cell can be determined from the change in the periodic directions. By means of the finite element method, the nodal forces of the unit cell, $\mathbf{f}$, can be expressed in terms of the nodal DoFs as:

$$\mathbf{f} = \mathbf{K}_{uc}\mathbf{d} \qquad (2.43)$$

where $\mathbf{K}_{uc}$ is the unit cell stiffness matrix, and $\mathbf{d}$, is the array of its nodal DoFs.
4. The microscopic stress in the solid material is then obtained via the Hooke's law.

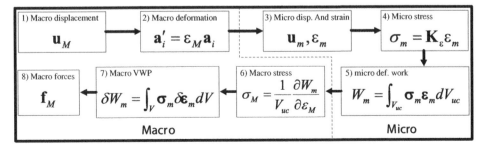

Figure 2.3 Multiscale scheme [51, 52]. VWP, virtual work principle.

5. The deformation work of the unit cell, W_m, is calculated by means of the finite element model of the unit cell, after applying periodic boundary conditions and solving the equilibrium problem of the unit cell:

$$W_m = \frac{1}{2A_{uc}} \boldsymbol{\varepsilon}^T \mathbf{D}_e^T \mathbf{K}_{uc} \mathbf{D}_e \boldsymbol{\varepsilon} \tag{2.44}$$

where A_{uc} is the area of the unit cell and $\mathbf{D}_e$ effectively links the components of the macroscopic strain to the DoFs of the unit cell nodes.

6. The macroscopic stress tensor is calculated as the gradient of the strain energy density with respect to the macroscopic strain;

7 and 8. The macroscopic forces are obtained by applying the virtual work principle at the component level.

This procedure can be used to account also for geometrical and material nonlinearity, and has no limitations in terms of the cell topology and range of relative density. The approach has been applied to both the linear and nonlinear analysis of lattice materials [2, 51, 52, 58, 59]. The model can also capture local bucking of cell struts under multiaxial macroscopic loading conditions, and thus is able to predict bifurcation points on the loading path. We recall that – in contrast to other approaches relying on the Taylor series of the displacement field or the Cauchy–Born rule for the approximation of the displacements within the repeating cell – this method does not make any kinematic assumption on the internal points; instead, conditions are applied to the boundary points of the unit cell only. In addition, this approach does not resort to micropolar theory to determine the lattice nodal rotations; the rotational DoFs of the cell nodes are evaluated by enforcing periodic equilibrium conditions on the unit cell. As with any RVE approach, however, this scheme cannot be used to model the impact of randomly distributed imperfections within the lattice.

2.10   Homogenization based on the Equation of Motion

The analysis of the dynamic response of a lattice can be used to calculate its equivalent mechanical properties [50, 138]. An example is given by the extraction of the elastostatic properties of a lattice from the analysis of their wave propagation and frequency band gaps. As recently demonstrated, the equivalent properties of a lattice material can be obtained through the analysis of the partial differential equations (PDEs) of motion associated with its homogenized continuum [50]. The discrete lattice equations can be formulated by an FE discretization of the unit cell, and then homogenized by expressing the discrete vector of DoFs as a continuous vector of the generalized displacements through a Taylor series expansion. Equivalent mechanical properties for the lattice can then be obtained by a direct comparison between the coefficients of the homogenized equations and the elasticity equations of an equivalent continuum material. The method enables the continuum equations to be obtained for periodic media with any periodicity direction. For this purpose, a "lattice (transformation) matrix" has been proposed; this maps lattice periodic vectors to a Cartesian reference frame. For a 2D periodic lattice, this process enables the quadrilateral geometry inscribing the unit cell to be mapped to

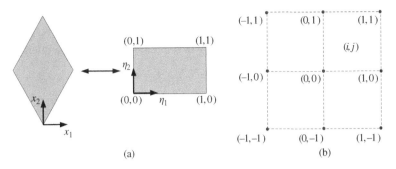

Figure 2.4 Lattice (transformation) matrix: (a) mapping a unit cell into the lattice space; (b) lattice topology in the lattice space.

a square of unit area, as shown in Figure 2.4. The components of the position vector in the two frames can be related to:

$$[x_1, x_2]^T = \mathbf{P}[\eta_1, \eta_2]^T \tag{2.45}$$

where x_1 and x_2 are the position of a material point in the RVE with respect to the 2D Cartesian reference frame, and η_1 and η_2 are the coordinates of a given material point with respect to a frame defined by the lattice vectors.

For the cell topology shown in Figure 2.4, the discretized equation of motion can be expressed in terms of mass and stiffness as [50]:

$$\sum_{m,n=-1}^{+1} \mathbf{K}^{(m,n)} \mathbf{u}^{(m,n)} + \mathbf{M}^{(m,n)} \ddot{\mathbf{u}}^{(m,n)} = \mathbf{f}^{(0,0)} \tag{2.46}$$

where $\mathbf{u}^{(m,n)}$ denotes the generalized displacements of a node, $\mathbf{f}^{(0,0)}$ is the vector of the generalized forces applied to node $(0,0)$, while $\mathbf{K}^{(m,n)}$ and $\mathbf{M}^{(m,n)}$ are proper partitions of the mass and stiffness matrices for the four-cell assembly in Figure 2.4 [139]. For the homogenization, the discrete vector of DoFs $\mathbf{u}^{(m,n)}$ is expressed by a continuous vector of generalized displacements $\mathbf{u} = \mathbf{u}(\eta_1, \eta_2)$ through the following Taylor series expansion:

$$\mathbf{u}^{(m,n)} \approx \mathbf{u}(\eta_1, \eta_2) + m\frac{\partial \mathbf{u}}{\partial \eta_1} + n\frac{\partial \mathbf{u}}{\partial \eta_2} + \frac{1}{2}m^2\frac{\partial^2 \mathbf{u}}{\partial \eta_1^2} + \frac{1}{2}n^2\frac{\partial^2 \mathbf{u}}{\partial \eta_2^2} + mn\frac{\partial^2 \mathbf{u}}{\partial \eta_1 \partial \eta_2} + \cdots \tag{2.47}$$

Substituting Eq. (2.47) into Eq. (2.46) yields a system of PDEs, whose dimension depends on the number of DoFs for the lattice nodes. Using the lattice (transformation) matrix – in other word, matrix $\mathbf{P}$ in Eq. (2.45) – the equivalent continuum equations can be expressed in the Cartesian frame x_1 and x_2. Direct comparison between the coefficients of the homogenized equations and the elasticity equations of an equivalent elastic domain allows the identification of the equivalent mechanical properties of a lattice. The homogenization procedure has been applied to estimate the in-plane properties for hexagonal and re-entrant (auxetic) lattices [50]. Equivalent Young's moduli, Poisson's ratios, and relative density were estimated and compared with the closed-form expressions available in literature.

One advantage of this method is that it can directly provide the inertial terms as well as other dynamic characteristics for vibration analysis. One downside is the effort that might be required to obtain closed-form equivalent properties for cells with complex geometries, although numeric PDE methods can be used. Another aspect of the method is that currently it has not been developed to calculate the microscopic stress distribution in the cells that results from a macroscopic strain applied to the lattice. This might reduce the scope of application to problems where no assessment of the microscopic behavior of the cell is required.

2.11   Case Study: Property Predictions for a Hexagonal Lattice

In this section, we apply the approaches described above to a planar lattice with regular hexagonal cells of uniform thickness. The aim of this basic case study is to offer a comparison between the predictions provided by each method, in particular their degree of closeness. Some approaches lose accuracy with complex cell geometries, but we select here a simple planar hexagonal lattice, aware that other cell topologies might better emphasize dissimilarities among the approaches. The focus of this study, nevertheless, is on the terms of the effective stiffness matrix and yield strength properties obtained at discrete increments of relative density.

As described in Section 2.3, the impact of applying a given boundary condition has been sometimes overlooked in the literature. We therefore start with this aspect in Figure 2.5, where we plot the results of alternative boundary conditions on the static response of a hexagon. The DBC and NBC provide, respectively, upper and lower bounds, within which lie the homogenized properties obtained with the PBC (and MBC). A uniform displacement (DBC) applied to the RVE constrains the cell to maintain its plane section, thereby over-constraining the RVE; as a result, the hexagon's properties are predicted to be stiffer than expected [60, 140, 141]. On the other hand,

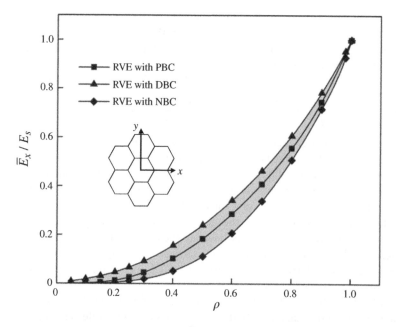

Figure 2.5 Effective Young's modulus, $\bar{E}_x$, as function of relative density for the hexagonal cell. Results obtained with the DBC and NBC bound those calculated with the PBC.

with the application of a uniform traction (NBC), we observe two factors. First, cell periodicity and displacement continuity are not guaranteed at the RVE boundaries. Second, the magnitude of the traction distributions applied to one side of the hexagon is dissimilar from that at the other side. These conditions result in an under-constrained RVE with lower stiffness. Besides impacting the stiffness properties, the DBC and NBC provide predictions that depend on the RVE size (and shape), in contrast to the application of the PBC. In general, it has been demonstrated that an increase in the number of cells defining the RVE drives the results obtained with the DBC and NBC to converge (respectively from above and below) to those obtained with the PBC [59, 90]. The reason for this is that the PBC guarantees the continuity of the boundary traction as well as the unicity of the solution.

Figure 2.6 shows the results obtained with the homogenization methods investigated here. The effective Young's modulus, shear modulus, and Poisson ratio are normalized with respect to the solid material and plotted as a function of relative density up to the value of 0.3. The reason for avoiding higher densities is that some methods assume the struts to behave as Euler–Bernoulli beams, which is accurate for low values of relative density only. See, for example, the results obtained with the force-based approach [27]: for relative density above 0.3, the elements are no longer slender and the results tend to lose accuracy. With AH and the surface- and volume-based approaches, on the other hand, the cell wall can be modelled with continuum (here meshed as planar (eight-node)) elements, without the range of relative density being restricted to below 0.3. Another advantage of AH, as well as surface- and volume-based methods is that FE analysis can be conveniently used to capture with high accuracy material deformation and stress localization throughout the cell walls and at the nodes.

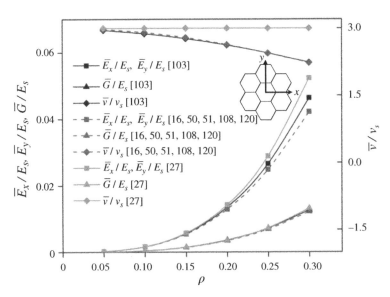

Figure 2.6 Effective elastic constants obtained with: force-based approach [27], asymptotic homogenization method [103], generalized continuum theory [16, 108], Bloch's theorem [120], matrix-based approach [51], and homogenization based on the equations of motion [50].

The properties plotted in Figure 2.6 confirm that the assumption of Euler–Bernoulli theory holds well for low relative densities. The results show that the effective mechanical properties converge to the values obtained with AH, as well as surface- and volume-based approaches with the PBC. We can also observe that for the hexagon no difference exists between these approaches. In contrast, dissimilarity in their ability to capture local phenomena would emerge in critical regions of high gradients, where macroscopic fields can rapidly vary, which is not the case for this simple study. For instance, if high variations of macroscopic strain were to occur, then AH would provide reasonable results, especially if higher-order terms are accounted for in the asymptotic expansion of a macroscopic field, such as displacement. In the example of the hexagon here, we considered only the first term, which is sufficient to define an effective constitutive relation with a zero-order approximation for the stress field [130, 142].

Figure 2.7 shows the relative percentage difference among the effective properties obtained with the selected methods. AH is the baseline for comparison and is used to normalize the properties predicted with the others. We chose AH as the benchmark because it is accurate with any cell geometry throughout the whole range of relative densities. Methods that assume cell elements to behave as Euler–Bernoulli beams, on the other hand, lose accuracy with increasingly higher values of relative density. This is the case for the properties obtained with a force-based approach, where the flexural deformation is considered as the main mode withstood by the cell elements. Other approaches, including generalized continuum homogenization and Bloch's theorem, account for both flexural and stretching deformation modes, thereby providing more accurate results. The difference in accuracy is evident in the computation of the effective

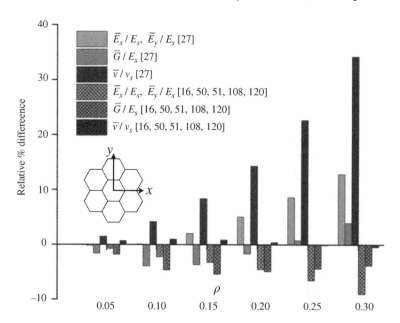

Figure 2.7 Relative difference of the effective elastic constants obtained with the force-based approach [27], generalized continuum theory [16, 108], Bloch's theorem [120], matrix-based approach [51], and homogenization based on equation of motion [50]. AH is used as the baseline.

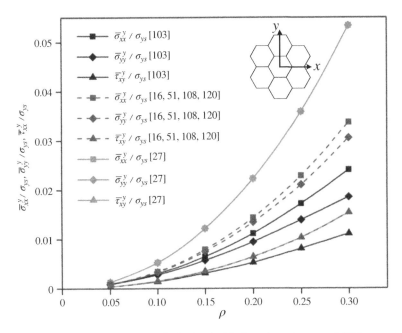

Figure 2.8 Effective yield strength obtained with: force-based approach [27], asymptotic homogenization method [103], generalized continuum theory [16, 108], Bloch's theorem [120], and matrix-based approach [51].

Poisson ratio. Compared to AH, the force-based approach yields a 34% difference at 30% relative density, while this percentage is below 1 with the other methods.

Similar to the effective stiffness, we now examine yield strength. Figure 2.8 illustrates the initial yield strength of the hexagonal cell under uniaxial and pure shear stresses. Among the homogenization methods described here, we notice that only the homogenization based on the equations of motion cannot capture the effective yield strength of the hexagonal lattice. Similar to the trends observed for stiffness, yield strength predictions diverge from the AH results as relative density increases. On the other hand, for low relative densities, the results converge to those of AH.

The assumptions used to model the deformation mechanisms in the hexagon members have an impact also on how the lattice responds to the direction of loading. For example, if cell members are assumed to undergo pure bending only [27], the yield strength would not change in the x and y directions, leading to an isotropic response [27]. In contrast, if the cell walls are allowed to deform axially (although the magnitude of axial deformation might be much lower than the bending mode), then the yield strength changes with the loading direction, resulting in an anisotropic response [16, 51, 103, 108, 120]. Since the computational model used with AH can generally capture all deformation modes, including bending, axial, and significant shear deformation at cell edges and vertices, and the resulting stress distribution reflects the actual response of the unit cell with a degree of accuracy higher than that provided by other methods, such as the force-based approach, generalized continuum theory, Bloch's theorem, and the matrix-based approach.

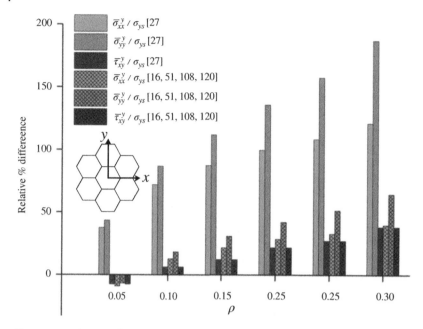

Figure 2.9 Relative difference of the yield strength obtained with force-based approach [27], generalized continuum theory [16, 108], Bloch's theorem [120], and matrix-based approach [51]. Effective property values are normalized by AH results.

Similar to Figure 2.7, Figure 2.9 shows the relative difference for yield strength. The results of AH theory are used as the benchmark to normalize the properties obtained with the other methods. We recall that this choice is motivated by the assumptions of treating cell members as continuum elements with sufficiently refined mesh density, so as to capture all significant deformation modes both at the edges and vertices of the unit cell. This is essential, especially for high values of relative density. The results for yield in Figure 2.9 resemble – to certain extent – those for stiffness. The force-based approach overpredicts the in-plane strength properties, since beam elements used to model the cell walls cannot capture any stress concentration at the cell nodes. For instance, at 30% relative density, the force-based approach overpredicts the uniaxial yield strength of a factor of 2, which is more than 100% off the results of AH. The accuracy can be improved if model refinements are implemented. For example, if axial forces, besides bending moments, are modeled in the cell members, the relative error for the prediction of uniaxial yield strength can decrease to about 50% for a relative density of 30%. However, for higher relative densities, the relative error rises, since the deformation of the solid material is neglected at the cell nodes. This observation is critical and cannot be overlooked if the lattice material is to be designed to withstand cyclic loading [93, 143, 144]

2.12 Conclusions

The study of the mechanics of periodic cellular materials can be considered to fall within the realm of composite materials, where the concept of RVE was first introduced. Several homogenization methods have been developed from a fairly wide range of disciplines. This chapter has provided a brief review of a set of methods that can be used to study the elastostatics of a lattice. We have started by examining the definition of RVE, along with the selection of the boundary conditions that can be applied to it. In particular, the focus has been on the impact of boundary conditions on the effective property bounds. Then for each method, we have discussed the assumptions, reviewed the main steps, and highlighted advantages and limitations.

Relative density and cell element assumptions play an important role in the elastostatic response of a lattice. It is thus critical to assume the proper model of the cell members with respect to the relative density of the lattice. For low relative densities, the force-based approach and generalized continuum theory, where slender beams or rod elements are generally used, provide accurate results, whereas continuum elements should be used for high values of relative density, usually done with AH. The case study closing the chapter has tested the degree of closeness of the methods in the analysis of a hexagonal lattice.

References

1 Ashby M. The properties of foams and lattices. *Philosophical Transactions A.* 2006;**364**(1838):15.
2 Vigliotti A, Deshpande VS, Pasini D. Non linear constitutive models for lattice materials. *Journal of the Mechanics and Physics of Solids.* 2014;**64**(0):44–60.
3 Deshpande V, Ashby M, Fleck N. Foam topology: bending versus stretching dominated architectures. *Acta Materialia.* 2001;**49**(6):1035–40.
4 Maxwell JC. L. on the calculation of the equilibrium and stiffness of frames. *The London, Edinburgh, and Dublin Philosophical Magazine and Journal of Science.* 1864;**27**(182):294–9.
5 Calladine C. Buckminster Fuller's "tensegrity" structures and Clerk Maxwell's rules for the construction of stiff frames. *International Journal of Solids and Structures.* 1978;**14**(2):161–72.
6 Pellegrino S, Calladine CR. Matrix analysis of statically and kinematically indeterminate frameworks. *International Journal of Solids and Structures.* 1986;**22**(4):409–28.
7 Pellegrino S. Structural computations with the singular value decomposition of the equilibrium matrix. *International Journal of Solids and Structures.* 1993;**30**(21):3025–35.
8 Gibson L, Ashby M. *Cellular solids: structure and properties: Cambridge University Press;* 1999.
9 Masters I, Evans K. Models for the elastic deformation of honeycombs. *Composite Structures.* 1996;**35**(4):403–22.
10 Christensen RM. Mechanics of cellular and other low-density materials. *International Journal of Solids and Structures.* 2000;**37**(1–2):93–104.
11 Wang A, McDowell D. In-plane stiffness and yield strength of periodic metal honeycombs. *Journal of Engineering Materials and Technology.* 2004;**126**(2):137–56.
12 Askar A, Cakmak A. A structural model of a micropolar continuum. *International Journal of Engineering Science.* 1968;**6**(10):583–9.

13 Chen J, Huang M. Fracture analysis of cellular materials: a strain gradient model. *Journal of the Mechanics and Physics of Solids*. 1998;**46**(5):789–828.

14 Bazant Z, Christensen M. Analogy between micropolar continuum and grid frameworks under initial stress. *International Journal of Solids and Structures*. 1972;**8**(3):327–46.

15 Kumar R, McDowell D. Generalized continuum modeling of 2-D periodic cellular solids. *International Journal of Solids and Structures*. 2004;**41**(26):7399–422.

16 Wang X, Stronge W. Micropolar theory for two-dimensional stresses in elastic honeycomb. *Proceedings: Mathematical, Physical and Engineering Sciences*. 1999:**2091–116**.

17 Warren W, Byskov E. Three-fold symmetry restrictions on two-dimensional micropolar materials. *European Journal of Mechanics-A/Solids*. 2002;**21**(5):779–92.

18 Hassani B, Hinton E. A review of homogenization and topology optimization I – Homogenization theory for media with periodic structure. *Computers and Structures*. 1998;**69**(707):707–17.

19 Hassani B, Hinton E. A review of homogenization and topology optimization I I – Analytical and numerical solution of homogenization equations. *Computers & Structures*. 1998;**69**(6):719–38.

20 Fang Z, Starly B, Sun W. Computer-aided characterization for effective mechanical properties of porous tissue scaffolds. *Computer-Aided Design*. 2005;**37**(1):65–72.

21 Wang W-X, Luo D, Takao Y, Kakimoto K. New solution method for homogenization analysis and its application to the prediction of macroscopic elastic constants of materials with periodic microstructures. *Computers & Structures*. 2006;**84**(15–16):991–1001.

22 Fang Z, Sun W, Tzeng J. Asymptotic homogenization and numerical implementation to predict the effective mechanical properties for electromagnetic composite conductor. *Journal of Composite Materials*. 2004;**38**(16):1371.

23 Andrews E, Gioux G, Onck P, Gibson L. Size effects in ductile cellular solids. Part II: Experimental results. *International Journal of Mechanical Sciences*. 2001;**43**(3):701–13.

24 Foo C, Chai G, Seah L. Mechanical properties of Nomex material and Nomex honeycomb structure. *Composite Structures*. 2007;**80**(4):588–94.

25 Cheng G-D, Cai Y-W, Xu L. Novel implementation of homogenization method to predict effective properties of periodic materials. *Acta Mechanica Sinica*. 2013;**29**(4):550–6.

26 Wang C, Feng L, Jasiuk I. Scale and boundary conditions effects on the apparent elastic moduli of trabecular bone modeled as a periodic cellular solid. *Journal of Biomechanical Engineering*. 2009;**131**(12):121008-.

27 Gibson LJ, Ashby MF. *Cellular solids: structure and properties*: Cambridge University Press; 1999.

28 Wang A, McDowell D. Yield surfaces of various periodic metal honeycombs at intermediate relative density. *International Journal of Plasticity*. 2005;**21**(2):285–320.

29 Sab K, Pradel F. Homogenisation of periodic Cosserat media. *International Journal of Computer Applications in Technology*. 2009;**34**(1):60–71.

30 Eringen AC. Linear theory of micropolar elasticity. *Journal of Mathematics and Mechanics*. 1966;**15**(6):909–23.

31 Lakes RS. Strongly Cosserat elastic lattice and foam materials for enhanced toughness. *Cellular Polymers.* 1993;**12**:17.

32 Hassani B, Hinton E. A review of homogenization and topology optimization I -- Homogenization theory for media with periodic structure. *Computers & Structures.* 1998;**69**(6):707–17.

33 Guedes J, Kikuchi N. Preprocessing and postprocessing for materials based on the homogenization method with adaptive finite element methods. *Computer methods in applied mechanics and engineering.* 1990;**83**(2):143–98.

34 Kalamkarov AL, Andrianov IV, Danishevs'kyy VV. Asymptotic homogenization of composite materials and structures. *Applied Mechanics Reviews.* 2009;**62**:030802.

35 Lin CY, Kikuchi N, Hollister SJ. A novel method for biomaterial scaffold internal architecture design to match bone elastic properties with desired porosity. *Journal of Biomechanics.* 2004;**37**(5):623–36.

36 Sturm S, Zhou S, Mai Y-W, Li Q. On stiffness of scaffolds for bone tissue engineering – a numerical study. *Journal of Biomechanics.* 2010;**43**(9):1738–44.

37 Hollister SJ. Porous scaffold design for tissue engineering. *Nature Materials.* 2005;**4**(7):518–24.

38 Hassani B, Hinton E. A review of homogenization and topology optimization I I I – Topology optimization using optimality criteria. *Computers & Structures.* 1998;**69**(6):739–56.

39 Sigmund O. Materials with prescribed constitutive parameters: an inverse homogenization problem. *International Journal of Solids and Structures.* 1994;**31**(17):2313–29.

40 Bendsøe M. Optimal shape design as a material distribution problem. *Structural and Multidisciplinary Optimization.* 1989;**1**(4):193–202.

41 Kikuchi N, Bendsoe M. Generating optimal topologies in structural design using a homogenization method. *Computer methods in applied mechanics and engineering.* 1988;**71**(2):197–224.

42 Takano N, Ohnishi Y, Zako M, Nishiyabu K. Microstructure-based deep-drawing simulation of knitted fabric reinforced thermoplastics by homogenization theory. *International Journal of Solids and Structures.* 2001;**38**(36):6333–56.

43 Takano N, Zako M, Kubo F, Kimura K. Microstructure-based stress analysis and evaluation for porous ceramics by homogenization method with digital image-based modeling. *International Journal of Solids and Structures.* 2003;**40**(5):1225–42.

44 Guinovart-Díaz R, Rodríguez-Ramos R, Bravo-Castillero J, Sabina FJ, Dario Santiago R, Martinez Rosado R. Asymptotic analysis of linear thermoelastic properties of fiber composites. *Journal of thermoplastic composite materials.* 2007;**20**(4):389–410.

45 Guinovart-Diaz R, Bravo-Castillero J, Rodriguez-Ramos R, Martinez-Rosado R, Serrania F, Navarrete M. Modeling of elastic transversely isotropic composite using the asymptotic homogenization method. Some comparisons with other models. *Materials Letters.* 2002;**56**(6):889–94.

46 Peng X, Cao J. A dual homogenization and finite element approach for material characterization of textile composites. *Composites Part B: Engineering.* 2002;**33**(1):45–56.

47 Visrolia A, Meo M. Multiscale damage modelling of 3D weave composite by asymptotic homogenisation. *Composite Structures*. 2013;**95**(0):105–13.

48 Hutchinson R, Fleck N. The structural performance of the periodic truss. *Journal of the Mechanics and Physics of Solids*. 2006;**54**(4):756–82.

49 Elsayed MSA, Pasini D. Analysis of the elastostatic specific stiffness of 2D stretching-dominated lattice materials. *Mechanics of Materials*. 2010;**42**(7):709–25.

50 Gonella S, Ruzzene M. Homogenization and equivalent in-plane properties of two-dimensional periodic lattices. *International Journal of Solids and Structures*. 2008;**45**(10):2897–915.

51 Vigliotti A, Pasini D. Linear multiscale analysis and finite element validation of stretching and bending dominated lattice materials. *Mechanics of Materials*. 2012;**46**(0):57–68.

52 Vigliotti A, Pasini D. Stiffness and strength of tridimensional periodic lattices. *Computer methods in Applied Mechanics and Engineering*. 2012;**229–232**(0):27–43.

53 Eshelby JD. The determination of the elastic field of an ellipsoidal inclusion, and related problems. *Proceedings of the Royal Society of London Series A: Mathematical and Physical Sciences*. 1957;**241**(1226):376–96.

54 Hill R. Elastic properties of reinforced solids: some theoretical principles. *Journal of the Mechanics and Physics of Solids*. 1963;**11**(5):357–72.

55 Hashin Z. *Theory of mechanical behavior of heterogeneous media*. DTIC Document, 1963.

56 Hashin Z. Analysis of composite materials. *Journal of Applied Mechanics*. 1983;**50**(2):481–505.

57 Nemat-Nasser S. *Overall Stresses and Strains in Solids with Microstructure*. Modelling Small Deformations of Polycrystals: Springer; 1986. p. 41–64.

58 Willis JR. Variational and related methods for the overall properties of composites. *Advances in Applied Mechanics*. 1981;**21**:1–78.

59 Yan J, Cheng G, Liu S, Liu L. Comparison of prediction on effective elastic property and shape optimization of truss material with periodic microstructure. *International Journal of Mechanical Sciences*. 2006;**48**(4):400–13.

60 Xia Z, Zhou C, Yong Q, Wang X. On selection of repeated unit cell model and application of unified periodic boundary conditions in micro-mechanical analysis of composites. *International Journal of Solids and Structures*. 2006;**43**(2):266–78.

61 Hohe J, Becker W. Effective stress-strain relations for two-dimensional cellular sandwich cores: Homogenization, material models, and properties. *Applied Mechanics Reviews*. 2002;**55**(1):61–87.

62 Bishop J, Hill R. XLVI. A theory of the plastic distortion of a polycrystalline aggregate under combined stresses. *Philosophical Magazine*. 1951;**42**(327):414–27.

63 Ponte Castaneda P, Suquet P. On the effective mechanical behavior of weakly inhomogeneous nonlinear materials. *European Journal of Mechanics A: Solids*. 1995;**14**(2):205–36.

64 Becker W. The in-plane stiffnesses of a honeycomb core including the thickness effect. *Archive of Applied Mechanics*. 1998;**68**(5):334–41.

65 Hohe J, Becker W. Determination of the elasticity tensor of non-orthotropic cellular sandwich cores. *Technische Mechanik*. 1999;**19**:259–68.

66 Voigt W. On the relation between the elasticity constants of isotropic bodies. *Annual Review of Physical Chemistry*. 1889;**274**:573–87.

67 Reuss A. Determination of the yield point of polycrystals based on the yield condition of sigle crystals. *Zeitschrift für Angewandte Mathematik und Mechanik.* 1929;**9**:49–58.

68 Pahr DH, Zysset PK. Influence of boundary conditions on computed apparent elastic properties of cancellous bone. *Biomechanics and Modeling in Mechanobiology.* 2008;**7**(6):463–76.

69 Warren W, Kraynik A, Stone C. A constitutive model for two-dimensional nonlinear elastic foams. *Journal of the Mechanics and Physics of Solids.* 1989;**37**(6):717–33.

70 Warren W, Kraynik A. Foam mechanics: the linear elastic response of two-dimensional spatially periodic cellular materials. *Mechanics of Materials.* 1987;**6**(1):27–37.

71 Overaker D, Cuitino A, Langrana N. Elastoplastic micromechanical modeling of two-dimensional irregular convex and nonconvex (re-entrant) hexagonal foams. *Journal of Applied Mechanics.* 1998;**65**(3):748–57.

72 Kelsey S, Gellatly R, Clark B. The shear modulus of foil honeycomb cores: A theoretical and experimental investigation on cores used in sandwich construction. *Aircraft Engineering and Aerospace Technology.* 1958;**30**(10):294–302.

73 Overaker DW, Cuitiño AM, Langrana NA. Effects of morphology and orientation on the behavior of two-dimensional hexagonal foams and application in a re-entrant foam anchor model. *Mechanics of Materials.* 1998;**29**(1):43–52.

74 Ueng CE, Kim TD. Shear modulus of core materials with arbitrary polygonal shape. *Computers & Structures.* 1983;**16**(1):21–5.

75 Gent A, Thomas A. The deformation of foamed elastic materials. *Journal of Applied Polymer Science.* 1959;**1**(1):107–13.

76 Lederman J. The prediction of the tensile properties of flexible foams. *Journal of Applied Polymer Science.* 1971;**15**(3):693–703.

77 Menges G, Knipschild F. Estimation of mechanical properties for rigid polyurethane foams. *Polymer Engineering & Science.* 1975;**15**(8):623–7.

78 Chan R, Nakamura M. Mechanical properties of plastic foams the dependence of yield stress and modulus on the structural variables of closed-cell and open-cell foams. *Journal of Cellular Plastics.* 1969;**5**(2):112–8.

79 Ko W. Deformations of foamed elastomers. *Journal of Cellular Plastics.* 1965;**1**(1):45–50.

80 Gibson L, Ashby M. The mechanics of three-dimensional cellular materials. *Proceedings of the Royal Society of London A Mathematical and Physical Sciences.* 1982;**382**(1782):43–59.

81 Gibson LJ, Ashby M, Schajer G, Robertson C. The mechanics of two-dimensional cellular materials. *Proceedings of the Royal Society of London A Mathematical and Physical Sciences.* 1982;**382**(1782):25–42.

82 Gent A, Thomas A. Mechanics of foamed elastic materials. *Rubber Chemistry and Technology.* 1963;**36**(3):597–610.

83 Simone A, Gibson L. Effects of solid distribution on the stiffness and strength of metallic foams. *Acta Materialia.* 1998;**46**(6):2139–50.

84 Sánchez-Palencia E., (ed). *Non-homogeneous Media and Vibration Theory.* Springer, 1980.

85 Bensoussan A, Lions J-L, Papanicolaou G. *Asymptotic Analysis for Periodic Structures: American Mathematical Society*; 2011.

86 Cioranescu D, Paulin JSJ. Homogenization in open sets with holes. *Journal of Mathematical Analysis and Applications*. 1979;**71**(2):590–607.

87 Sixto-Camacho LM, Bravo-Castillero J, Brenner R, Guinovart-Díaz R, Mechkour H, Rodríguez-Ramos R, et al. Asymptotic homogenization of periodic thermo-magneto-electro-elastic heterogeneous media. *Computers & Mathematics with Applications*. 2013;**66**(10):2056–74.

88 Zhang H, Zhang S, Bi J, Schrefler B. Thermo-mechanical analysis of periodic multiphase materials by a multiscale asymptotic homogenization approach. *International Journal for Numerical Methods in Engineering*. 2007;**69**(1):87–113.

89 Andrianov IV, Bolshakov VI, Danishevskyy VV, Weichert D. Higher order asymptotic homogenization and wave propagation in periodic composite materials. *Proceedings of the Royal Society A: Mathematical, Physical and Engineering Science*. 2008;**464**(2093):1181–201.

90 Hollister S, Kikuchi N. A comparison of homogenization and standard mechanics analyses for periodic porous composites. *Computational Mechanics*. 1992;**10**(2):73–95.

91 Hassani B. A direct method to derive the boundary conditions of the homogenization equation for symmetric cells. *Communications in Numerical Methods in Engineering*. 1996;**12**(3):185–96.

92 Jansson S. Homogenized nonlinear constitutive properties and local stress concentrations for composites with periodic internal structure. *International Journal of Solids and Structures*. 1992;**29**(17):2181–200.

93 Masoumi Khalil Abad E, Arabnejad Khanoki S, Pasini D. Fatigue design of lattice materials via computational mechanics: Application to lattices with smooth transitions in cell geometry. *International Journal of Fatigue*. 2013;**47**:126–36.

94 Matsui K, Terada K, Yuge K. Two-scale finite element analysis of heterogeneous solids with periodic microstructures. *Computers & Structures*. 2004;**82**(7–8):593–606.

95 Dumontet H. Study of a boundary layer problem in elastic composite materials. *RAIRO Modélisation Mathématique et Analyse Numérique*. 1986;**20**(2):265–86.

96 Ghosh S, Lee K, Raghavan P. A multi-level computational model for multi-scale damage analysis in composite and porous materials. *International Journal of Solids and Structures*. 2001;**38**(14):2335–85.

97 Lefik M, Schrefler B. F E modelling of a boundary layer corrector for composites using the homogenization theory. *Engineering Computations*. 1996;**13**(6):31–42.

98 Raghavan P, Ghosh S. Concurrent multi-scale analysis of elastic composites by a multi-level computational model. *Computer Methods in Applied Mechanics and Engineering*. 2004;**193**(6–8):497–538.

99 Takano N, Zako M, Okuno Y. Multi-scale finite element analysis for joint members of heterogeneous dissimilar materials with interface crack. *Zairyo*. 2003;**52**(8):952–7.

100 Takano N, Zako M, Okuno Y. Multi-scale finite element analysis of porous materials and components by asymptotic homogenization theory and enhanced mesh superposition method. *Modelling and Simulation in Materials Science and Engineering*. 2003;**11**(2):137–56.

101 Yuan F, Pagano N. Size Scales for Accurate Homogenization in the presence of severe stress gradients. *Mechanics of Advanced Materials and Structures.* 2003;**10**(4):353–65.

102 Kruch S. Homogenized and relocalized mechanical fields. *Journal of Strain Analysis for Engineering Design.* 2007;**42**(4):215–26.

103 Arabnejad S, Pasini D. Mechanical properties of lattice materials via asymptotic homogenization and comparison with alternative homogenization methods. *International Journal of Mechanical Sciences.* 2013;**77**(0):249–62.

104 Onck P, Andrews E, Gibson L. Size effects in ductile cellular solids. Part I: Modeling. *International Journal of Mechanical Sciences.* 2001;**43**(3):681–99.

105 Tekoglu C. Size effect in cellular solids: PhD thesis, Rijksunivesiteit, Netherlands, 2007.

106 Cosserat E, Cosserat F, Brocato M, Chatzis K. *Théorie des corps déformables: A.* Hermann Paris; 1909.

107 Eringen AC. *Microcontinuum Field Theories I: Foundations and Solids.* Springer; 1999.

108 Dos Reis F, Ganghoffer J-F. *Construction of micropolar continua from the homogenization of repetitive planar lattices.* Mechanics of Generalized Continua: Springer; 2011. p. 193–217.

109 Spadoni A, Ruzzene M. Elasto-static micropolar behavior of a chiral auxetic lattice. *Journal of the Mechanics and Physics of Solids.* 2012;**60**(1):156–71.

110 Onck P. Cosserat modeling of cellular solids. *Comptes Rendus Mécanique.* 2002;**330**:717–22.

111 Hutchinson RG. *Mechanics of lattice materials.* PhD Thesis, University of Cambridge; 2004.

112 Phani AS, Woodhouse J, Fleck N. Wave propagation in two-dimensional periodic lattices. *The Journal of the Acoustical Society of America.* 2006;**119**:1995.

113 Bloch F. Über die quantenmechanik der elektronen in kristallgittern. *Zeitschrift für Physik.* 1929;**52**(7–8):555–600.

114 Cornwell JF. *Group Theory in Physics: An Introduction: Academic Press*; 1997.

115 Lomont JS. *Applications of Finite Groups: Academic Press*; 1959.

116 Yang W, Li Z, Shi W, Xie B, Yang M. Review on auxetic materials. *Journal of Materials Science.* 2004;**39**(10):3269–79.

117 Sutradhar A, Paulino G, Miller M, Nguyen T. Topological optimization for designing patient-specific large craniofacial segmental bone replacements. *Proceedings of the National Academy of Sciences.* 2010;**107**(30):13222.

118 Liu Y, Hu H. A review on auxetic structures and polymeric materials. *Scientific Research and Essays.* 2010;**5**(10):1052–63.

119 Zhang T. A general constitutive relation for linear elastic foams. *International Journal of Mechanical Sciences.* 2008;**50**(6):1123–32.

120 Elsayed MS. *Multiscale mechanics and structural design of periodic cellular materials.* PhD Thesis, McGill University; 2012.

121 Terada K, Kikuchi N. Nonlinear homogenization method for practical applications. *ASME Applied Mechanics Division Publications.* 1995;**212**:1–16.

122 Ghosh S, Lee K, Moorthy S. Multiple scale analysis of heterogeneous elastic structures using homogenization theory and Voronoi cell finite element method. *International Journal of Solids and Structures.* 1995;**32**(1):27–62.

123 Ghosh S, Moorthy S. Elastic-plastic analysis of arbitrary heterogeneous materials with the Voronoi cell finite element method. *Computer Methods in Applied Mechanics and Engineering.* 1995;**121**(1–4):373–409.

124 Ghosh S, Lee K, Moorthy S. Two scale analysis of heterogeneous elastic-plastic materials with asymptotic homogenization and Voronoi cell finite element model. *Computer Methods in Applied Mechanics and Engineering.* 1996;**132**(1–2):63–116.

125 Smit R, Brekelmans W, Meijer H. Prediction of the mechanical behavior of nonlinear heterogeneous systems by multi-level finite element modeling. *Computer Methods in Applied Mechanics and Engineering.* 1998;**155**(1–2):181–92.

126 Miehe C, Schotte J, Schröder J. Computational micro-macro transitions and overall moduli in the analysis of polycrystals at large strains. *Computational Materials Science.* 1999;**16**(1–4):372–82.

127 Miehe C, Schroder J, Schotte J. Computational homogenization analysis in finite plasticity simulation of texture development in polycrystalline materials. *Computer Methods in Applied Mechanics and Engineering.* 1999;**171**(3–4):387–418.

128 Michel J, Moulinec H, Suquet P. Effective properties of composite materials with periodic microstructure: a computational approach. *Computer Methods in Applied Mechanics and Engineering.* 1999;**172**(1–4):109–43.

129 Chaboche J-L, Kruch S, Pottier T. Micromechanics versus macromechanics: a combined approach for metal matrix composite constitutive modelling. *European Journal of Mechanics - A/Solids.* 1998;**17**(6):885–908.

130 Kanouté P, Boso D, Chaboche J, Schrefler B. Multiscale methods for composites: a review. *Archives of Computational Methods in Engineering.* 2009;**16**(1):31–75.

131 Terada K, Kikuchi N. A class of general algorithms for multi-scale analyses of heterogeneous media. *Computer Methods in Applied Mechanics and Engineering.* 2001;**190**(40–41):5427–64.

132 Kouznetsova V, Brekelmans W, Baaijens F. An approach to micro-macro modeling of heterogeneous materials. *Computational Mechanics.* 2001;**27**(1):37–48.

133 Geers MGD, Kouznetsova VG, Brekelmans WAM. Multi-scale computational homogenization: Trends and challenges. *Journal of Computational and Applied Mathematics.* 2010;**234**(7):2175–82.

134 Kouznetsova V, Geers M, Brekelmans W. Multi scale constitutive modelling of heterogeneous materials with a gradient enhanced computational homogenization scheme. *International Journal for Numerical Methods in Engineering.* 2002;**54**(8):1235–60.

135 Suquet P. Local and global aspects in the mathematical theory of plasticity. *Plasticity Today: Modelling, Methods and Applications.* 1985:**279–310**.

136 Vigliotti A, Pasini D. Mechanical properties of hierarchical lattices. *Mechanics of Materials.* 2013;**62**(0):32–43.

137 Asaro RJ, Lubarda VA. Mechanics of Solids And Materials: Cambridge University Press; 2006.

138 Fish J, Chen W. Higher-order homogenization of initial/boundary-value problem. *Journal of Engineering Mechanics.* 2001;**127**(12):1223–30.

139 Brown G, Byrne K. Determining the response of infinite, one-dimensional, non-uniform periodic structures by substructuring using waveshape coordinates. *Journal of Sound and Vibration.* 2005;**287**(3):505–23.

140 Needleman A, Tvergaard V. Comparison of crystal plasticity and isotropic hardening predictions for metal-matrix composites. *Journal of Applied Mechanics.* 1993;**60**(1):70–6.

141 Sun C, Vaidya R. Prediction of composite properties from a representative volume element. *Composites Science and Technology.* 1996;**56**(2):171–9.

142 Schrefler B, Lefik M, Galvanetto U. Correctors in a beam model for unidirectional composites. *Mechanics of Composite Materials and Structures.* 1997;**4**(2):159–90.

143 Arabnejad Khanoki S, Pasini D. Fatigue design of a mechanically biocompatible lattice for a proof-of-concept femoral stem. *Journal of the Mechanical Behavior of Biomedical Materials.* 2013;**22**:65–83.

144 Arabnejad S, Pasini D. The fatigue design of a bone preserving hip implant with functionally graded cellular material. *ASME Journal of Medical Devices.* 2013;**7**(2):020908.

3

Elastodynamics of Lattice Materials

A. Srikantha Phani

Department of Mechanical Engineering, University of British Columbia, Vancouver, British Columbia, Canada

3.1 Introduction

Elastic wave propagation in lattice materials and structures is the central theme of this chapter. As we have already seen in Chapter 1, present interest in lattice materials, viewed as periodic composite materials, lies in tailoring their effective thermoelastic properties as well as their elastodynamic responses at macro [1], micro [2], and nano [3] length scales. Given the fundamental nature of the current topic, a very brief overview of wave transmission phenomena in periodic media in general, and their practical applications in several disciplines is provided first. This sets the context – both historical and modern – for subsequent development of the topics presented in this chapter.

In 1877, Rayleigh considered the propagation of waves in a medium endowed with a periodic structure [4] and specifically the case of transverse vibrations of a stretched string with point masses (loads) attached uniformly along its length. He showed that for:

> ... the propagation of waves in an infinite laminated medium (where, however, the properties are supposed to vary continuously according to the harmonic law) however slight the variation, reflexion is ultimately total, provided the agreement be sufficiently close between the wave-length of the structure and the half wave-length of the vibration.

Rayleigh observed that when

> ... the wave-length of a train of progressive waves be approximately equal to the *double* interval between the loads[1], the partial reflexions from the various loads

---

1 It should be observed that the frequency of the transverse wave that suffers total reflection in the mass-loaded string is given by $f \approx \frac{c}{2L}$ Hz, where c is the speed of sound (m/s), and L is the interval between two masses or the length of the unit cell (m). This frequency coincides with the pinned–pinned resonant frequency, which constitutes the edge of the pass band. For a detailed discussion of the relationship between the resonant modes of a unit cell and band gap edges, refer to pp. 10–12 in [5] in the context of Kelvin's resolution of Cauchy's paradox concerning refractive dispersion using a diatomic lattice, to [5–7] for Bragg band gap edges, and to [8] for sub-Bragg band gap edges.

Dynamics of Lattice Materials, First Edition. Edited by A. Srikantha Phani and Mahmoud I. Hussein.
© 2017 John Wiley & Sons Ltd. Published 2017 by John Wiley & Sons Ltd.

will all occur in phase, and the result must be a powerful aggregate reflexion, even though the effect of an individual load may be insignificant.

Rayleigh later extended these acoustic ideas to optical domain in explaining Stoke's observation of the remarkable reflection of colour by crystals of chlorate of potash [9], attributing the total reflection and iridescent properties of crystals to the presence of layers of twinned and counter-twinned crystals of small but irregular thickness. For a complete account on this topic see Section 40 in the *Baltimore Lectures* by Lord Kelvin [10]. These earlier works by Rayleigh anticipated Bragg's law and offered insights into what emerged much later – through the efforts of researchers in solid-state and other branches of physics – as *band gap* phenomena in photonic and phononic crystals. For our purposes here, Rayleigh's pioneering studies can be taken as a useful starting point, since many of the fundamental insights concerning wave propagation in periodic media emanated from his works. Since the time of Rayleigh, wave phenomena in periodic media have received attention from physicists as well as engineers, sometimes simultaneously, mostly independently. It is worth examining, very briefly, the development of ideas to synthesize useful insights from diverse fields of study.

A comprehensive summary of wave propagation in periodic structures can be found in the definitive work of Brillouin [5] from the perspective of solid-state physics. Here, we take an engineering perspective, set in the context of applications of periodic structures, materials, and devices. Practical acoustic applications of periodic structures, exploiting band gap phenomena, appeared as early as the 1920s in the studies by Stewart [11, 12]. He designed acoustic filters to control acoustic wave transmission through a periodic arrangement of acoustic volumes of contrasting impedance mismatch, an arrangement analogous to a diatomic crystal lattice. Such designs have also been used in constructing *finite* acoustic filters to suppress low-frequency pressure disturbances in hydraulic loops [13]. These studies illustrate the potential of spatially periodic design principles in realizing low-pass, high-pass, and band-pass filters for acoustic wave propagation. Filtering characteristics of periodic systems have also been extensively used in the telecommunications industry [14, 15], including in transceivers used in mobile phones and wireless communications [16].

Structural engineering interest in periodic media, particularly from aerospace engineers in the period after the early 1960s, originates in the development of periodic rib–skin structures, stiffened panels, and grillage and space-frame structures used in the aerospace industry [17–26]. Initial studies used a transfer matrix formalism [19] in mostly 1D systems, and propagation constant [17] formalisms in 1D, 2D, and 3D. Links between wave propagation in periodic structures and solid-state physics were firmly established in the 1980s, and the mechanical analogues of phenomena such as Anderson localization in disordered, or nearly periodic, structures and material systems have also been uncovered [27]. The central role of Floquet–Bloch theory has been recognized in the context of periodic structures [20, 21, 25] and composite materials [6, 28–33] widely used in aerospace industry.

More recent developments in composite materials [34–38], cellular solids [39], hybrid materials [40], and lattice materials [1] utilize periodic architectures to design ultralight, multifunctional lattice-structure materials. Periodic material architectures are also exploited in phononic crystals for elastic wave tailoring [41], acoustic metamaterials with negative index for focusing and cloaking applications [42–46] and sonic crystals with acoustic transmission levels above the mass law [47].

The field of phononic crystals and metamaterials is rapidly expanding, and interested readers are referred to recent research monographs [48–50]. The developments in phononic crystals closely parallel similar developments in photonic crystals, although fundamental differences exist between electromagnetic and elastic wave propagation. In the field of biomedical engineering, implants with periodic architecture are used in stents [51, 52], for example. Here, the mechanical deformation characteristics can be tailored by tuning the geometry of the unit cell [53]. The foregoing illustrate the emerging avenues for applications of lattice materials and structures.

This chapter's focus is on wave transmission in lattice materials and structures. A lattice structure is defined here to be any structure comprising a periodic assembly of 1D *structural* elements such as beams and bars, independent of the length scale at which such periodicity manifests itself. Lattice material is defined as a spatially periodic network of beams or shells whose constituent length scale is much larger than the load-deformation length scales. Such materials possess a spatially ordered pattern specified by a unit cell and associated tessellation directions (lattice basis vectors). The unit cell itself is an interconnected network of beams. Material constituents of each beam can be a single homogeneous isotropic material (such as steel or aluminium) or a hierarchical anisotropic composite [1]. Thus lattice materials, in the form of an interconnected spatially periodic network of composite beams, can be viewed as discrete multiscale materials with a hierarchy.

3.2 One-dimensional Lattices

This section deals with 1D lattice structures. Repetition of a single unit at regular spatial intervals in one direction is a characteristic feature of these structures. Some illustrative examples are monoatomic and diatomic lattice models for crystalline materials at atomic scale, Bragg gratings, mechanical filters used in the telecommunication industry, cardiovascular stents, rib–skin aerospace structures, reinforced cylinders in submarines, turbines in power generation, and railway tracks in transportation. It should be remarked that even though a lattice structure can be 2D, as in a turbine, it can be regarded as a 1D lattice once a unit cell (a blade) and the single direction of repetition (along the circumference of a circle) are identified. Propagation of free waves in such 1D lattices is considered in this section. Here, only elastic and inertial forces and their mutual interactions are considered. Dissipative forces and their influence on wave propagation in lattice structures will be described in Chapter 4.

Consider the monoatomic, spring–mass lattice model shown in Figure 3.1. It is apparent that two unit cells (Figures 3.1b,c) are equally attractive for the purpose of studying *free* wave propagation along the lattice. If it is preferred that the chosen unit cell reflect the symmetries of the parent lattice, then the symmetric unit cell in Figure 3.1c is a good choice. Alternate symmetric unit cells can be readily constructed, for example by surrounding one mass on either side with springs of equal stiffness. Similarly, other choices for asymmetric unit cells also exist. Similar remarks apply to the diatomic lattice model shown in Figure 3.2.

We will consider wave propagation in the diatomic lattice in detail, since it is more general, in the sense that it supports two types of wave motion. In the first type, both masses within the unit cell move in phase, while they move out of phase in the second

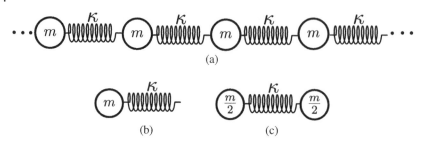

Figure 3.1 A monoatomic lattice is shown in (a), along with two possible unit cell choices in (b) and (c). Note that other choices exist for both symmetric and asymmetric unit cells. All masses are equidistant.

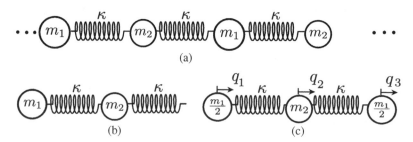

Figure 3.2 A diatomic lattice is shown in (a), along with two possible unit cell choices in (b) and (c). Note that other choices exist for both symmetric and asymmetric unit cells. The spacing between all masses is equal. $m_1 = \frac{1}{2}, m_2 = m$.

type. In this chapter, a lattice is taken to be infinite in extent (or sufficiently damped near the boundaries) so that attention can be focussed on the *direct* wave field.

For the choice of the unit cell in Figure 3.2c, the explicit scalar differential equations governing the displacements of the unit-cell masses are:

$$\frac{m_1}{2}\ddot{q}_1 + \kappa(q_1 - q_2) = f_1,$$
$$m_2\ddot{q}_2 + \kappa(2q_2 - q_1 - q_3) = f_2, \tag{3.1}$$
$$\frac{m_1}{2}\ddot{q}_3 + \kappa(q_3 - q_2) = f_3,$$

where $f_1, f_2,$ and f_3 are external forces associated with displacement degrees of freedom $q_1, q_2,$ and q_3, respectively. The above scalar equations can be expressed in a compact matrix form

$$M\ddot{q} + Kq = f, \tag{3.2}$$

where

$$M = \begin{bmatrix} \frac{m_1}{2} & 0 & 0 \\ 0 & m_2 & 0 \\ 0 & 0 & \frac{m_1}{2} \end{bmatrix}, \quad K = \begin{bmatrix} \kappa & -\kappa & 0 \\ -\kappa & 2\kappa & -\kappa \\ 0 & -\kappa & \kappa \end{bmatrix}, \quad q = \begin{bmatrix} q_1 \\ q_2 \\ q_3 \end{bmatrix}, \text{ and } f = \begin{bmatrix} f_1 \\ f_2 \\ f_3 \end{bmatrix}. \tag{3.3}$$

It is useful to partition this matrix equation into three parts, corresponding to the left edge q_l, the right edge q_r and the interior displacements q_i. Since each mass in Figure 3.2

has only one displacement degree of freedom, the partitioned vectors contain only one displacement:

$$q = \begin{bmatrix} q_l \\ q_r \\ q_i \end{bmatrix}, \quad q_l = q_1, \quad q_r = q_3, \quad \text{and} \quad q_i = q_2. \tag{3.4}$$

At this stage, the above partitioning may seem unnecessary in the case of a diatomic lattice, but it is of immense value when we consider more complicated lattices that require a finite-element (FE) numerical model of the unit cell. One can eliminate the consideration of q_i using the internal equilibrium conditions.

3.2.1 Bloch's Theorem

This section serves as an introduction to Bloch's theorem. Elementary concepts from solid-state physics are necessary in a 1D setting. These will be generalized to two and higher dimensions in Section 3.3.1.

The matrix differential equations of motion of the *entire* lattice can be viewed as linear ordinary differential equations with spatially *periodic* coefficients. The spatial periodicity of the coefficient matrices is due to *intrinsic* periodicity of a lattice structure. Studies of 1D differential equations with periodic coefficients, such as Hill's equation[2] and Mathieu's equation, have resulted in the establishment of Floquet's theorem. Further generalization of Floquet's theorem [5] to 3D problems, specifically of electron waves in a crystalline lattice, was done by Felix Bloch in 1929 [54]; see Chapter 8 of Ashcroft and Mermin [55] for a formal proof of Bloch's theorem.

Before proceeding to Bloch's theorem, it is worth reviewing relevant concepts from solid-state physics. A lattice can be visualized as a collection of points, called lattice points, and these points are specified by basis vectors, not necessarily of unit length. For example, the location of masses in the infinite monoatomic chain in Figure 3.1 can be specified by a set of vectors $r_n = ne_1, n = -\infty \ldots - 1, 0, 1, 2, \ldots \infty$, and the magnitude of the basis vector is $|e_1| = l$, l being the distance between the masses in the rest state. The lattice point system together with the basis is usually referred to as a direct lattice. A 2D lattice is specified by two basis vectors. In general, one needs N basis vectors for an N-dimensional lattice.

Upon selecting a suitable unit cell, the entire direct lattice can be obtained by repeating the unit cell along the basis vectors e_i. Denote the lattice points in a unit cell by $r_j = je_1$: these will correspond to the two masses in the unit cell shown in Figure 3.2c or, more generally, a subset of the nodes of an FE model of the unit cell in the case of 2D lattices. Let $q(r_j)$ denote the displacement of a lattice point in the reference unit cell. If a plane-wave solution is admitted, then $q(r_j)$ is of the form

$$q(r_j) = q_j e^{(i\omega t - k \cdot r_j)}, \quad i = \sqrt{-1} \tag{3.5}$$

where q_j is the amplitude, ω is frequency, k is the wavevector of the plane wave, and the symbol $\cdot$ denotes the scalar or dot product. With reference to the chosen unit cell,

---

2 A vibrating string subjected to periodically varying tension, by attaching it to a tuning fork axially, is an acoustic analog of truncated Hill's equation, which in fact was studied as such by Rayleigh in 1887 [4].

let the integer n identify any other cell obtained by n translations along the e_1 direction. The point in the cell n, corresponding to the jth point in the reference unit cell, is denoted by the vector $r = r_j + ne_1$. According to Bloch's theorem, the displacement at the jth point in any cell, identified by the integer n in the direct lattice basis, is given by

$$q(r) = q(r_j)e^{k \cdot (r - r_j)} = q(r_j)e^{(kn)}. \tag{3.6}$$

Here, $k = \delta + i\mu$ represents the component of the wavevector k along the e_1 vector; that is, $k = k \cdot e_1$. The real part δ and the imaginary part μ are called the *attenuation* and *phase* constants, respectively. The real part is a measure of the attenuation (or growth) of a wave as it progresses from one unit cell to the next. For waves propagating without attenuation, the real part is zero and the components of the wavevector reduce to $k = i\mu$. The imaginary part or the phase constant is a measure of the phase change across one unit cell.

It is convenient to define a reciprocal lattice in the wavevector space (k-space) such that the basis vectors of the direct and reciprocal lattices satisfy:

$$e_i \cdot e_j^* = \delta_{ij} \tag{3.7}$$

where e_i denote the basis vectors of the direct lattice and e_j^* denote the bases of reciprocal lattice, while δ_{ij} is the Kronecker delta function. For a 1D lattice, the subscripts i and j take the integer value 1. The magnitude of the reciprocal basis vector e_1^* is $|e_1^*| = \frac{1}{l}$. Thus for an N-dimensional lattice, the volumes of the unit cell in the direct and reciprocal lattices are reciprocal to each other; that is, their product is identity, and hence the name 'reciprocal' lattice.

The wavevectors can be expressed in terms of the reciprocal lattice basis e_i^*. Since the reciprocal lattice is also periodic, one can restrict the wavevectors to certain regions in the reciprocal lattice called Brillouin zones [5]. The wavevectors are restricted to the edges of the irreducible part of the first Brillouin zone (IBZ) to explore band gaps, since the band extrema almost always occur along the boundaries of the irreducible zone [5, 55, 56]. The first Brillouin zone is defined as a Wigner–Seitz or primitive unit cell of the reciprocal lattice, and it can be constructed as follows:

1. Select any lattice point in the reciprocal lattice as the origin and connect it to neighbouring points.
2. Construct the perpendicular bisectors of these lines. The region of intersection is the first Brillouin zone.

The first Brillouin zone for 1D monoatomic and diatomic lattices is defined by the interval $-\pi \leq \mu \leq \pi$ and the irreducible Brillouin zone is defined by $0 \leq \mu \leq \pi$. The phase change in displacements of two lattice points n units apart is μn from Eq. (3.6). Thus μ can be visualized as the phase change across *one* unit cell. Hence, at the edges of the IBZ, corresponding to $\mu = 0$ and $\mu = \pi$, the motion of the left- and right-hand edges of the unit cell are completely in phase or out of phase. They correspond to normal *modes* or standing waves with zero group velocity.

To summarize, Bloch's theorem (or Floquet's theorem in the case of 1D periodic structures) states that, for any structure with repetitive identical units, the change in complex wave amplitude across a unit cell due to a propagating wave without attenuation does not depend upon the location of the unit cell within the structure. By virtue of this theorem, one can understand wave propagation through the *entire* lattice by considering wave motion within a *single* unit cell. Bloch's theorem thus leads to enormous savings in the analysis of wave propagation in periodic structures. Born–von Kármán periodic boundary conditions are expedient for infinite lattices.

3.2.2 Application of Bloch's Theorem

By virtue of Bloch's theorem and the Born–von Kármán periodic boundary conditions, the following relationships between the displacements, q, and forces, f, are obtained for a 1D lattice:

$$q_r = e^k q_l, \quad f_r = -e^k f_l \tag{3.8}$$

where the subscripts l, and r respectively denote the left- and right-hand edges of the unit cell.

Using the above relationships one can define the following transformation to project the displacement vector q to a reduced vector $\tilde{q}$ for a propagating wave without attenuation; that is, $k = i\mu$:

$$q = T\tilde{q},$$

$$q = \begin{bmatrix} q_l \\ q_r \\ q_i \end{bmatrix}, \quad T = \begin{bmatrix} I & 0 \\ Ie^{i\mu} & 0 \\ 0 & I \end{bmatrix}, \quad \tilde{q} = \begin{bmatrix} q_l \\ q_i \end{bmatrix} \tag{3.9}$$

where $\tilde{q}$ denote the displacements of the nodes in the Bloch reduced coordinates, and I is an identity matrix of appropriate size.

Inserting the Bloch transformation given by Eq. (3.9) into the governing equations of motion in Eq. (3.2) and pre-multiplying the resulting equation with T^H to enforce force equilibrium [57], one obtains the following governing equations in the reduced coordinates:

$$\tilde{D}\tilde{q} = \tilde{f}, \quad \tilde{D} = T^H DT, \quad \tilde{f} = T^H f, \quad D \equiv K - \omega^2 M \tag{3.10}$$

where the superscript ^{H} denotes the Hermitian transpose. Harmonic motion with $e^{i\omega t}$ time dependence has been assumed to define the dynamic stiffness matrix D. The wave vector $k = \delta + i\mu$ is complex in general. For a plane wave propagating without attenuation along the lattice ($\delta = 0$), the propagation constant is $k = i\mu$. For a harmonic *free* wave motion ($f = 0$), the Eq. (3.10) can be rewritten as an eigenvalue problem,

$$\tilde{D}(\mu, \omega)\tilde{q} = 0 \tag{3.11}$$

Any pair (μ, ω) obtained by solving the eigenvalue problem in Eq. (3.11) represents a plane wave propagating at frequency ω. A plot of ω against μ is called a dispersion curve, whose tangent slope indicates group velocity and secant slope indicates phase velocity. We shall now illustrate dispersion curves for selected 1D lattices. Of interest here is the significance of resonant frequencies of the unit cell for lattice structures with a symmetric unit cell.

3.2.3 Dispersion Curves and Unit-cell Resonances

Dispersion curves for monoatomic and diatomic lattices are illustrated in Figure 3.3. These curves are obtained by specifying the phase constant for each point in the first

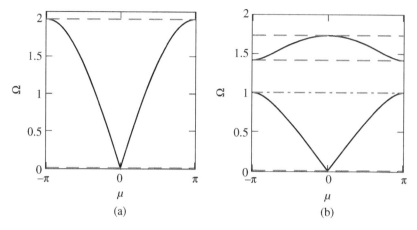

Figure 3.3 Dispersion curves for (a) monoatomic and (b) diatomic discrete spring–mass lattices. Natural frequencies of the unit cell under free–free boundary conditions and pinned–pinned boundary conditions are shown by the horizontal dashed lines, and dash-dot lines, respectively. Note that the resonances of the unit cell bound the band edges.

Brillouin zone, defined by the interval $-\pi \leq \mu \leq \pi$, and solving the eigenvalue problem given in Eq. (3.11) for the frequency ω. Propagating wave frequency ω is normalized according to $\Omega = \frac{\omega}{\omega_0}$ where $\omega_0 = \sqrt{\frac{\kappa}{m}}$, see Figure 3.1 for κ and m. For the diatomic lattice $m_1 = m$, and $m_2 = 2m$.

Consider the monoatomic lattice in Figure 3.3a. Two features can be immediately identified: there is a cut-off or an upper frequency limit to propagating waves, and at the edges of the Brillouin zone the dispersion curve becomes flat, indicating zero group velocity $c_g = \frac{d\omega}{dk} = 0$; in other words, a standing wave. These standing waves are the normal modes of the unit cell shown in Figure 3.1c when both end masses are allowed to move freely. There is a rigid body or zero- frequency line that forms the lower bound for the dispersion curve. Thus we observe that the resonances of the *unit cell* under free boundary conditions define the lower and upper bounds of the dispersion curve for a monoatomic lattice. The fixed boundary conditions for the unit cell in Figure 3.1c do not have any modes.

Next, the diatomic lattice in Figure 3.3b exhibits two dispersion curves. The lower and upper branches are respectively called the *acoustic* and *optic* branches, presumably to signify that light waves are of higher frequencies than sound waves. In addition to the cut-off we also observe that there is a gap between the acoustic and optic branches. In this gap there are no solutions for the Bloch-wave eigenvalue problem in Eq. (3.11), indicating that there are no propagating solutions. This frequency interval is dubbed the *band gap*, a frequency interval in which waves are forbidden to propagate. More importantly, the unit cell resonances again bound the dispersion curves! However, some care is required, since unlike the monoatomic case, the *symmetric* unit cell for the diatomic lattice in Figure 3.2c has nontrivial resonant modes under pinned (locked-end masses) boundary condition. In fact, once we clamp the end masses, it is immediately clear that this pinned–pinned resonance of the unit cell is at $\omega = \sqrt{\frac{2\kappa}{m_2}} = \sqrt{\frac{\kappa}{m}}$ or $\Omega = 1$, shown by the dash-dotted line in Figure 3.3b. Under free–free boundary conditions we expect

Table 3.1 Unit-cell resonant frequencies under free and fixed boundary conditions that bound the pass-band edges of discrete 1D lattices with a symmetric unit cell.

Lattice	Unit cell	Resonant frequencies (free)	Resonant frequencies (fixed)
Monoatomic	Symmetric Figure 3.1c	$0, \sqrt{\dfrac{4k}{m}}$	0
Diatomic	Symmetric Figure 3.2c	$0, \sqrt{\dfrac{2k}{m_1}}, \sqrt{\dfrac{2k(m_1+m_2)}{m_1 m_2}}$	$\sqrt{\dfrac{2k}{m_2}}$

three resonant modes: one at $\Omega = 0$ (rigid-body mode) and the remaining two nonzero resonant frequencies shown by dashed horizontal lines in Figure 3.3b. The unit-cell resonances are summarized in Table 3.1.

The fact that, for a *symmetric* unit cell, the edges of dispersion curves are bounded by the resonances of the unit cell is of significant practical engineering interest, since these edges define the band gap width. For example, if one wants to maximize the band gap width, one can place the resonances of the pinned-pinned and free-free unit-cell resonances far apart. band gap design thus becomes an *inverse* structural dynamics problem, where the unit-cell resonances under two boundary conditions are specified and the geometry and material properties are to be optimized to meet these objectives. This avenue of *band gap* design using inverse structural dynamics approaches needs further study. This approach can be used in existing topology optimization [58, 59] and other frameworks for optimization [60] of elastodynamic response properties of periodic materials and structures.

3.2.4 Continuous Lattices: Local Resonance and sub-Bragg Band Gaps

Continuous beam lattice systems differ from discrete monoatomic and diatomic lattices in one crucial aspect: the unit cell has an infinite number of resonances because the governing equations for wave propagation in a beam are partial differential equations. Consequently, dispersion branches extend to infinite frequency. Nonetheless, within a specified frequency of interest, it is possible to capture the elastodynamic response by using a finite-element numerical model to accurately discretize the partial differential equations [61–63]. We consider three continuous periodic systems, as shown in Figure 3.4. An elastic flexural beam is subjected to periodic changes in mass distribution, a flexural beam with periodically arranged resonators, and a beam which is a combination of the previous two. These systems are chosen to illustrate the band gap formation mechanism: through Bragg scattering, which was seen earlier in discrete monoatomic and diatomic lattices, and a new local resonance phenomenon arising from the interaction between propagating waves in the beam with a resonator arranged periodically or otherwise. Here we consider a periodic arrangement, but periodicity is not necessary to observe local resonance effects used, for example, in acoustic metamaterials.

An FE model of the unit cell provides a set of matrix equations of motion of the form in Eq. (3.2). Note that the mass and stiffness matrices contain the contributions from

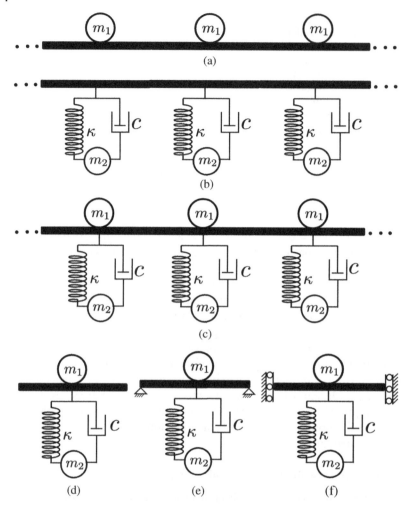

Figure 3.4 Continuous beam lattices: (a) beam with uniformly spaced masses, (b) beam with uniformly spaced resonators, (c) beam with uniformly spaced masses and resonators, (d) unit cell for the general case (c), with pinned and guided boundary conditions shown in (e) and (f), respectively.

the inertial and elastic properties of the beam, as well as any additional inertial and stiffness contributions from the periodic masses and resonators. Invoking Bloch's theorem in Eq. (3.6) we once again arrive at the Hermitian Bloch eigenvalue problem in Eq. (3.11), which can be solved numerically to obtain the dispersion curves. The dispersion curves for the three beam lattices are shown in Figure 3.5. It is useful to examine these dispersion curves in detail to glean the significance of the unit-cell resonant frequencies.

3.2.5 Dispersion Curves of a Beam Lattice

The computed dispersion curves for the three lattices in Figures 3.4a–c are shown in Figures 3.5a–c, respectively. Figures 3.5d–f are the magnified versions in the low-frequency region of Figures 3.5a–c, respectively. Consider the band gaps in

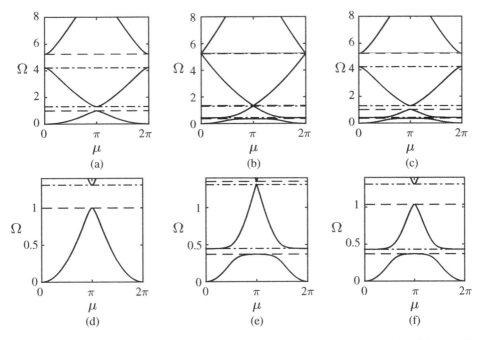

Figure 3.5 Dispersion curves for beam lattices in Figure 3.4. The top row corresponds to (a) beam with periodic masses, (b) beam with periodic resonators, and (c) beam with periodic masses and resonators. The bottom row is a magnified version of the top-row figures in a lower frequency range. The sub-Bragg band gap in (e) around $\Omega \approx 0.4$ is due to local resonance. Notice how the width of the Bragg band around $\Omega \approx 1$ decreases in (e) compared to (d), but is restored in (f) when periodic masses are introduced in the beam with periodic resonators. Ω is the nondimensional frequency $\Omega = \frac{\omega}{\omega_0}$, where ω_0 is the pinned–pinned resonance frequency of the unit cell of the beam with periodic masses. μ is the phase constant. The dashed and dashed-dotted horizontal lines are the resonant frequencies of the unit cell under pinned–pinned and free–free (guided) boundary conditions shown in Figures 3.4e and f, respectively.

Figure 3.5a due to the introduction of periodic masses in a uniform beam. These are Bragg band gaps. This can be verified by observing in Figure 3.5d that the first band gap occurs at $\Omega \approx 1$ which in turn is associated with the first pinned–pinned mode of the unit cell. The first pinned–pinned resonant mode has wavelength $\lambda = 2l$, where l is the length of the generic unit cell in Figure 3.4d, with $m_2 = \kappa = c = 0$. In contrast to the periodic mass system, a low-frequency sub-Bragg band gap is evident in the case of a beam with periodically attached resonators in Figure 3.5b and e. This band gap is well below the frequencies at which the wavelength approaches the unit-cell size, and it arises from the interaction of a propagating wave (lower dispersion branch in Figure 3.5a with the resonance of the attached oscillator at $\Omega = \frac{\omega_a}{\omega_0}$, where

$\omega_a = \sqrt{\frac{\kappa}{m_2}}$ and ω_0 is the natural frequency of the unit cell of the beam with periodic masses under pinned–pinned conditions. It may be noticed that this interaction is local, hence the term 'local resonance'. Remarkably, the Bragg band gap width at $\Omega \approx 1$ decreases compared to Figure 3.5d. A combination of periodic masses and resonators in Figure 3.4c has the dispersion curves shown in Figures 3.5c and f, and has the desirable features of a local resonance sub-Bragg band gap and Bragg band gaps. Further details on optimization of the sub-Bragg band gap width can be found in the literature [8].

A feature that reappears and is consistent with earlier observations is that the band gap edges are coincident with the resonant frequencies of the unit cell. It may be noticed again in Figure 3.5 that the edges of the band gaps coincide with a dashed line; the resonant frequencies of the unit cell under a pinned–pinned boundary condition, as shown in Figure 3.4e, or, a dash-dotted line; the resonant frequencies of the unit cell under guided support boundary conditions, as shown in Figure 3.4f. This property is true for both Bragg and sub-Bragg band gaps. Thus *one can calculate the band gap widths, Bragg or sub-Bragg, without Bloch-wave analysis by simply computing the resonant frequencies of the unit cell under free and fixed boundary conditions.* This remarkable property of periodic systems with a symmetric unit cell has enormous implications for design optimization of band gap phenomena in acoustic metamaterials and phononic crystals and remains to be explored. Raghavan and Phani [8] have elucidated a receptance coupling method to predict the band gap widths, given the receptances of the individual components of the unit cell. We shall briefly review this procedure in the next section.

3.2.6 Receptance Method

Receptance, or dynamic compliance, is defined as the ratio of the steady-state displacement response of a linear system to the harmonic input force to which it is subjected. Receptance functions depend on the forcing frequency and the spatial location of the force and response points. Of particular interest here are point receptance functions for which the force and response points coincide. The inverse Fourier transform of the receptance leads to an impulse response function in the time domain [64, 65]. The receptance and analogous Green's function based techniques are widely applied in structural acoustics. Tables of receptances exist for a range of structural components [64]. Receptance functions expressed in terms of normal modes of the system in a *modal series* form are extremely powerful. Receptance functions can also be defined in terms of characteristic Bloch waves for periodic structures [20, 24, 25]. Note that receptance functions can be measured or calculated from the governing equations of motion using FE models. Their usefulness lies in analysing the response of coupled systems. Receptance methods were pioneered to synthesize the dynamic response of an assembled component from the individual dynamic characteristics of its constituent components. Coupling a local resonator to a medium can be achieved by enforcing the necessary force equilibrium and displacement compatibility conditions on the receptance functions of the individual subsystems. Simple rules exist for coupling point receptances of two systems, each evaluated at the point of coupling. For two systems with point receptances H_1 and H_2, the following results can be derived:

$$\frac{1}{H} = \frac{1}{H_1} + \frac{1}{H_2} \quad \text{parallel connection}$$

$$H = H_1 + H_2 \quad \text{series connection}$$

(3.12)

These rules provide the characteristic equations, whose roots are the natural frequencies of the coupled system. These coupling procedures can be applied to 1D wave guides.

For our present purposes, we are interested in the natural frequencies of the unit cell under guided and pinned boundary conditions, as shown in Figure 3.6. Recognizing

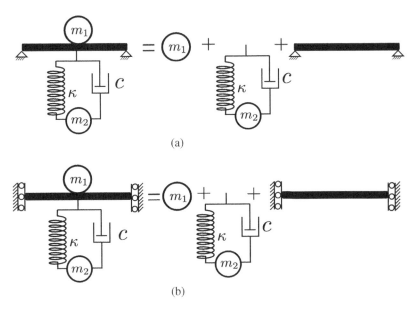

Figure 3.6 The concept of receptance coupling: the vertical displacement of the beam, mass, and resonator are coupled to obtain the respective unit cells for (a) the pinned and (b) the guided boundary conditions. Here coupling is viewed as a parallel coupling, in which the displacements of the connected units are the same but the forces add up.

that at the point of coupling, the mass m_1 and the resonator $m_2 - \kappa - c$ are attached in parallel with the flexural beam under pinned or guided boundary conditions, one can apply the receptance method as follows. The dynamics of mass m_1 are governed by Newton's second law $f = m_1 a$, where f is the force acting on m_1 and a is its acceleration. Using this equation of motion one defines the receptance of m_1 in the Fourier domain as

$$H_{m_1}(\omega) = -\frac{1}{m_1 \omega^2} \tag{3.13}$$

Similarly, from the governing equations of motion of the resonator $m_2 - \kappa - c$ at the point of coupling, one can obtain the receptance function of the resonator H_a. Point receptance of the damped resonator at the point of attachment can be shown to be:

$$H_a = -\frac{\kappa - m_2 \omega^2 + ic\omega}{(\kappa + ic\omega)m_2 \omega^2} \tag{3.14}$$

where, κ, m_2, and c correspond to the stiffness, mass and damping coefficient of the resonator respectively.

Point receptance of a beam at any location, x, can be expressed in terms of the normal modeshapes of the beam using the modal series expansion [64, 65] as:

$$H_b(x) = \sum_{r=1}^{n} \frac{\Phi_r^2(x)}{a_r(\omega_r^2 - \omega^2 + i2\zeta_r\omega\omega_r)} \tag{3.15}$$

where $\Phi_r(x)$ is the modeshape associated with the normal mode r with natural frequency ω_r, and ζ_r is the associated damping ratio. The mass normalization constant is a_r. H_b

depends on the boundary conditions. A table of receptance function for various dynamic components can be found in the literature [64].

We can now use the systems-in-parallel coupling rule in Figure 3.6, since all of the sub-systems (mass, resonator, and beam) have the same displacement at the point of coupling:

$$\frac{1}{H} = \frac{1}{H_{m_1}} + \frac{1}{H_a} + \frac{1}{H_b} \tag{3.16}$$

The natural frequencies are the roots of the denominator of H. A graphical method, for point coupling, is illustrated in Figure 3.7b. Here H_g and H_p are the receptance function of the beam with central mass m_1 under guided and pinned boundary conditions, and H_a is as defined in Eq. (3.14). The frequency, $\Omega \approx 0.37$, corresponding to the point of intersection of H_p with H_a, defines the resonances of the unit cell shown in Figure 3.4e. This frequency is indeed the lower edge of the sub-Bragg band gap in Figure 3.5e. Similarly, the frequency, $\Omega \approx 0.448$, associated with the point of intersection of H_g with H_a, defines the resonances of the unit cell shown in Figure 3.4f, which forms the upper edge of the sub-Bragg pass band in Figure 3.5e.

A full exposition of the receptance method can be found in the literature [8]. The fact that the resonant frequencies, with appropriate boundary conditions, define the widths of Bragg or sub-Bragg band gaps highlights the importance of vibration modes in designing 1D band gap materials with a symmetric unit cell. It should be remarked that Rayleigh's principle, in particular the Cauchy–Rayleigh interlacing theorem, may

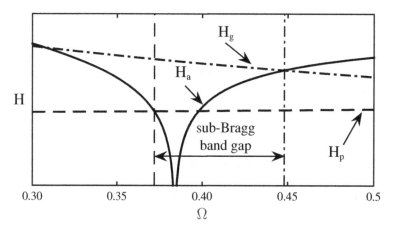

Figure 3.7 Sub-Bragg band gap viewed from a receptance perspective. The lower and upper band edges are, respectively, the resonances of the unit cell under pinned and guided boundary conditions. These resonant frequencies are shown as the vertical dashed (pinned–pinned supports) and dashed-dotted lines (guided supports). Resonant frequencies of the pinned–pinned unit cell (lower edge of the sub-Bragg band gap) lies at the intersection of receptance of the resonator (H_a) and the pinned–pinned beam (H_p). The same holds for the guided boundary conditions: the resonance (upper edge of the band gap) is the intersection of H_g and H_a. The deep anti-resonance of the attached resonator is evident in the minimum of H_a. Notice the agreement with Figure 3.5e for the sub-Bragg band gap.

be relevant here if one views the addition of periodic mass to a uniform beam as a constraint. Pursuing this further is outside the scope of this work. Interested readers are referred to the literature [8] for a complete treatment of the receptance method applied to 1D lattice models and its experimental verification.

3.2.7 Synopsis of 1D Lattices

The essential phenomena that we have observed in wave propagation through 1D lattices, in the discrete (mass-spring) or continuous (flexural beam with masses, and/or resonators) types, can be summarized as follows:

1. Spatially repetitive regular changes in impedance of a material or a structure will lead to the emergence of forbidden frequency intervals for wave propagations, or band gaps, through Bragg reflections and/or through the sub-Bragg local resonance phenomenon.
2. Regardless of the underlying mechanism, Bragg or sub-Bragg, resonant frequencies of the unit cell under free and locked boundary conditions define the band edges, provided the unit cell is symmetric. For asymmetric unit cells no such general principles exist [66, 67].
3. It has been demonstrated using discrete and continuous lattices that one can infer the band gap widths, without having to compute the dispersion curves, for lattices with a symmetric unit cell. Thus band gap design becomes a natural frequency design problem, for which many techniques have been outlined in the structural dynamics literature.
4. Receptance coupling methods simplify the band gap analysis of complex unit cells of continuous lattices.

We now move on to 2D lattice materials.

3.3 Two-dimensional Lattice Materials

Two-dimensional lattices are considered here. Three-dimensional lattices are studied in Chapter 12. Plane-wave propagation in 2D planar lattice materials, producing displacements within the plane of the lattice, can be studied using several techniques [41]. Given its versatility to accommodate complex geometries and the availability of efficient numerical algorithms, an FE setting is chosen here as the modeling framework. The discrete matrix equations of motion obtained from the FE models of a lattice are efficiently solved by invoking Bloch's theorem.

3.3.1 Application of Bloch's Theorem to 2D Lattices

The Bloch theory used in this section has been widely employed in a number of research fields: in solid-state physics to investigate wave propagation in crystal structures, Bragg gratings and photonic crystals [5, 56] and in mechanical systems such as stiffened panels to study their in-plane and out-of-plane vibration responses [20, 68].

We have already seen Bloch's theorem in the context of 1D lattices in Section 3.2.1. Here, we introduce appropriate notation for 2D lattices. With reference to the chosen unit cell of a 2D lattice, let the integer pair (n_1, n_2) identify any other cell obtained by n_1 translations along the e_1 direction and n_2 translations along the e_2 direction. The point in the cell (n_1, n_2), corresponding to the jth point in the reference unit cell, is denoted by the vector $r = r_j + n_1 e_1 + n_2 e_2$. According to Bloch's theorem, the displacement at the jth point in any cell identified by the integer pair (n_1, n_2) in the direct lattice basis of a 2D lattice is given by

$$q(r) = q(r_j)e^{k \cdot (r - r_j)} = q(r_j)e^{(k_1 n_1 + k_2 n_2)} \tag{3.17}$$

Here, $k_1 = \delta_1 + i\mu_1$ and $k_2 = \delta_2 + i\mu_2$ represent the components of the wavevector k along the e_1 and e_2 vectors; that is, $k_1 = k \cdot e_1$ and $k_2 = k \cdot e_2$. The real part δ and the imaginary part μ are called the *attenuation* and *phase* constants, respectively. The real part is a measure of the attenuation of a wave as it progresses from one unit cell to the next. For waves propagating without attenuation, the real part is zero and the components of the wavevector reduce to $k_1 = i\mu_1$ and $k_2 = i\mu_2$. The imaginary part or the phase constant is a measure of the phase change across one unit cell.

The reciprocal lattice can be defined in a similar way as for 1D lattices, according to Eq. (3.7). For a 2D lattice the subscripts i and j take the integer values 1 and 2. The wavevectors can be expressed in terms of the reciprocal lattice basis e_i^* and they are restricted to the edges of the irreducible part of the first IBZ for exploring band gaps [5, 55, 56].

By virtue of Bloch's theorem and the Born–von-Kármán periodic boundary conditions, the following relationships between the displacements q and forces f are obtained:

$$\begin{aligned}
q_r &= e^{k_1} q_l, \quad q_t = e^{k_2} \\
q_b q_{rb} &= e^{k_1} q_{lb}, \quad q_{rt} = e^{k_1 + k_2} q_{lb}, \quad q_{lt} = e^{k_2} q_{lb} \\
f_r &= -e^{k_1} f_l, \quad f_t = -e^{k_2} f_b \\
f_{rt} &+ e^{k_1} f_{lt} + e^{k_2} f_{rb} + e^{k_1 + k_2} f_{lb} = 0
\end{aligned} \tag{3.18}$$

where the subscripts l, r, b, t, and i respectively denote the displacements corresponding to the left, right, bottom, top, and internal nodes of a generic unit cell, as shown in Figure 3.8. The displacements of the corner nodes are denoted by double subscripts: for example, lb denotes the left bottom corner.

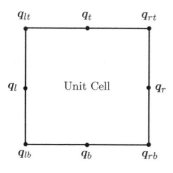

Figure 3.8 A generic unit cell for a 2D periodic structure, showing the degrees of freedom shared with the neighboring unit cells.

Using the above relationships one can define the following transformation:

$$q = T\tilde{q},$$

$$
T =
\begin{bmatrix}
I & 0 & 0 & 0 \\
Ie^{k_1} & 0 & 0 & 0 & 0 \\
0 & I & 0 & 0 & 0 \\
0 & Ie^{k_2} & 0 & 0 & 0 \\
0 & 0 & I & 0 \\
0 & 0 & Ie^{k_1} & 0 \\
0 & 0 & Ie^{k_2} & 0 \\
0 & 0 & Ie^{(k_1+k_2)} & 0 \\
0 & 0 & 0 & I
\end{bmatrix},
\quad
\tilde{q} =
\begin{bmatrix}
q_l \\
q_b \\
q_{lb} \\
q_i
\end{bmatrix}
\tag{3.19}
$$

where $\tilde{q}$ denotes the displacements of the nodes in the Bloch reduced coordinates. Now the transformation given by Eq. (3.19) can be substituted into the governing FE equations of motion in Eq. (3.2) and the resulting equation can be multiplied by T^H to enforce force equilibrium [57]. This gives the following governing equations in the reduced coordinates:

$$\tilde{D}\tilde{q} = \tilde{f}, \quad \tilde{D} = T^H D T, \quad \tilde{f} = T^H f \tag{3.20}$$

where the superscript ^{H} denotes the Hermitian transpose. For a plane wave propagating without attenuation in the $x–y$ plane, the propagation constants along the x and y directions are $k_1 = i\mu_1$ and $k_2 = i\mu_2$. For free wave motion ($f = 0$), the above equation can be written in the frequency domain to give the following Hermitian eigenvalue problem,

$$\tilde{D}(k_1, k_2, \omega)\tilde{q} = 0 \tag{3.21}$$

Any triad (k_1, k_2, ω) obtained by solving the eigenvalue problem in Eq. (3.21) represents a plane wave propagating at frequency ω.

In the eigenvalue problem defined by Eq. (3.21) there exist three unknowns: the two propagation constants k_1, k_2, which are complex in general, and the frequency of wave propagation ω, which is real since the matrix $\tilde{D}$ in the eigenvalue problem is Hermitian. At least two of the three unknowns have to be specified in order to obtain the third. For wave motion without attenuation the propagation constants are purely imaginary, and of the form $k_1 = i\mu_1$ and $k_2 = i\mu_2$. In this case one obtains the frequencies of wave propagation as a solution to the linear algebraic eigenvalue problem defined in Eq. (3.21) for each pair of phase constants (μ_1, μ_2). In contrast to the 1D lattices considered in earlier sections, solutions of the eigenvalue problem in Eq. (3.21) form nested 3D surfaces in the $\omega - k_1 - k_2$ coordinates. These are called the *dispersion surfaces*. There are as many surfaces as there are eigenvalues of the problem in Eq. (3.21). If two surfaces do not overlap each other then there is a gap along the ω-axis in which no wave motion occurs. This gap between the dispersion or phase constant surfaces (which as analogues of the Fermi surfaces in solid-state physics) is called the band gap in the solid-state physics literature [56] and the stop band in structural dynamics [20, 68]. For all frequencies on the phase constant surface, wave motion can occur and hence the frequency range occupied by these surfaces is a pass band. Furthermore, the normal to the phase constant surface at any point gives the Poynting vector or group velocity, and this indicates the speed and direction of energy flow [5].

A complete description of dispersion characteristics of 2D lattices requires one to sketch dispersion surfaces with propagating frequency, and two wavevector components forming the axes. However, a common practice emanating from solid-state physics is to plot dispersion curves by stitching together slices of dispersion surfaces, obtained by traveling along a path in the reciprocal space known as the edges of the first Brillouin zone. This conventional practice can sometimes lead to a misleading dispersion portrait. We illustrate this in the context of a discrete square lattice of masses and springs in the next section.

3.3.2 Discrete Square Lattice

Consider the out-of-plane wave propagation in a discrete square lattice, as shown in Figure 3.9a, where the linear spring k resists the relative out-of-plane displacements of the masses m at its ends. In the interests of simplicity, all springs and masses have equal values, k and m, respectively. Among the three possible choices for the unit cell indicated in Figure 3.9b–d, the unit cell shown in Figure 3.9b is the smallest symmetric unit cell and hence it is used in the Bloch-wave analysis. The first Brillouin zone of a square lattice is specified by the interval $-\pi \leq \mu_1, \mu_2 \leq \pi$, where μ_1 and μ_2 are phase constants. Strictly speaking, one should calculate propagating frequencies by solving the eigenvalue problem in Eq. (3.21) for *every* point (μ_1, μ_2) in the first Brillouin zone. Such a dispersion surface, or equivalently phase constant surface, is shown in Figure 3.10a as a contour map. The frequency of wave propagation ω in nondimensional form $\Omega = \frac{\omega}{\sqrt{\frac{k}{m}}}$ is shown as a function of μ_1 and μ_2 for *all* points within the interval $-2\pi \leq \mu_1, \mu_2 \leq 2\pi$. Propagating wave directions at a given frequency, or frequencies of propagating waves in a given direction, can be inferred from the dispersion surface. There is only one dispersion surface for the monoatomic square lattice. Further, one may see in Figure 3.9a that the dispersion surface repeats periodically outside the first Brillouin zone $-1 \leq \frac{\mu_1}{\pi}, \frac{\mu_2}{\pi} \leq 1$. In solid-state physics, and subsequent literature on periodic structures, it is a common practice to draw 2D curves stitched together by following a locus in the Brillouin zone, typically along the edge of the irreducible part of the first IBZ. Two such loci, $\Gamma - X - M - \Gamma$ and $Y - \Gamma - X - Y$, are shown in Figure 3.10a. The points $\Gamma = (0, 0)$, $X = (\pi, 0)$, $Y = (0, \pi)$, and $M = (\pi, \pi)$ indicate the directions of symmetry emanating from the center of zone $\Gamma = (0, 0)$. They coincide with the edges of the IBZ. Dispersion curves calculated along these two path are shown in Figure 3.10b for the $\Gamma - X - M - \Gamma$ path, and in Figure 3.10c for the path $Y - \Gamma - X - Y$. It is remarkable that the barely evident zero group velocity branch at $\Omega = 2$ in Figure 3.10b is strikingly clear in Figure 3.10c but absent in Figure 3.10d. Further, the maximum frequency $\Omega = \sqrt{8}$ is missed in Figure 3.10c. Such pitfalls plague the 2D representation of 3D dispersion surfaces and hence one must exercise caution.

The resonant frequencies of the unit cell in Figure 3.9b under the boundary condition of all the masses being free to oscillate are expected to be four in total. And, they are calculated to be at $\omega = 0$, $\omega = 2\sqrt{\frac{k}{m}}$ (this frequency repeats twice), and $\omega = \sqrt{\frac{8k}{m}}$,

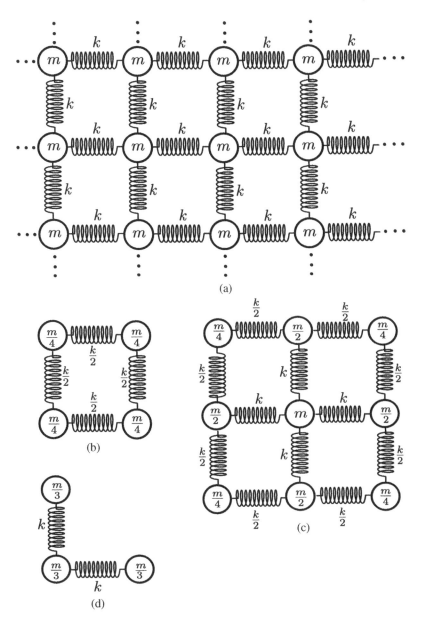

Figure 3.9 A square lattice with possible unit-cell choices. Note that other choices exist for both symmetric and asymmetric unit cells.

or in the nondimensional form $\Omega = 0$, $\Omega = 2$, and $\Omega = \sqrt{8}$. Table 3.2 summarizes the resonant frequencies of the unit cells for the discrete square lattices. It can be seen that $\Omega = 2$ forms the upper bounds of the dispersion curves along the $\Gamma - Y$ and $\Gamma - X$ directions in Figure 3.10c. Similarly, $\Omega = 0$ forms the lower bound of the dispersion curve. The significance of $\Omega = \sqrt{8}$ is evident from Figure 3.10b; it forms the upper bound of the dispersion curve along the $\Gamma - M$ path. It is remarkable that

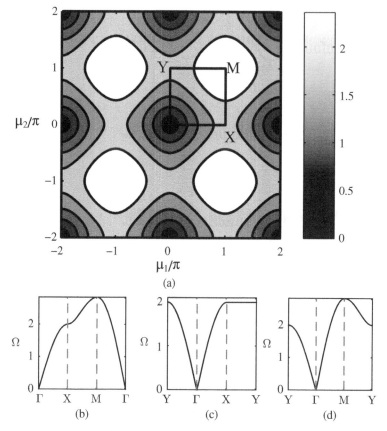

Figure 3.10 Out-of-plane wave dispersion in a square lattice: (a) dispersion surface with the IBZ shown as a square in the interval $0 \leq \mu_1, \mu_2 \leq \pi$; notice the repetition of the dispersion surface outside the full Brillouin zone (not shown) defined by the interval $-\pi \leq \mu_1, \mu_2 \leq \pi$. The dispersion surface in (a) can also be represented as dispersion curves obtained by following different paths along the edges of the IBZ. Shown in the second row are the dispersion curves for the path $\Gamma - X - M - \Gamma$ in (b), $Y - \Gamma - X - Y$ in (c) and $Y - \Gamma - M - Y$ in (d). Notice how the normal mode (zero group velocity) associated with $\Omega = 2$ is clearly evident in (c) but not so in (d). Also, note that the curve in (c) misses the maximum frequency $\Omega = \sqrt{8}$ evident in (b) and (d). It should be observed that the band edges $\Omega = 0, 2, \sqrt{8}$ are the natural frequencies of the unit cell under free and fixed boundary conditions.

the normal modes of the unit cell define the bounds for a square lattice. Whether these properties hold for lattices of other symmetries, such as a planar hexagonal lattice in two dimensions, and other lattices in three dimensions, remains to be explored.

3.4 Lattice Materials

Lattice materials can be visualized as a periodic network of beams, when their relative density $\bar{\rho}$ is low. We define relative density according to $\bar{\rho} \equiv \rho^*/\rho$, where ρ and ρ^* are the densities of the solid material and lattice material, respectively. Lattice materials are

Table 3.2 Resonant frequencies of the unit cell in Figure 3.9b. These are $\omega_1 = 0$, $\omega_2 = \omega_3 = \sqrt{\frac{4k}{m}}$ (repeated twice), $\omega_4 = \sqrt{\frac{8k}{m}}$ for free boundary condition, and $\omega_0 = 0$ for fixed boundary condition. Note that the natural frequencies of the unit cell are normalized using $\Omega_n = \omega_n / \sqrt{\frac{k}{m}}$.

Symmetry path/point	Calculated bound using Eq. (3.21)	Predicted bound	Unit-cell boundary condition
Γ	$\Omega = 0$	$\Omega_1 = 0$	Fixed-fixed
At X in $\Gamma - X$	$\Omega = 2$	$\Omega_2 = 2$	Free–free
At M in $\Gamma - M$	$\Omega = \sqrt{8}$	$\Omega_4 = \sqrt{8}$	Free–free
At Y in $\Gamma - Y$	$\Omega = 2$	$\Omega_3 = 2$	Free–free

spatially discrete and cellular at macroscopic dimensions, but at the constituent-element level, their dynamic response is governed by that of a *continuous* structural element, namely, a beam or a shell. The reader is referred to Chapter 1 for background on these materials and their innovative application as lightweight multifunctional materials, with extraordinary stiffness and strength that is not obtainable using conventional composite or homogeneous materials. Given that a beam has infinitely many resonances it is conceivable that lattice materials are the most general locally resonant metamaterials. The reader is referred to Chapter 8, on pentamode lattice materials, for the application of lattices as a 'meta fluid', which has low shear resistance but can resist compression. Other lattice topologies, inspired by origami and chiral architectures [69] and topological order effects [70], are also the subject of growing interest.

We next explore wave propagation phenomena, such as band gaps and wave directionality, in a set of planar lattice materials; see Figure 3.11. Three regular honeycombs – hexagonal, square, and triangular – and a semi-regular Kagome lattice are considered. Additionally, we consider a pentamode lattice geometry. The wavevectors are restricted to the edges of the irreducible part of the first IBZ, shown by the shaded region $\Gamma - X - M$. The basis vectors for the direct lattice and reciprocal lattice and the Brillouin zone points are given in the literature [33] for the first four lattice geometries. The pentamode lattice in Figure 3.11e has the same symmetries as a hexagonal latttice and hence the wavevector trajectory along the first IBZ remain the same as that of a hexagonal lattice.

In all these geometries, the constituent beams are of uniform length L and depth d, and 2D prismatic topologies of unit thickness into the page are considered. The hexagonal and triangular honeycombs and the Kagome lattice have isotropic in-plane effective properties, while the square honeycomb is strongly anisotropic [32, 39]. The effective elastic properties of these microstructures are summarized in Table 3.3. It is anticipated that the results based upon effective medium theory are the asymptotes to the dispersion curves in the long-wavelength limit at zero frequency. Clearly, the lattices have different long-wavelength deformation limits. It is of particular interest to know whether the finite-frequency short-wavelength deformation behavior of the lattices shows similar differences. To explore this, dispersion curves of the five lattices are computed using Floquet–Bloch theory.

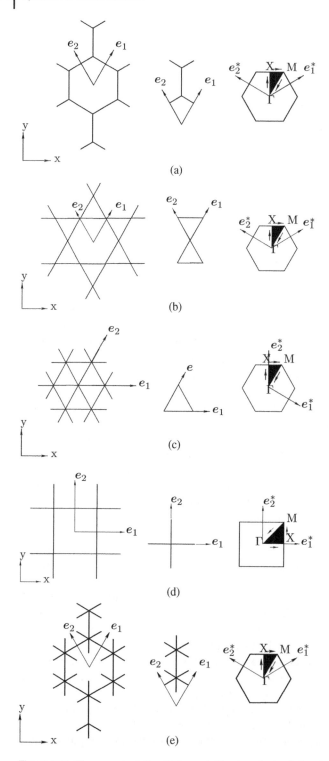

Figure 3.11 Five representative 2D beam lattice topologies (left column), along with the unit cell (middle column) and the first Brillouin zone (right column): (a) hexagonal lattice; (b) Kagome lattice; (c) triangular lattice; (d) square lattice; (e) pentamode lattice.

Table 3.3 Effective elastic properties of the microstructures in Figure 3.11.

Topology	$\bar{\rho}$		$\bar{K} = \dfrac{K^*}{E}$	$\bar{G} = \dfrac{G^*}{E}$	v^*	Isotropic
Triangular honeycomb	$2\sqrt{3}\left(\dfrac{d}{L}\right) = \dfrac{12}{\lambda}$		$\dfrac{1}{4}\bar{\rho}$	$\dfrac{1}{8}\bar{\rho}$	$\dfrac{1}{3}$	Yes
Hexagonal honeycomb	$\dfrac{2}{\sqrt{3}}\left(\dfrac{d}{L}\right) = \dfrac{4}{\lambda}$		$\dfrac{1}{2}\bar{\rho}$	$\dfrac{3}{8}\bar{\rho}^3$	1	Yes
Kagome lattice	$\sqrt{3}\left(\dfrac{d}{L}\right) = \dfrac{6}{\lambda}$		$\dfrac{1}{4}\bar{\rho}$	$\dfrac{1}{8}\bar{\rho}$	$\dfrac{6 - \bar{\rho}^2}{18 + \bar{\rho}^2} \approx \dfrac{1}{3}$	Yes
Square honeycomb	$2\left(\dfrac{d}{L}\right) = \dfrac{4\sqrt{3}}{\lambda}$		$\dfrac{1}{4}\bar{\rho}$	$\dfrac{1}{16}\bar{\rho}^3$		No

E, Young's modulus of solid material; K^*, G^*, effective bulk modulus and shear modulus of cellular solid, respectively; $\bar{\rho} \equiv \rho^*/\rho$, relative density of cellular solid; ρ, ρ^*, densities of the solid and lattice material, respectively; *d*, *L*, thickness and length of the cell walls, respectively; λ, slenderness ratio $\lambda = 2\sqrt{3}L/d$

3.4.1 Finite Element Modelling of the Unit Cell

Each lattice is considered to be a rigid-jointed network of beams with no pre-stress. The unit cell is discretized into a network of Timoshenko beams. Each beam is assumed to have three degrees of freedom at each end: two translations (u, v) in the (x, y) plane and a rotation θ_z about the z-axis. The continuous variation of these displacements within a typical beam element, shown in Figure 3.12, is approximated by:

$$u(x, t) = \sum_{r=1}^{6} a_r(x)q_r(t) \quad v(x, t) = \sum_{r=1}^{6} b_r(x)q_r(t) \quad \theta_z(x, t) = \sum_{r=1}^{6} c_r(x)q_r(t) \quad (3.22)$$

where *x* is measured along the axis of the beam, and the degrees of freedom q_r consist of the six nodal degrees of freedom $(u_1, v_1, \theta_{z1}, u_2, v_2, \theta_{z2})$. The shape functions a_r, b_r, and c_r, $(r = 1 \ldots 6)$ for the six nodal displacements can be found in the literature [63, 71].

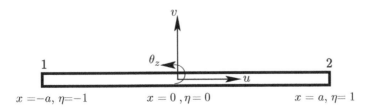

Figure 3.12 Beam element with nodes numbered 1 and 2. The three nodal degrees of freedom are shown, together with the local element coordinate axes with origin located at the middle of the beam. The nondimensional coordinate is $\eta = x/a = 2x/L$, where *L* is the length of the beam.

The kinetic and potential energies per unit thickness of the beam into the page are given by:

$$T = \frac{1}{2}\int_{-L/2}^{L/2}\rho d\dot{u}^2 dx + \frac{1}{2}\int_{-L/2}^{L/2}\rho d\dot{v}^2 dx + \frac{1}{2}\int_{-L/2}^{L/2}\rho I_z\dot{\theta}_z^2 dx$$

$$U = \frac{1}{2}\int_{-L/2}^{L/2}Ed\left(\frac{du}{dx}\right)^2 dx + \frac{1}{2}\int_{-L/2}^{L/2}EI_z\left(\frac{d\theta_z}{dx}\right)^2 dx$$

$$+ \frac{1}{2}\int_{-L/2}^{L/2}\kappa_s dG\left(\frac{dv}{dx} - (\theta_z)\right)^2 dx \tag{3.23}$$

where ρ is the density of the material used to make the lattice. L and I_z denote the length and second moment of area of the beam, respectively, and κ_s denotes the shear correction factor used in Timoshenko beam theory [61]. Substituting Eq. (3.22) into Eq. (3.23) gives:

$$T = \frac{1}{2}\sum_{r=1}^{6}\sum_{s=1}^{6}\dot{q}_r\dot{q}_s\int_{-L/2}^{L/2}(\rho da_r a_s + \rho db_r b_s + \rho I_z c_r c_s)dx$$

$$U = \frac{1}{2}\sum_{r=1}^{6}\sum_{s=1}^{6}q_r q_s\int_{-L/2}^{L/2}(Eda_r' a_s' + EI_z b_r'' b_s'' + \kappa_s Gd(b_r' - c_r)(b_s' - c_s))dx \tag{3.24}$$

where the primes denote differentiation with respect to the axial coordinate x. The equations of motion are obtained by applying Hamilton's variational principle

$$\delta\int \mathcal{L}dt = 0, \quad \mathcal{L} = T - U + W_e \tag{3.25}$$

where δ denotes the first variation, $\mathcal{L}$ is the Lagrangian of the dynamical system as defined above, and W_e denotes the work done by the external forces. Upon evaluating the first variation in Eq. (3.25), the Euler–Lagrangian equations of motion for the dynamics of the beam element are obtained as:

$$\frac{d}{dt}\left(\frac{\partial\mathcal{L}}{\partial\dot{q}_r}\right) - \frac{\partial\mathcal{L}}{\partial q_r} = f_r \tag{3.26}$$

where f_r is the force corresponding to the degree of freedom q_r. The above equations of motion can be written for each beam element, and the assembled equation of the motion for the unit cell takes the form:

$$M\ddot{q} + Kq = f. \tag{3.27}$$

where the matrices M, K denote the assembled global mass and stiffness matrix of the unit cell, respectively. The vectors q, $\ddot{q}$ and f respectively denote the nodal displacements, accelerations and forces. For any jth node, the nodal displacement vector is given by $q_j = [u_j\ v_j\ \theta_{zj}]^T$.

Having thus obtained the governing equations of motion in Eq. (3.27), which are identical to Eq. (3.2), we can formulate the Hermitian algebraic Bloch eigenvalue problem in Eq. (3.21) by following the steps already outlined in Section 3.3.1.

3.4.2 Band Structure of Lattice Topologies

The computational procedure adopted to calculate the dispersion surfaces (band structure) for the selected lattices in Figure 3.11 is as follows:

1. Select a primitive unit cell of the lattice.
2. Construct the mass and stiffness matrices of the unit cell using the FE technique described in Section 3.4.1.
3. Apply Bloch's theorem to the equations of motion of the unit cell and form the eigenvalue problem in Eq. (3.21).
4. Specify the phase constants (μ_1, μ_2) by restricting the wavevector to the edges of the irreducible part of the first Brillouin zone.
5. Solve the resulting linear algebraic eigenvalue problem in Eq. (3.21) for the wave propagation frequencies.

To explore the band gaps it is sufficient to choose wavevectors along the edges of the first irreducible Brillouin zone [56]. Instead of solving the eigenvalue problem in Eq. (3.21) for each pair (μ_1, μ_2) over the shaded region $\Gamma - X - M$ in the first Brillouin Zone, one need only explore the edges of the triangle $\Gamma - X - M$. The parameter s is introduced as the arc length along the perimeter $\Gamma - X - M - \Gamma$ of the shaded region $\Gamma - X - M$ in the first Brillouin Zone. It is a scalar path-length parameter in k-space and is used to denote the location of any point on the perimeter. Thus the extremes of the frequency on a 3D dispersion surface can be represented by a 2D dispersion curve, with the wavevector as the abscissa and the frequency as ordinate. The band gaps now correspond to regions along the ordinate wherein no dispersion branch is present. Within this frequency band wave motion cannot occur and hence these are stop bands. In contrast, the frequency values for which there is at least one dispersion curve correspond to the pass bands. *Partial* band gaps can exist, whereby waves in selected directions are forbidden to propagate. The long-wavelength limit corresponds to the origin, as denoted by point Γ in the first Brillouin zone.

The band structure of each lattice is computed by solving the eigenvalue problem in Eq.(3.21) for wavevectors along the closed locus $\Gamma - X - M - \Gamma$ in k-space. Results are presented for the slenderness ratio $(\lambda = 2\sqrt{3}L/d)$, equal to 50. Recall that the relative density scales directly with the slenderness ratio, and the scaling factor for each geometry is summarized in Table 3.3. Pentamode geometry has symmetry similar to a hexagonal lattice, hence only the dispersion curves for this geometry are discussed; see Chapter 8 for the elastostatic properties. The dispersion curves for slenderness ratios of 50 is shown in Figure 3.13 for a hexagonal lattice. A nondimensional wave-propagation frequency is plotted on the vertical axis as a function of the wavevector locus $\Gamma - X - M - \Gamma$ along the edges of the first IBZ, using the arc length parameter s in k-space. The nondimensional frequency Ω is defined as:

$$\Omega = \frac{\omega}{\omega_1} \qquad (3.28)$$

where ω is the frequency of the plane wave obtained by solving the eigenvalue problem of Eq.(3.21), and $\omega_1 = \pi^2 \sqrt{(EI/\rho dL^4)}$ is the first pinned–pinned flexural resonance frequency of a lattice beam. Consequently, at $\Omega = 1$ the cell deformation exhibits the first pinned–pinned flexural mode of the beam. All of the parameters are as defined in Section 3.4.1.

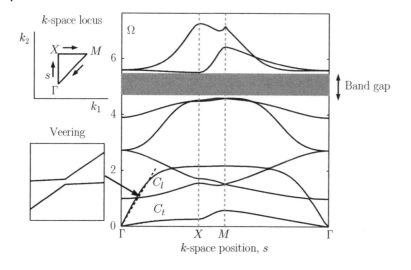

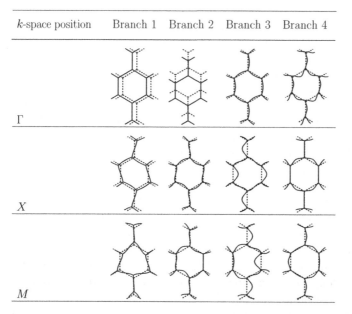

Figure 3.13 Band structure of a hexagonal honeycomb with slenderness ratio equal to 50. The eigenwaves of a typical cell are shown in tabular form. The three rows correspond to the three points Γ, *X* and *M* in *k*-space, while the four columns correspond to the first four dispersion branches in ascending order.

The point Γ corresponds to the long-wavelength limit at which the effective medium representation of the lattice remains valid. Two branches of the dispersion curve emanate from the origin Γ: these are the longitudinal waves (also known as irrotational or dilatational waves) and transverse waves (also known as distortional, shear or equivoluminal waves). Recall that the tangent to the dispersion curve at any point gives the group velocity, while the secant slope of the line connecting the origin Γ to the point of interest on the dispersion curve gives the phase velocity. The two group velocities corresponding to the dispersion branches in the long-wavelength limit correspond to those of an effective elastic medium with modulus as given in Table 3.3 for the hexagonal honeycomb. For an isotropic medium with Young's modulus E^*, shear modulus G^*, bulk modulus K^* and density ρ^* the group velocities are given by [32, 72]:

$$C_l = \sqrt{\frac{K^* + G^*}{\rho^*}}, \quad C_t = \sqrt{\frac{G^*}{\rho^*}}, \tag{3.29}$$

Thus the two lines starting at O and with slopes corresponding to the group velocities of C_l and C_t are a best approximation to the dispersion curves in the long-wavelength limit. The deviation of these long-wavelength asymptotes from the dispersion curves indicates the range of validity of effective medium theories: at higher frequencies of wave propagation the effective-medium results in Table 3.3 do not apply. At these frequencies the detailed geometry of the lattice has a profound influence upon the wave-bearing properties.

A common feature of the dispersion curves for the hexagonal honeycomb (and for the other topologies) is the phenomenon of veering of frequencies (or repulsion of the dispersion branches), where the dispersion curves approach one another very closely but avoid crossing each other along the locus $\Gamma - X - M - \Gamma$ in k-space. Such veering can be observed in Figure 3.13 between the second and third branches of the dispersion curve along the locus $\Gamma - X$. From the magnified picture of the veering zone, it can be noted that the eigenvalues do not cross, but veer away from each other. Avoided crossing is a ubiquitous feature of eigenvalue problems of weakly coupled systems [73, 74]. In lattice structures at micro and nano length scales, the phenomenon has potential applications in ultrasensitive sensors [75].

The eigenwaves of a typical honeycomb cell for each dispersion curve at the points Γ, X, and M in k-space are summarized in Figure 3.13. The three rows in each table correspond to the three points Γ, X, and M, while the four columns correspond to the first four dispersion branches in ascending order. Consider the four eigenwaves in the first column of Figure 3.13. At the point Γ, the unit cell exhibits rigid-body translation. Transverse wave motion occurs along the first branch, with increasing wavenumber; the transverse nature of the eigenwaves is clear from the cell deformation at point X. The second column of the table gives the second branch of the dispersion curve and this comprises a longitudinal wave. In higher branches, the eigenwave exhibits combined transverse and longitudinal motion. A complete band gap (shaded space) between the sixth and the seventh branches of the dispersion curves can be observed in Figure 3.13.

The dispersion responses of the Kagome lattice and of the triangular honeycomb and square honeycomb are similarly shown in a plot of Ω versus length parameter s in k-space in Figure 3.14, for slenderness ratios of 50. The long-wavelength asymptotes, as calculated from the group velocities using Eq. (3.29), are superimposed on the plots and they agree with the dispersion curves. The Kagome lattice does not exhibit any complete band gaps for the slenderness ratio considered. It is instructive to compare the long-wavelength behavior of transverse waves in the Kagome and honeycomb lattices. For the honeycomb, the slope of the dispersion curve associated with the transverse wave speed decreases with an increase in slenderness ratio, while the dispersion curves of the Kagome lattice do not show this behavior. This can be explained as follows. The transverse wave speed depends upon the ratio of effective shear modulus G^* to density

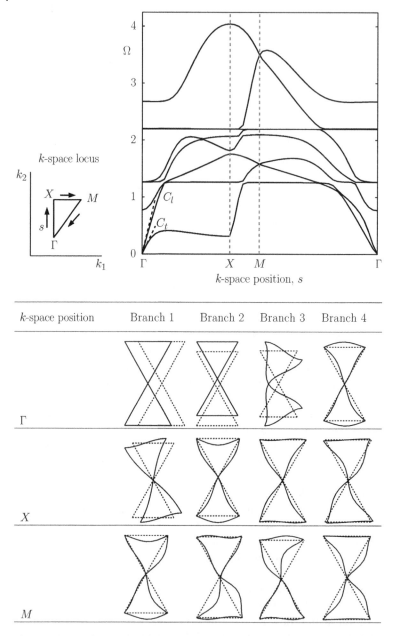

Figure 3.14 Band structure of a Kagome lattice with slenderness ratio equal to 50. The eigenwaves of a typical cell are shown in tabular form. The three rows correspond to the three points Γ, X and M in k-space, while the four columns correspond to the first four dispersion branches in ascending order.

according to Eq.(3.29). Now G^* scales as ρ^{*^3} for the hexagonal honeycomb, whereas G^* scales as ρ^* for the Kagome lattice. Consequently, the transverse wave decreases with an increase in slenderness ratio for the hexagonal honeycomb but not for the Kagome lattice.

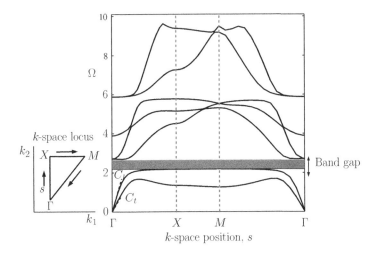

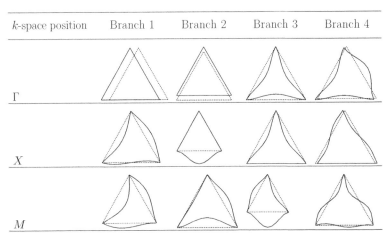

Figure 3.15 Band structure of a triangular honeycomb with slenderness ratio equal to 50. The eigenwaves of a typical cell are shown in tabular form. The three rows correspond to the three points Γ, X and M in k-space, while the four columns correspond to the first four dispersion branches in ascending order.

The dispersion results for the triangular lattice are plotted in Figure 3.15, for slenderness ratio 50. Again, the long-wavelength asymptotes agree with the dispersion curve. A complete band gap (shaded region) is present. But the location of the band gap depends upon slenderness ratio. Refer to the literature [33] for a full discussion on the influence of the slenderness ratio on the band gap.

The dispersion results for the square lattice are provided in Figure 3.16. The group velocities for the longitudinal and shear waves in the long-wavelength limit are given by [32, 72]:

$$C_l = \sqrt{\frac{2K^*}{\rho^*}}, \quad C_t = \sqrt{\frac{G^*}{\rho^*}} \tag{3.30}$$

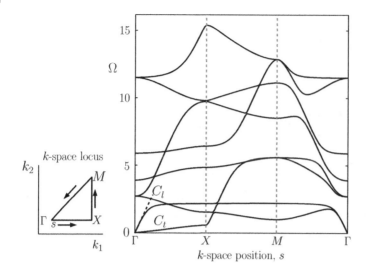

Figure 3.16 Band structure of a hexagonal honeycomb with slenderness ratio equal to 50. The eigenwaves of a typical cell are shown in tabular form. The three rows correspond to the three points Γ, X and M in k-space, while the four columns correspond to the first four dispersion branches in ascending order.

where the effective bulk modulus (K^*) and shear modulus (G^*) are as given in Table 3.3. As before, the long-wavelength asymptotes, as calculated from the group velocities via Eq. (3.30), agree well with the dispersion curve. The general features are similar to those already discussed for the other lattices: the first and second branches comprise shear waves and longitudinal waves, respectively. Pinned–pinned eigenmodes exist at $\Omega = 1$. Above this frequency value the dispersion curves tend to cluster. No complete band gaps are observed.

A remarkable feature of *anamolous* dispersion, in the form of a negative group velocity at sonic speed, can be observed in the pentamode lattice's dispersion curves along the $\Gamma - X$ locus for the fifth branch; see Figure 3.17. This negative group velocity (or negative index) property enables exotic wave manipulations, such as wave focusing; in other words, an acoustic lens behavior in a selected frequency region. Such metamaterial concepts are of widespread ongoing interest to materials scientists and engineers alike.

Finally, a comment on the connection between band gap and nodal connectivity may be worth making here. From the previous discussion, it would appear that a nodal connectivity of four leads to band gap emergence. However, this is not entirely correct. For example, by placing a point mass on the lattice beams it is possible to alter sub-Bragg or Bragg band gaps.

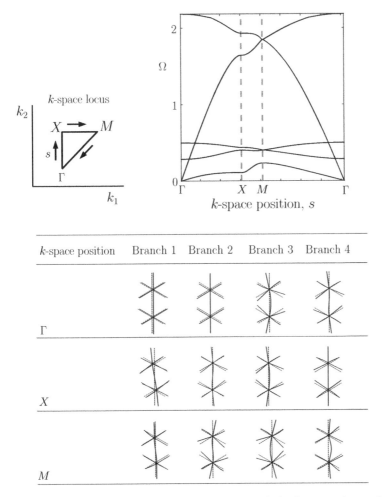

Figure 3.17 Band structure of a pentamode lattice with slenderness ratio equal to 50 in the interests of brevity. The eigenwaves of a typical cell are shown in tabular form. The three rows correspond to the three points Γ, X and M in k-space, while the four columns correspond to the first four dispersion branches in ascending order.

3.4.3 Directionality of Wave Propagation

The directionality of waves indicates the degree to which a medium is isotropic with respect to wave propagation. In an isotropic medium no preferred directions exist, and waves propagate equally in all directions. A common technique for displaying directionality is to construct iso-frequency contours of the dispersion surface and to plot these contours in a Cartesian reference frame in k-space. The physical coordinate system (x, y) has already been introduced for each lattice in Figure 3.11. The wavevectors (k_x, k_y) are aligned with these physical orthonormal vectors, and form the axes of the iso-frequency plots shown in Figure 3.18 for each lattice at a slenderness ratio of 50. Contours are given for selected nondimensional frequencies Ω.

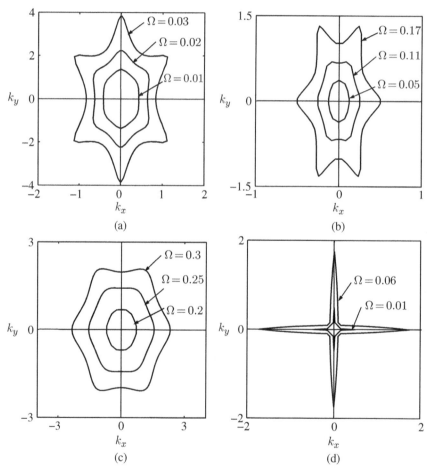

Figure 3.18 Directionality of plane-ave propagation in the four lattice topologies with slenderness ratio equal to 50: (a) hexagonal honeycomb; (b) Kagome lattice; (c) triangular honeycomb; (d) square honeycomb. The nondimensional frequency (Ω) associated with each contour is labelled. At each frequency, energy flow direction is given by the normal to the contour and is in the direction of maximum rate of change of frequencies.

The relative shapes of the iso-frequency contours are compared in Figure 3.18 for the four topologies of interest. The symmetries of these contours show the rotational and reflective symmetries of the parent lattices. At low frequencies, the triangular and hexagonal honeycomb and Kagome lattice are isotropic, whereas the square honeycomb is strongly anisotropic. At high frequencies, the lattice symmetries have a strong influence upon directionality.

3.5   Tunneling and Evanescent Waves

Much of the discussion so far has been confined to *purely* propagating waves without attenuation, where the wave components supplied to the Bloch eigenvalue problem in Eq. (3.21) are real, and the frequency is real for propagating waves lying on the dispersion surface. Evanescent waves, or inhomogeneous waves, with complex wavenumbers exhibit the property that they propagate along a certain direction but their amplitudes decay in one of the perpendicular directions. These waves lie within the partial band gaps of a lattice material. The long-wavelength zero-frequency evanescent waves are associated with Saint Venant elastic boundary layers [76]. Surface waves in lattice materials have been studied in the long-wavelength limit [76] and for a finite frequency based on the solutions of quadratic and linear eigenvalue problems [77].

An inhomogeneous plane wave is mathematically represented as $A \exp(-i\mathbf{k}.\mathbf{r})$, where $\mathbf{k}$ is complex. It is possible to find real frequency solutions for the Bloch eigenvalue problem in Eq. (3.21) for complex values of k_1 and k_2. Recall that a complex wave number k_1 or k_2 would represent a wave with a spatially oscillating and decaying amplitudes. One has to distinguish between pure evanescent waves, associated with no spatial oscillation but decay only, identified by purely real k_1 and k_2 values, and spatially oscillating and decaying (leaky) waves. Such *leaky* waves can carry energy., but while a pure evanescent wave, with exponentially decaying spatial amplitude, cannot carry energy on its own, it can interact with other evenescent waves to transmit energy. For example, a left-going evanescent wave incident on the free boundary of a beam generates a right-going reflected evanescent wave and a left-going transmitted wave in the beam that is propagating, see the literature for an example [78, 79]. The existence of such a propagating wave that carries energy away from the boundary into the medium can be argued from Snell's law or the reciprocity principle [80].

A more quantitative argument is obtained by Snell's law; see Graff [81], p. 316 and Brekhovskikh [82], Section 32 and p. 5 for example. Consider the interface between two isotropic elastic media. Let c_1 and c_2 denote the phase velocities in the first and the second medium respectively. A wave, originating in medium 1 and incident at an angle θ_1 on the interface separating the media 1 and 2, will generate a refracted (transmitted) wave in medium 2 at angle θ_r and a reflected wave in medium 1. Our attention is focused on the refracted wave. For the refracted wave, by Snell's law, one has $n \sin \theta_r = \sin \theta_1$, where $n \equiv \frac{c_1}{c_2}$ is an elastic equivalent of 'refractive index'[3]. For $n < 1$ and $\sin \theta_1 > n$, we have,

---

3 It must be observed that elastic wave propagation exhibits significant differences to electromagnetic wave propagation. These arise from the fact that there is only one wave velocity for an electromagnetic wave propagating in an isotropic medium, whereas an elastic wave has two velocities (pressure (P) and shear (S) velocities) in any given direction. Hence elastic waves exhibit birefringence, even in isotropic media,

from Snell's law, $\sin\theta_r > 1$; that is, the refracted wave is inhomogeneous (evanescent) since θ_r is complex. This situation corresponds to total internal reflection. This total internal reflection can be frustrated by an evanescent wave; see Figure 33-11 in Volume II of Feynman's lectures on physics [83] for a beautiful illustration for centimetre waves. Frustrated total internal reflection is a generic phenomenon relevant to light and electromagnetic waves as well. Here, an incident inhomogenous (evanescent) wave generates a homogeneous (propagating) wave in a medium. This can be explained by $\sin\theta_1 < n$ with $n > 1$, which gives $\sin\theta_r < 1$. Thus evanescent waves emanating from medium 1 will convert into propagating refracted waves in the medium 2, and carry energy away from the interface. Thus the incident wave energy is tunneled through many such conversions in a layered elastic medium comprising alternating high- and low-speed media. Refer to the literature [84, 85] for more recent studies on wave tunneling in phononic crystals

To further consolidate the preceding ideas in the context of structural dynamics, consider an evanescent wave $w_i(x) = Ae^{kx}$ incident on the free boundary $x = 0$ of a semi-infinite, prismatic flexural beam extending from $x = 0$ to $x = \infty$. This is a simple 1D system, but nonetheless it captures the essential feature of evanescent waves. The fourth-order governing partial differential equation for the flexural beam:

$$EI\frac{\partial^4 w}{\partial x^4} + \rho A\frac{\partial^2 w}{\partial t^2} = 0 \tag{3.31}$$

has the wave-field solution comprising the incident wave Ae^{kx} and two reflected waves: $r_1 e^{ikx}$ propagating energy to the right and an evanescent wave $r_2 e^{-kx}$. Thus, the total wave field with *complex* wave amplitude r_1 and r_2 is

$$w(x, t) = Ae^{kx} + r_1 e^{-ikx} + r_2 e^{-kx}. \tag{3.32}$$

The reflected wave amplitudes r_1 and r_2 (complex numbers) must be found by enforcing free boundary conditions on the wave field at the free boundary ($x = 0$): $\frac{\partial^2 w}{\partial x^2} = 0$ for zero net bending moment and $\frac{\partial^3 w}{\partial x^3} = 0$ for zero net shear force. These two conditions yield $r_1 = -iA$ and $r_2 = A(1 - i)$, which gives the wave field in the beam as

$$w(x, t) = Ae^{kx} + A(1 - i)e^{ikx} - iAe^{-kx}. \tag{3.33}$$

From the above it is clear that the two evanescent waves (incident and reflected), decaying in mutually opposite directions, interact to produce a propagating wave that carries energy to the right. The energy for the propagating wave arises from the interactions of the two evanescent waves, ensuring the conservation of energy. Specifically, the bending moments and shear forces created within the beam by the incident wave Ae^{kx} do net nonzero work on the displacements produced by the reflected wave $r_2 e^{-kx}$, which gives rise to the propagating wave $r_1 e^{-ikx}$. Admittedly, this is an idealized example. It is not difficult to foresee that such phenomena can exist in lattice materials and structures made of beam assemblies. Since evanescent waves fall within the band gap, one needs to be aware of these in studying energy transmission in finite and semi-infinite lattice materials and structures. For infinite systems one cannot have this phenomenon, in order to preserve the requirement of boundedness of the wave field at infinity [82].

---

whereby two refracted (transmitted) waves are generated by a single incident wave. In anisotropic solid, trirefringence is possible.

While recent work has explored the tunneling phenomenon in phononic crystals [84, 85], much remains to be discovered in the case of truss lattice materials and structures. To conclude this discussion, one must exercise care in interpreting the experimentally measured transmission spectra of finite (or semi-infinite) lattice materials and structures, since tunneling is possible.

3.6 Concluding Remarks

This chapter has reviewed essential phenomena present in the elastodynamic responses of 1D and 2D lattices. Bloch theorem forms the cornerstone to compute dispersion relations. 1D and 2D lattices exhibit band gap phenomena, whereby free waves are forbidden from propagating over a certain frequency interval. Thus, lattice materials act as filters for waves in the frequency domain. In 2D lattices, an additional feature – directivity – has been observed, suggesting that 2D and 3D lattices act as filters in both the frequency and wavenumber (spatial) domain. The significance of unit-cell resonances, and their use in designing band gap materials with symmetric unit cells is underscored. Surface waves and evanescent waves are briefly discussed. This chapter has focused on linear undamped waves in one and 2D lattices. Damped lattices will be studied in Chapter 4. Chapters 5 and 6 will address the nonlinear and stability aspects of lattices, respectively. Given the rapid strides in manufacturing, theoretical development, and applications, it is safe to surmise that future studies on lattice materials will yield significant insights on wave phenomena at a theoretical level and lead to wide-ranging multidisciplinary applications, from molecular- to machine-scale lattice materials, structures and devices.

3.7 Acknowledgments

Parts of this work are funded by Discovery Grant program and Canada Research Chairs program of the National Sciences and Engineering Research Council (NSERC) of Canada. Assistance with figures from my present and past graduate students – Ehsan Moosavimehr, Behrooz Yousefzadeh and Lalitha Raghavan – is gratefully acknowledged. Discussions with Profs. J. Woodhouse, R.S. Langley, and N.A. Fleck from Cambridge University are acknowledged.

References

1 N. A. Fleck, V. S. Deshpande, and M. F. Ashby, "Micro-architectured materials: past, present and future," *Proceedings of the Royal Society A: Mathematical, Physical and Engineering Science*, vol. 466, no. 2121, pp. 2495–2516, 2010.

2 T. A. Schaedler, A. J. Jacobsen, A. Torrents, A. E. Sorensen, J. Lian, J. R. Greer, L. Valdevit, and W. B. Carter, "Ultralight metallic microlattices," *Science*, vol. 334, no. 6058, pp. 962–965, 2011.

3 L. R. Meza, S. Das, and J. R. Greer, "Strong, lightweight, and recoverable three-dimensional ceramic nanolattices," *Science*, vol. 345, no. 6202, pp. 1322–1326, 2014.

4 Strutt, J.W. (Lord Rayleigh), "On the maintenance of vibrations by forces of double frequency, and on the propagation of waves through a medium endowed with a periodic structure," *Philosophical Magazine Series 5*, vol. 24, no. 147, pp. 145–159, 1887.

5 L. Brillouin, *Wave Propagation in Periodic Structures*. Dover Publications, 2nd ed., 1953.

6 E. H. Lee and W. H. Yang, "On waves in composite materials with periodic structure," *SIAM Journal on Applied Mathematics*, vol. 25, no. 3, pp. 492–499, 1973.

7 D. J. Mead, "Wave propagation and natural modes in periodic systems: I. Mono-coupled systems," *Journal of Sound and Vibration*, vol. 40, no. 1, pp. 1–18, 1975.

8 L. Raghavan and A. S. Phani, "Local resonance bandgaps in periodic media: Theory and experiment," *The Journal of the Acoustical Society of America*, vol. 134, no. 3, pp. 1950–1959, 2013.

9 Strutt, J.W. (Lord Rayleigh), "On the remarkable phenomenon of crystalline reflexion described by Prof. Stokes," in *Scientific Papers*, vol. 3, pp. 204–212, Cambridge University Press, 2009.

10 W. Thomson, *Baltimore Lectures on Molecular Dynamics and the Wave Theory of Light*. Cambridge University Press, 2010.

11 G. W. Stewart, "Acoustic wave filters," *Physical Review*, vol. 20, pp. 528–551, 1922.

12 G. W. Stewart, "Acoustic wave filters; an extension of the theory," *Physical Review*, vol. 25, pp. 90–98, 1925.

13 M. Paidoussis, "High-pass acoustic filters for hydraulic loops," *Journal of Sound and Vibration*, vol. 14, no. 4, pp. 433–437, 1971.

14 R. Johnson, M. Borner, and M. Konno, "Mechanical filters - A review of progress," *Sonics and Ultrasonics, IEEE Transactions on*, vol. 18, no. 3, pp. 155–170, 1971.

15 R. Johnson, *Mechanical Filters in Electronics*. Wiley Series on Filters: Design, Manufacturing and Applications, John Wiley & Sons, 1983.

16 C.-C. Nguyen, "Frequency-selective MEMSmead1996turized low-power communication devices," *Microwave Theory and Techniques, IEEE Transactions on*, vol. 47, no. 8, pp. 1486–1503, 1999.

17 M. A. Heckl, "Investigations on the vibrations of grillages and other simple beam structures," *The Journal of the Acoustical Society of America*, vol. 36, no. 7, pp. 1335–1343, 1964.

18 E. E. Ungar, "Steady state responses of one dimensional periodic flexural systems," *The Journal of the Acoustical Society of America*, vol. 39, no. 5A, pp. 887–894, 1966.

19 Y. Lin and B. Donaldson, "A brief survey of transfer matrix techniques with special reference to the analysis of aircraft panels," *Journal of Sound and Vibration*, vol. 10, no. 1, pp. 103–143, 1969.

20 D. Mead, "Wave propagation in continuous periodic structures: research contributions from Southampton 1964–1995," *Journal of Sound and Vibration*, vol. 190, no. 3, pp. 495–524, 1996.

21 R. M. Orris and M. Petyt, "A finite element study of harmonic wave propagation in periodic structures," *Journal of Sound and Vibration*, vol. 33, no. 2, pp. 223–236, 1974.

22 W. Zhong and F. Williams, "On the direct solution of wave propagation for repetitive structures," *Journal of Sound and Vibration*, vol. 181, no. 3, pp. 485–501, 1995.

23 B. Mace, "Wave reflection and transmission in beams," *Journal of Sound and Vibration*, vol. 97, no. 2, pp. 237–246, 1984.

24 R. Langley, "A variational principle for periodic structures," *Journal of Sound and Vibration*, vol. 135, no. 1, pp. 135–142, 1989.

25 R. Langley, N. Bardell, and H. Ruivo, "The response of two-dimensional periodic structures to harmonic point loading: a theoretical and experimental study of beam grillage," *Journal of Sound and Vibration*, vol. 207, no. 4, pp. 521–535, 1997.

26 J. M. Renno and B. R. Mace, "Vibration modelling of structural networks using a hybrid finite element/wave and finite element approach," *Wave Motion*, vol. 51, no. 4, pp. 566–580, 2014.

27 C. H. Hodges and J. Woodhouse, "Theories of noise and vibration transmission in complex structures," *Reports on Progress in Physics*, vol. 49, no. 2, p. 107, 1986.

28 M. S. Kushwaha, P. Halevi, L. Dobrzynski, and B. Djafari-Rouhani, "Acoustic band structure of periodic elastic composites," *Physical Review Letters*, vol. 71, pp. 2022–2025, 1993.

29 C. T. Sun, J. D. Achenbach, and G. Herrmann, "Continuum theory for a laminated medium," *Journal of Applied Mechanics*, vol. 35, no. 3, pp. 467–475, 1968.

30 S. Mukherjee and E. H. Lee, "Dispersion relations and mode shapes for waves in laminated viscoelastic composites by finite difference methods," *Computers & Structures*, vol. 5, pp. 279–285, 1975.

31 W. Warren and K. A.M., "Foam mechanics: the linear elastic response of two-dimensional spatially periodic cellular materials," *Mechanical of Materials*, vol. 6, pp. 27–37, 1987.

32 S. Torquato, L. V. Gubiansky, M. J. Silva, and L. J. Gibson, "Effective mechanical and transport properties of cellular solids," *International Journal of Mechanical Sciences*, vol. 40, no. 1, pp. 71–82, 1998.

33 A. S. Phani, J. Woodhouse, and N. A. Fleck, "Wave propagation in two-dimensional periodic lattices," *The Journal of the Acoustical Society of America*, vol. 119, no. 4, pp. 1995–2005, 2006.

34 R. M. Chsitensen, "Mechanics of cellular and other low density materials," *International Journal of Solids and Structures*, vol. 37, no. 1, pp. 93–104, 2000.

35 N. Wicks and J. W. Hutchinson, "Optimal truss plates," *International Journal of Solids and Structures*, vol. 38, pp. 5165–5183, 2001.

36 A. Evans, J. Hutchinson, N. Fleck, M. Ashby, and H. Wadley, "The topological design of multifunctional cellular metals," *Progress in Materials Science*, vol. 46, no. 3–4, pp. 309–327, 2001.

37 M. I. Hussein, G. M. Hulbert, and R. A. Scott, "Dispersive elastodynamics of 1D banded materials and structures: Analysis," *Journal of Sound and Vibration*, vol. 289, no. 4–5, pp. 779–806, 2006.

38 M. I. Hussein, G. M. Hulbert, and R. A. Scott, "Dispersive elastodynamics of 1D banded materials and structures: Design," *Journal of Sound and Vibration*, vol. 307, no. 3–5, pp. 865–893, 2007.

39 L. J. Gibson and M. F. Ashby, *Cellular Solids: Structure and Properties*. Cambridge University Press, 2nd ed., 1997.

40 M. Ashby, "Hybrid materials to expand the boundaries of material-property space," *Journal of the American Ceramic Society*, vol. 94, pp. s3–s14, 2011.

41 M. I. Hussein, M. J. Leamy, and M. Ruzzene, "Dynamics of phononic materials and structures: Historical origins, recent progress, and future outlook," *Applied Mechanics Reviews*, vol. 66, no. 4, p. 040802, 2014.

42 G. W. Milton, M. Briane, and J. R. Willis, "On cloaking for elasticity and physical equations with a transformation invariant form," *New Journal of Physics*, vol. 8, no. 10, p. 248, 2006.

43 A. N. Norris, "Acoustic cloaking theory," *Proceedings of the Royal Society of London A: Mathematical, Physical and Engineering Sciences*, vol. 464, no. 2097, pp. 2411–2434, 2008.

44 J. B. Pendry and J. Li, "An acoustic metafluid: Realizing a broadband acoustic cloak," *New Journal of Physics*, vol. 10, no. 11, p. 115032, 2008.

45 T. Bückmann, M. Thiel, M. Kadic, R. Schittny, and M. Wegener, "An elasto-mechanical unfeelability cloak made of pentamode metamaterials," *Nat Commun*, vol. 5, 2014.

46 S. Yang, J. H. Page, Z. Liu, M. L. Cowan, C. T. Chan, and P. Sheng, "Focusing of sound in a 3D phononic crystal," *Physical Review Letters*, vol. 93, p. 024301, 2004.

47 Z. Y. Liu, X. X. Zhang, Y. W. Mao, Y. Y. Zhu, Z. Y. Yang, C. T. Chan, and P. Sheng, "Locally resonant sonic crystals," *Science*, vol. 289, no. 5485, pp. 1734–1736, 2000.

48 P. Deymier, *Acoustic Metamaterials and Phononic Crystals*. Springer, 2013.

49 R. V. Craster and S. Guenneau, *Acoustic Metamaterials: Negative Refraction, Imaging, Lensing and Cloaking*. Springer, 2012.

50 A. Khelif and A. Adibi, *Phononic Crystals: Fundamentals and Applications*. Springer, 2015.

51 T. W. Tan, G. R. Douglas, T. Bond, and A. S. Phani, "Compliance and longitudinal strain of cardiovascular stents: Influence of cell geometry," *Journal of Medical Devices*, vol. 5, no. 4, p. 041002, 2011.

52 E. M. K. Abad, D. Pasini, and R. Cecere, "Shape optimization of stress concentration-free lattice for self-expandable nitinol stent-grafts," *Journal of Biomechanics*, vol. 45, no. 6, pp. 1028–1035, 2012.

53 G. R. Douglas, A. S. Phani, and J. Gagnon, "Analyses and design of expansion mechanisms of balloon expandable vascular stents," *Journal of Biomechanics*, vol. 47, no. 6, pp. 1438–1446, 2014.

54 F. Bloch, "Über die quantenmechanik der elektronen in kristallgittern," *Zeitschrift für Physik*, vol. 52, no. 7–8, pp. 555–600, 1929.

55 N. Ashcroft and N. Mermin, *Solid State Physics*. Holt, Rinehart and Winston, 1976.

56 C. Kittel, *Elementary Solid State Physics: A Short Course*. John Wiley & Sons, 1st ed., 1962.

57 R. S. Langley, "A note on the forced boundary conditions for two-dimensional periodic structures with corner freedoms.," *Journal of Sound and Vibration*, vol. 167, no. 2, pp. 377–381, 1993.

58 O. Sigmund, "Tailoring materials with prescribed elastic properties," *Mechanics of Materials,* vol. 20, no. 4, pp. 351–368, 1995.

59 O. Sigmund and J. Søndergaard Jensen, "Systematic design of phononic band gap materials and structures by topology optimization," *Philosophical Transactions of the Royal Society of London A: Mathematical, Physical and Engineering Sciences,* vol. 361, no. 1806, pp. 1001–1019, 2003.

60 M. Hussein, K. Hamza, G. Hulbert, R. Scott, and K. Saitou, "Multiobjective evolutionary optimization of periodic layered materials for desired wave dispersion characteristics," *Structural and Multidisciplinary Optimization,* vol. 31, no. 1, pp. 60–75, 2006.

61 W. Weaver and P. Johnston, *Structural Dynamics by Finite Elements.* Prentice-Hall, 1987.

62 M. Petyt, *Introduction to Finite Element Vibration Analysis.* Cambridge University Press, 1998.

63 T. Hughes, *The Finite Element Method: Linear Static and Dynamic Finite Element Analysis.* Dover Publications, 2012.

64 R. E. D. Bishop and D. C. Johnson, *The Mechanics of Vibration.* Cambridge University Press, 1960.

65 D. J. Ewins, *Modal Testing: Theory, Practice and Application.* Wiley Publishers, 2nd ed., 2009.

66 D. J. Mead, "Wave propagation and natural modes in periodic systems: II. Multi-coupled systems, with and without damping," *Journal of Sound and Vibration,* vol. 40, no. 1, pp. 19–39, 1975.

67 N. S. Bardell, R. S. Langley, J. M. Dunsdon, and T. Klein, "The effect of period asymmetry on wave propagation in periodic beams," *Journal of Sound and Vibration,* vol. 197, no. 4, pp. 427–445, 1996.

68 D. Mead, "A general theory of harmonic wave propagation in linear periodic systems with multiple coupling," *Journal of Sound and Vibration,* vol. 27, no. 2, pp. 235–260, 1973.

69 A. Spadoni, M. Ruzzene, S. Gonella, and F. Scarpa, "Phononic properties of hexagonal chiral lattices," *Wave Motion,* vol. 46, no. 7, pp. 435–450, 2009.

70 R. Süsstrunk and S. D. Huber, "Observation of phononic helical edge states in a mechanical topological insulator," *Science,* vol. 349, no. 6243, pp. 47–50, 2015.

71 A. Bazoune, Y. A. Khulief, and N. G. Stephen, "Shape functions of three-dimensional Timoshenko beam element," *Journal of Sound and Vibration,* vol. 259, no. 2, pp. 473–480, 2003.

72 J. W. Eischen and S. Torquato, "Determining the elastic behevior of composites by the boundary element method," *Journal of Applied Physics,* vol. 74, no. 1, pp. 159–170, 1993.

73 N. Perkins and C. Mote Jr., "Comment on curve veering in eigenvalue problems.," *Journal of Sound and Vibration,* vol. 106, no. 3, pp. 451–463, 1986.

74 B. R. Mace and E. Manconi, "Wave motion and dispersion phenomena: Veering, locking and strong coupling effects," *Journal of the Acoustical Society of America.,* vol. 131, no. 2, pp. 1015–1028, 2012.

75 M. Manav, G. Reynen, M. Sharma, E. Cretu, and A. S. Phani, "Ultrasensitive resonant MEMS transducers with tuneable coupling," *Journal of Micromechanics and Microengineering,* vol. 24, no. 5, p. 055005, 2014.

76 A. S. Phani and N. A. Fleck, "Elastic boundary layers in two-dimensional isotropic lattices," *Journal of Applied Mechanics*, vol. 75, p. 021020, 2008.

77 A. S. Phani, "On elastic waves and related phenomena in lattice materials and structures," *AIP Advances*, vol. 1, no. 4, p. 041602, 2011.

78 Y. Bobrovnitskii, "On the energy flow in evanescent waves," *Journal of Sound and Vibration*, vol. 152, no. 1, pp. 175–176, 1992.

79 D. J. Mead, "Waves and modes in finite beams: Application of the phase-closure principle," *Journal of Sound and Vibration*, vol. 171, no. 5, pp. 695–702, 1994.

80 J. Achenbach, *Reciprocity in Elastodynamics*. Cambridge University Press, 2003.

81 K. Graff, *Wave Motion in Elastic Solids*. Dover Publications, 1975.

82 L. Brekhovskikh, *Waves in Layered Media*. Elsevier Science, 2012.

83 R. Feynman, R. Leighton, and M. Sands, *The Feynman Lectures on Physics, Desktop Edition Volume II: The New Millennium Edition*. Feynman Lectures on Physics, Basic Books, 2013.

84 S. Yang, J. H. Page, Z. Liu, M. L. Cowan, C. T. Chan, and P. Sheng, "Ultrasound tunneling through 3D phononic crystals," *Physical Review Letters*, vol. 88, p. 104301, 2002.

85 P. Peng, C. Qiu, Y. Ding, Z. He, H. Yang, and Z. Liu, "Acoustic tunneling through artificial structures: From phononic crystals to acoustic metamaterials," *Solid State Communications*, vol. 151, no. 5, pp. 400–403, 2011.

4

Wave Propagation in Damped Lattice Materials

Dimitri Krattiger[1], A. Srikantha Phani[2] and Mahmoud I. Hussein[1]

[1]Department of Aerospace Engineering Sciences, University of Colorado Boulder, Boulder, Colorado, USA
[2]Department of Mechanical Engineering, University of British Columbia, Vancouver, British Columbia, Canada

4.1 Introduction

There is a trade-off between a material's stiffness, quantified by the Young's modulus, and its capacity to dissipate energy, quantified by the loss factor or damping ratio. On the one hand, ceramic materials such as diamond possess high modulus but poor damping capacity, while on the other hand materials such as foam possess high damping capacity but low material modulus. Naturally evolved hierarchical composite materials such as bone possess high stiffness and damping [1]. Many engineered materials, including metal alloys and composites, exhibit this trade-off between damping capacity and stiffness. Lattice materials – or more generally phononic materials – with their tuneable hierarchical geometries, offer a unique opportunity to simultaneously maximize stiffness and damping capacity. For example, viscous or viscoelastic metamaterials possessing internal resonating bodies have been shown to demonstrate enhanced dissipation under certain conditions, as seen in observed rises in the wavenumber-dependent damping ratios (in other words, a net dissipation beyond the level that would be deduced from the rule of mixtures according to Voigt or Reuss rules) [2–5]. This property is beneficial in applications where enhanced dissipation is desired, but not at the expense of stiffness or load-bearing capacity. Furthermore, lattice materials, as well as lattice-based metamaterials, provide a unique opportunity to precisely control the level and characteristics of dissipation. In this chapter, we approach lattice materials from the general perspective of phononic materials and elucidate dissipation-driven dispersion effects by considering wave propagation in the presence of damping. This is in contrast to Chapter 3, where damping forces are ignored.

There are numerous approaches for the treatment of damping in material or structural models, including those representing phononic materials. A common approach is to consider viscous damping, for which a simple version is known as Rayleigh [6], or proportional, damping, whereby the matrix of damping coefficients is assumed to be proportional to the mass and/or stiffness matrices [7–10]. If the proportionality condition is not met, then the model is described as *generally damped* [11, 12]. Experiments are used to determine an appropriate damping model for a given material or structure [13].

Dynamics of Lattice Materials, First Edition. Edited by A. Srikantha Phani and Mahmoud I. Hussein.
© 2017 John Wiley & Sons Ltd. Published 2017 by John Wiley & Sons Ltd.

Beyond the choice of the damping model, an important consideration is whether the frequency is selected to be real or the wavenumber is selected to be real, and, consequently, which of the two is permitted to be complex. There are two classes of problems dealing with damped phononic materials. In one class, the frequencies are assumed *a priori* to be real, thus allowing the effects of damping to manifest only in the form of complex wavenumbers. Physically, this represents a medium experiencing wave propagation due to a sustained driving frequency (implying that a harmonic frequency is externally imposed) and dissipation taking effect in the form of spatial attenuation only. This approach follows a $\kappa = \kappa(\omega)$ formulation (where κ and ω denote wavenumber and frequency, respectively) resulting from either a linear [14–19] or a quadratic [20–22] eigenvalue problem (EVP). In the other class, the frequencies are permitted to be complex, thus allowing dissipation to take effect in the form of temporal attenuation. Physically, this represents a medium admitting free dissipative wave motion, for example due to impulse loading. Here, a $\omega = \omega(\kappa)$ formulation leading to a linear EVP is the common route (in some cases with the aid of a state-space transformation) [19, 23–27].

In the "driven-waves" case, a real frequency is prescribed and the underlying EVP is solved for the complex wavenumber solutions, the real and imaginary components of which represent propagation and attenuation constants, respectively. All modes are described by complex wavenumbers due to the dissipation. In the "free-waves" case, on the other hand, a real wavenumber is specified, and complex frequencies emerge as the solution (the real and imaginary parts respectively provide the damping ratio and the frequency for each mode). Because of the common association of the driven-waves problem to an EVP for which the frequency is the independent variable and, in contrast, the free-waves problem to an EVP for which the wavenumber is the independent variable, it is often viewed that the two only available options are: real frequencies and complex wavenumbers versus real wavenumbers and complex frequencies [28–30]. However, if the medium permits spatial attenuation in its undamped state – which is the case for phononic materials within band gap frequencies – then, in principle, there should be an imaginary wavenumber component (in addition to the real wavenumber component) even when the frequencies are complex. This, in fact, represents a more complete picture of the dispersion curves for damped free-wave motion in media that contain inherent mechanisms for spatial attenuation, such as Bragg scattering and local resonance. Since this scenario pertains only to free waves, one expects to see complex frequencies for bands admitting only spatial propagation as well as bands admitting evanescent waves (with the real part of the wavenumber being either zero or π divided by the lattice spacing; the two values that represent the limits of the irreducible Brillouin zone (IBZ)). For a proportionally damped problem, a solution that permits both the frequencies and wavenumbers to be complex has been obtained using the transfer matrix method [31]. This gives a $\kappa = \kappa(\omega)$ linear EVP. For a generally damped problem, however, an all-complex solution cannot be obtained from a linear EVP, nor from directly solving a quadratic EVP; however, in principle, it may be obtained by an algorithm that searches for unique and compatible pairs of complex frequencies and wavenumbers [32].

In this chapter, mathematical formulations are presented for both types of damped-wave propagation problem – the free and the driven – and examples of band structures are shown for each using a simple 1D mass–spring–damper model and a 2D finite-element model of a damped lattice plate.

4.2 One-dimensional Mass–Spring–Damper Model

4.2.1 1D Model Description

The simplest model for a 1D damped lattice, or phononic material, consists of a chain of masses connected by springs and viscous dampers, as shown in Figure 4.1a. Let us consider the case where each unit cell contains two masses and has the following properties:

$$m_1 = 1 \text{ kg}, \qquad m_2 = 9 \text{ kg} \tag{4.1}$$

$$k_1 = 2 \times 10^5 \text{ N/m}, \quad k_2 = 2 \times 10^5 \text{ N/m} \tag{4.2}$$

$$c_1 = r\omega_0 \overline{m}, \qquad c_2 = 2c_1 \tag{4.3}$$

where

$$\overline{m} = \frac{m_1 m_2}{m_1 + m_2}, \qquad \omega_0 = \sqrt{\frac{k_1 + k_2}{\overline{m}}} \tag{4.4}$$

and r is a tuneable parameter to adjust the damping.

In order to model the mass–spring–damper chain, we begin by modeling the unit cell as though it were cut out from the surrounding chain, as shown in Figure 4.1b. Note that we borrow from an adjacent unit cell a placeholder "phantom" mass that will be used to apply the appropriate boundary conditions. This is similar to the process used for the undamped mass-spring chain in Chapter 3. Applying Newton's law to each mass in the freed unit cell, the equations of motion for the mass–spring–damper chain are as follows:

$$K u + C \dot{u} + M \ddot{u} = 0 \tag{4.5}$$

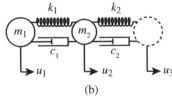

(a)

(b)

Figure 4.1 (a) One-dimensional mass–spring–damper chain with unit cell highlighted, and (b) freed unit cell with position of the first mass from an adjacent unit cell indicated.

where

$$K = \begin{bmatrix} k_1 & -k_1 & 0 \\ -k_1 & k_1 + k_2 & -k_2 \\ 0 & -k_1 & k_2 \end{bmatrix}$$

$$C = \begin{bmatrix} c_1 & -c_1 & 0 \\ -c_1 & c_1 + c_2 & -c_2 \\ 0 & -c_1 & c_2 \end{bmatrix} \qquad (4.6)$$

$$M = \begin{bmatrix} m_1 & 0 & 0 \\ 0 & m_2 & 0 \\ 0 & 0 & 0 \end{bmatrix}, \qquad u = \begin{Bmatrix} u_1 \\ u_2 \\ u_3 \end{Bmatrix}$$

Bloch's theorem dictates that wave fields in periodic media are composed of a plane-wave term multiplied by a periodic function [33]. In 1D, the Bloch displacement vector can be written as follows:

$$u = \bar{u}\gamma, \qquad \gamma = e^{i\kappa x} \qquad (4.7)$$

where $\bar{u}$ is the periodic field, γ is the plane wave term, and a is the lattice constant or unit cell length. By definition, the periodic field must satisfy $\bar{u}(x) = \bar{u}(x + a)$. Using this description it is possible to derive the Bloch-periodic equations of motion from the free unit cell equations of motion given above. Practically, this is accomplished by setting $u_3 = u_1\gamma$ (see Chapter 3 for more details regarding implementation). This gives

$$\overline{K}\bar{u} + \overline{C}\dot{\bar{u}} + \overline{M}\ddot{\bar{u}} = 0 \qquad (4.8)$$

where

$$\overline{K} = \begin{bmatrix} k_1 + k_2 & -k_1 - k_2\gamma \\ -k_1 - k_2\gamma^{-1} & k_1 + k_2 \end{bmatrix}$$

$$\overline{C} = \begin{bmatrix} c_1 + c_2 & -c_1 - c_2\gamma \\ -c_1 - c_2\gamma^{-1} & c_1 + c_2 \end{bmatrix} \qquad (4.9)$$

$$\overline{M} = \begin{bmatrix} m_1 & 0 \\ 0 & m_2 \end{bmatrix}, \qquad \bar{u} = \begin{Bmatrix} u_1 \\ u_2 \end{Bmatrix}$$

The bar denotes that Bloch periodicity has been enforced on the equations of motion. We now set $\bar{u}(t) = \bar{q}e^{\lambda t}$ so that the resulting characteristic equation is

$$(\overline{K} + \lambda\overline{C} + \lambda^2\overline{M})\bar{q} = 0 \qquad (4.10)$$

where $\bar{q}$ has no time dependence.

4.2.2 Free-wave Solution

Free waves are solutions to Eq. (4.10) wherein the wavenumber is specified and the frequency is solved for. Two methods for computing the free-wave dispersion are presented herein. The first method uses a linearization of the quadratic EVP via a state-space transformation, and then uses a linear eigenvalue solver to obtain the damped dispersion frequencies and mode shapes. This method can be computationally demanding for unit cells with many degrees of freedom because the dimension of the EVP is doubled. In the second method, an approximate free-wave solution is obtained using Bloch–Rayleigh perturbation analysis where the undamped solution is computed first, and then perturbation formulae are used to respectively obtain the damped dispersion frequencies and mode shapes [27].

State-space Wave Calculation

The state-space method allows us to take the quadratic EVP of Eq. (4.10) that results when damping is present in the equations of motion, and turn it into a linear EVP, although at the expense of doubling the problem size:

$$\left(\begin{bmatrix} \overline{K} & 0 \\ 0 & I \end{bmatrix} - \lambda \begin{bmatrix} -\overline{C} & -\overline{M} \\ I & 0 \end{bmatrix} \right) \begin{Bmatrix} \overline{q} \\ \lambda \overline{q} \end{Bmatrix} = \begin{Bmatrix} 0 \\ 0 \end{Bmatrix} \tag{4.11}$$

There are several forms for state-space transformation of the quadratic EVP [34]. These forms are nominally equivalent, but have different numerical properties. The formalism presented above gives well-conditioned state-space matrices for material and structural dynamics problems.

The damped eigenvalue can be decomposed in terms of the frequency ω and the damping ratio ζ,

$$\lambda = -\zeta \omega \pm i \omega \sqrt{1 - \zeta^2} \tag{4.12}$$

from which the damped frequency, $\omega_d = |\text{imag}(\lambda)|$, and the damping ratio, $\zeta = -\text{real}(\lambda)/|\lambda|$, are extracted. The quantity $\omega = |\lambda| = \sqrt{\text{real}(\lambda)^2 + \text{imag}(\lambda)^2}$ is referred to as the *resonant frequency*. For the special case of proportional viscous damping, the resonant frequency is equal to the undamped frequency.

Bloch–Rayleigh Perturbation Method

This section gives a brief overview of the computational steps required for the Bloch–Rayleigh perturbation analysis, which is a technique that allows us to obtain the dispersion solution for a damped periodic medium using the undamped solution as a basis. A more complete description of the theory is available in the literature [27]. The following orthogonality relations, expressed in Bloch coordinates, hold for the undamped system:

$$\phi_m^{\text{T}} \overline{M} \phi_n = \delta_{mn}, \qquad \phi_m^{\text{T}} \overline{K} \phi_n = \delta_{mn} \omega_n^2 \tag{4.13}$$

where ϕ_n and ω_n^2 are respectively the nth eigenvector and eigenvalue of the undamped system, and where

$$\delta_{mn} = \begin{cases} 1, & \text{if } n = m \\ 0, & \text{if } n \neq m \end{cases} \tag{4.14}$$

The matrix $\overline{C'}_{mn} = \phi_m^{\text{T}} \overline{C} \phi_n$ will generally not be diagonal, but it will be diagonally dominant as long as the damping values are not too large. The damped eigenvalues can be approximated as

$$\lambda_n \approx \omega_n^2 + \frac{i}{2} \sum_{\substack{j=1 \\ j \neq n}}^{N} \overline{C'}_{nn} \tag{4.15}$$

and the damped eigenvectors can be approximated as

$$z_n \approx \phi_n + \frac{i}{2} \sum_{\substack{j=1 \\ j \neq n}}^{N} \frac{\overline{C}'_{jn}\omega_n}{\omega_j^2 - \omega_n^2}\phi_j \tag{4.16}$$

4.2.3 Driven-wave Solution

Driven waves are solutions to Eq. (4.10) wherein the frequency is specified and the wavenumber is solved for. Note that in this approach the frequency is forced to be real, so no damping ratio can be obtained. The stiffness and damping matrices can be decomposed in terms of γ as follows:

$$\begin{aligned}
\overline{K} &= \gamma^{-1}\overline{K}_0 + \overline{K}_1 + \gamma\overline{K}_2 \\
\overline{C} &= \gamma^{-1}\overline{C}_0 + \overline{C}_1 + \gamma\overline{C}_2
\end{aligned} \tag{4.17}$$

Using this decomposition and substituting $\lambda = i\omega$, Eq. (4.10) is expressed as

$$\left[\underbrace{(\overline{K}_0 + i\omega\overline{C}_0 - \omega^2\overline{M})}_{\overline{D}_0} + \gamma\underbrace{(\overline{K}_1 + i\omega\overline{C}_1)}_{\overline{D}_1} + \gamma^2\underbrace{(\overline{K}_2 + i\omega\overline{C}_2)}_{\overline{D}_2} \right]\overline{q} = 0 \tag{4.18}$$

This equation can now be linearized using the state-space approach, yielding

$$\left(\begin{bmatrix} \overline{D}_0 & 0 \\ 0 & I \end{bmatrix} - \gamma \begin{bmatrix} -\overline{D}_1 & -\overline{D}_2 \\ I & 0 \end{bmatrix} \right) \left\{ \begin{matrix} \overline{q} \\ \gamma\overline{q} \end{matrix} \right\} = \left\{ \begin{matrix} 0 \\ 0 \end{matrix} \right\} \tag{4.19}$$

Using a linear eigenvalue solver, the eigenvalues γ are obtained. The wavenumbers are then obtained as $\kappa = \ln(\gamma)/ia$.

4.2.4 1D Damped Band Structures

Figure 4.2 shows a collection of damped band structure plots for the 1D spring–mass–damper system of Figure 4.1. Three different values of r are evaluated in order to demonstrate the effects of damping on the band structure. The damping levels are unrealistically high for practical systems, but they help to more clearly demonstrate the unique nature of these effects. The plots on the left-hand side of the figure show the free-wave band structures. In each dispersion plot, the acoustic branch is shown as a solid line and filled circles, and the optical branch is shown as a dashed line and hollow circles. Excellent agreement is seen between the state-space and Bloch–Rayleigh solutions, except for branches exhibiting exceedingly high damping. An interesting phenomenon that can be seen when damping is very high is *branch overtaking*, which involves the optical branches dropping significantly, to the extent of becoming fully, or partially, lower than the acoustic branches [25–27]. Furthermore, these drops in the branches may lead to a *wavenumber cut-off* or a *wavenumber cut-in* [25–27].

The plot on the right of the figure shows a band structure obtained using the driven-wave solution, wherein the wavenumber is solved for as a function of frequency. The left-hand side of the driven-waves plot shows the imaginary components of the wavenumber, and the right-hand side shows the real components. In the undamped

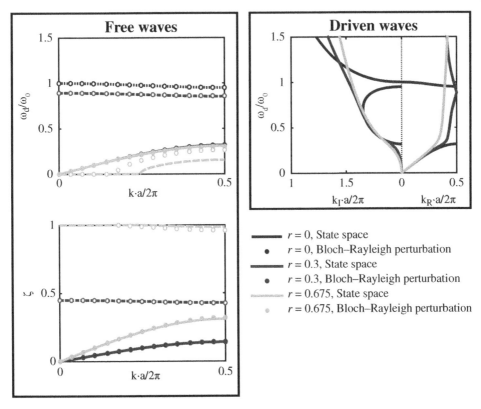

Figure 4.2 Damped dispersion for 1D spring–mass–damper chain: free-wave frequency band structure (top left), corresponding damping ratio diagram (bottom left), and driven-wave band structure (top right).

system, the wavenumber solutions typically have only nonzero imaginary components when the real component is either 0 or π/a. However, when damping is added, the wavenumber solutions are generally complex. Another noticeable, and very important, phenomenon in the driven-waves case is the closing up of band gaps due to damping.

4.3 Two-dimensional Plate–Plate Lattice Model

4.3.1 2D Model Description

In this section, the techniques developed for the 1D model are extended to 2D. The extension to 3D from 2D follows easily and is not covered. For demonstration purposes, a 2D damped plate lattice material is considered. The unit cell for this system is shown in Figure 4.3. The physical and material properties of the unit cell are summarized as follows:

$E = 2.4\,\mathrm{GPa}$
$\rho = 1040\,\mathrm{kg/m^3}$
$\nu = 0.33$

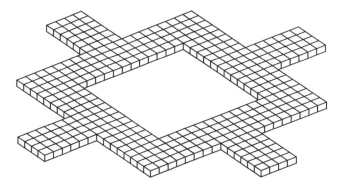

Figure 4.3 Plate lattice unit cell with finite-element mesh shown.

$L_x = L_y = 10\,\text{cm}$
$h = 1\,\text{mm}.$

where E is the elastic modulus, ρ is the density, v is the Poisson's ratio, L_x and L_y are the lattice constants, and h is the plate thickness. The plate is modeled using Mindlin–Reissner plate theory [35, 36], which allows for shear stresses to be included in plate models. The governing equation of Mindlin–Reissner plate theory is

$$\left(\nabla^2 - \frac{\rho}{\mu\alpha}\frac{\partial^2}{\partial t^2}\right)\left(D\nabla^2 - \frac{\rho h^3}{12}\frac{\partial^2}{\partial t^2}\right)w + \rho h\frac{\partial^2 w}{\partial t^2} = 0 \tag{4.20}$$

where μ is the shear modulus, α is the shear correction factor, $D = (Eh^3)/(12 - 12v^2)$ is the plate stiffness, and w is the transverse displacement of the plate. The plate is discretized into elements and the mass and stiffness matrices are formed using Mindlin–Reissner plate elements, as outlined in Chapter 9. The damping matrix is formed in exactly the same way as the stiffness matrix, but with modified elastic modulus, E_d, and Poisson ratio, v_d,

$$E_d = q\mu_d\frac{3\lambda_d + 2\mu_d}{\lambda_d + \mu_d}, \qquad v_d = \frac{\lambda_d}{2(\lambda_d + \mu_d)} \tag{4.21}$$

where

$$\mu_d = \mu = \frac{E}{2(1+v)}, \qquad \lambda_d = \frac{1}{2}\lambda = \frac{1}{2}\frac{Ev}{(1+v)(1-2v)} \tag{4.22}$$

The scaling parameter q is used to adjust the level of damping. Note that modeling the damping in this way is not necessarily physically meaningful, but it does create a general damping matrix.

With the mass, stiffness, and damping matrices formed for a free unit cell, Bloch-periodicity is enforced by application of Bloch boundary conditions.

4.3.2 Extension of Driven-wave Calculations to 2D Domains

Once $\overline{M}, \overline{K}$, and $\overline{C}$, are set up for the plate model as a function of the wave vector components, κ_x, and κ_y, the free-wave dispersion calculations proceed exactly as for the 1D problem (see Section 4.2.2). Thus, these calculations need not be discussed further here.

For the driven-wave case, there are two approaches. If we form the free equations of motion and apply Bloch boundary conditions we will end up with a polynomial

EVP whose order depends on the direction of wave propagation. This makes it a very difficult problem to solve numerically [37]. The second approach is to introduce the Bloch conditions as an operator applied to the governing equations [38] to form the damped equations of motion [39]. For 2D problems, this produces a stiffness matrix of the following form:

$$\overline{K} = \overline{K}_0 + \kappa_x \overline{K}_x + \kappa_y \overline{K}_y + \kappa_x^2 \overline{K}_{xx} + \kappa_y^2 \overline{K}_{yy} + \kappa_x \kappa_y \overline{K}_{xy} \tag{4.23}$$

Depending on how the damping matrix is constructed, it will have the same form as well. The mass matrix, however, is typically not a function of the wave vector in this formulation. If we specify a wave propagation direction we can write the wave vector as $\kappa = \begin{bmatrix} \kappa_x & \kappa_y \end{bmatrix}^T = \kappa \begin{bmatrix} a & b \end{bmatrix}^T$, where a and b are known constants. Substituting into Eq. (4.23), we obtain the following form for the stiffness matrix:

$$\overline{K} = \overline{K}_0 + \kappa \underbrace{\left(a\overline{K}_x + b\overline{K}_y \right)}_{\overline{K}_1} + \kappa^2 \underbrace{\left(a^2 \overline{K}_{xx} + b^2 \overline{K}_{yy} + ab\overline{K}_{xy} \right)}_{\overline{K}_2} \tag{4.24}$$

Assuming that the damping matrix has the same form as the stiffness matrix, the equations of motion can be described as follows:

$$[\underbrace{(\overline{K}_0 + i\omega\overline{C}_0 - \omega^2\overline{M})}_{\overline{A}_0} + \kappa\underbrace{(\overline{K}_1 + i\omega\overline{C}_1)}_{\overline{A}_1} + \kappa^2\underbrace{(\overline{K}_2 + i\omega\overline{C}_2)}_{\overline{A}_2}]\overline{q} = 0 \tag{4.25}$$

As before, this quadratic EVP can be linearized to a standard generalized form, yielding

$$\left(\begin{bmatrix} \overline{A}_0 & 0 \\ 0 & I \end{bmatrix} - \kappa \begin{bmatrix} -\overline{A}_1 & -\overline{A}_2 \\ I & 0 \end{bmatrix} \right) \begin{Bmatrix} \overline{q} \\ \kappa\overline{q} \end{Bmatrix} = \begin{Bmatrix} 0 \\ 0 \end{Bmatrix} \tag{4.26}$$

4.3.3 2D Damped Band Structures

Figure 4.4 shows a collection of damped band structure plots for the 2D plate lattice material of Figure 4.3. The plots on the left of the figure show free-wave band structures. As in the 1D case, we notice that the Bloch–Rayleigh technique becomes less accurate when the damping is exceptionally high. The highest characteristic frequencies computed from finite-element models are not typically physical. Rather, they are spurious modes that arise due to the spatial discretization. For undamped models we simply ignore all modes above a certain frequency of interest. However, with certain types of damping, high-frequency modes are heavily affected, causing the frequencies to fall. This makes the spurious modes fall into the frequency range of interest and makes it difficult to separate the physical modes from the nonphysical modes. The free-wave frequency band structure shown in Figure 4.4 shows several branches that have collapsed to zero in some parts of the Brillouin zone (BZ). Most likely these branches are spurious and should be filtered out. One way to perform this filtering is by removing all modes with damping ratio above some threshold.

The driven-wave plot (on the right side of Figure 4.4) shows complex-wavenumber band structures for waves traveling along the Γ–X direction. In the free-wave case, the branches mostly shift downward as damping is added but retain their shape. In contrast, for driven waves the change in the band structure seems much more significant as branches take on entirely new characteristics.

Figure 4.5 shows iso-frequency and iso-damping contours for the first branch. When using the driven-wave approach to compute the complex wavenumbers, curve sorting becomes very important. At first glance it appears that the acoustic branches from adjacent BZs intersect at the BZ edge, creating the folding pattern seen in the driven-wave plot of Figure 4.4. If the branch is viewed in the complex plane, however, the acoustic branches do not intersect but rather veer around each other and continue into the next BZ. Thus we follow the acoustic branch into the neighboring BZ for the driven-wave plots in Figure 4.5. The BZ edges are shown in the top-right plot with dashed lines. The driven-wave iso-frequency plots may be difficult to follow, so we plot the same contours against frequency in Figure 4.6 in order to separate them. An interesting feature that we observe from both Figures 4.5 and 4.6 is the smoothing of the iso-frequency contours with the addition of damping. This implies that damping reduces the degree of frequency-dependent elastodynamic anisotropy in lattice materials (or periodic materials in general).

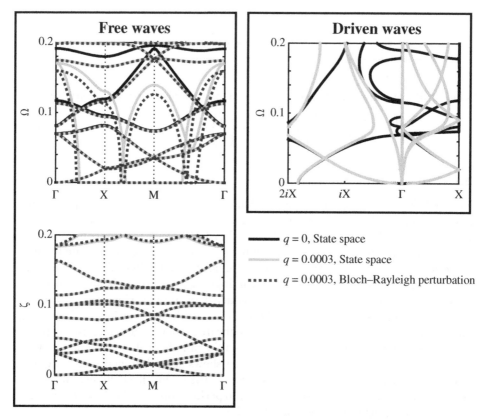

Figure 4.4 Damped dispersion for plate lattice material: free-wave frequency band structure (top left), corresponding damping ratio diagram (bottom left), and driven-wave band structure (top right). Frequencies are normalized as follows: $\Omega = \omega_d L_x \sqrt{\rho/E}$.

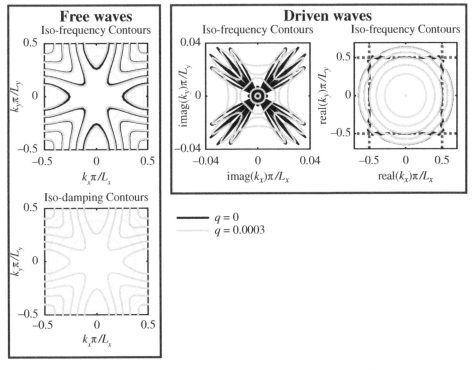

Figure 4.5 Damped dispersion for plate lattice material: iso-frequency plot (top left) and iso-damping plot (bottom left) for free waves; iso-frequency plots for imaginary (top middle) and real (top right) wavenumbers for driven waves.

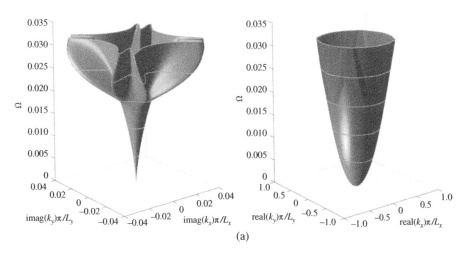

(a)

Figure 4.6 Iso-frequency plots for imaginary (left) and real (right) wavenumbers for driven waves for the 2D plate lattice. The top figures show the damped wavenumber surfaces, the middle plots show the undamped wavenumber surfaces, and the bottom plots show both damped and undamped wavenumbers at select frequencies.

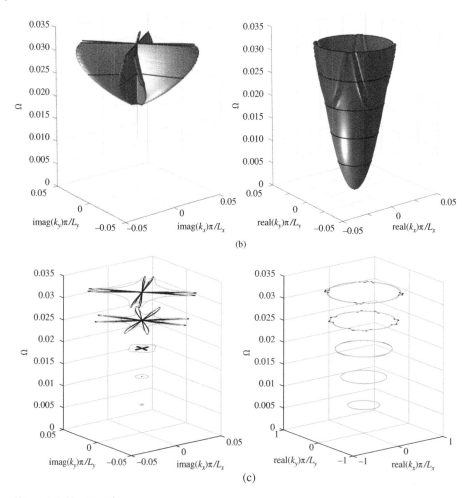

(b)

(c)

Figure 4.6 (*Continued*)

References

1 R. S. Lakes, "High damping composite materials: Effect of structural hierarchy," *Journal of Composite Materials*, vol. 36, no. 3, pp. 287–297, 2002.

2 M. I. Hussein and M. J. Frazier, "Metadamping: An emergent phenomenon in dissipative metamaterials," *Journal of Sound and Vibration*, vol. 332, pp. 4767–4774, 2013.

3 I. Antoniadis, D. Chronopoulos, V. Spitas, and D. Koulocheris, "Hyper-damping properties of a stiff and stable linear oscillator with a negative stiffness element," *Journal of Sound and Vibration*, vol. 346, pp. 37–52, 2015.

4 Y. Y. Chen, M. V. Barnhart, J. K. Chen, G. K. Hu, C. Sun, and G. L. Huang, "Dissipative elastic metamaterials for broadband wave mitigation at subwavelength scale," *Composite Structures*, vol. 136, pp. 358–371, 2016.

5 M. J. Frazier and M. I. Hussein, "Viscous-to-viscoelastic transition in phononic crystal and metamaterial band structures," *Journal of the Acoustical Society of America*, vol. 138, pp. 3169–3180, 2015.

6 J. W. S. Rayleigh, *The Theory of Sound*, vol. 1. Macmillan and Co., 1877.

7 T. K. Caughey and M. E. J. O'Kelly, "Classical normal modes in damped linear dynamic systems," *Journal of Applied Mechanics – Transactions of the ASME*, vol. 32, pp. 583–588, 1965.

8 A. S. Phani, "On the necessary and sufficient conditions for the existence of classical normal modes in damped linear dynamic systems," *Journal of Sound and Vibration*, vol. 264, pp. 741–745, 2003.

9 S. Adhikari, "Damping modelling using generalized proportional damping," *Journal of Sound and Vibration*, vol. 293, pp. 156–170, 2005.

10 S. Adhikari and A. S. Phani, "Experimental identification of generalized proportional viscous damping matrix," *Journal of Vibration and Acoustics*, vol. 131, p. 011008, 2009.

11 J. Woodhouse, "Linear damping models for structural vibration," *Journal of Sound and Vibration*, vol. 215, pp. 547–569, 1998.

12 S. Adhikari and J. Woodhouse, "Identification of Damping: Part 1, Viscous damping," *Journal of Sound and Vibration*, vol. 243, pp. 43–61, 2001.

13 A. S. Phani and J. Woodhouse, "Viscous damping identification in linear vibration," *Journal of Sound and Vibration*, vol. 303, pp. 475–500, 2007.

14 E. Tassilly, "Propagation of bending waves in a periodic beam," *International Journal of Engineering Science*, vol. 25, pp. 85–94, 1987.

15 R. S. Langley, "On the forced response of one-dimensional periodic structures: Vibration localization by damping," *Journal of Sound and Vibration*, vol. 178, pp. 411–428, 1994.

16 V. Laude, Y. Achaoui, S. Benchabane, and A. Khelif, "Evanescent Bloch waves and the complex band structure of phononic crystals," *Physical Review B*, vol. 80, p. 092301, 2009.

17 V. Romero-García, J. V. Sánchez-Pérez, and L. M. Garcia-Raffi, "Propagating and evanescent properties of double-point defects in sonic crystals," *New Journal of Physics*, vol. 12, p. 083024, 2010.

18 R. P. Moiseyenko and V. Laude, "Material loss influence on the complex band structure and group velocity in phononic crystals," *Physical Review B*, vol. 83, p. 064301, 2011.

19 E. Andreassen and J. S. Jensen, "Analysis of phononic bandgap structures with dissipation," *Journal of Vibration and Acoustics*, vol. 135, p. 041015, 2013.

20 D. J. Mead, "A general theory of harmonic wave propagation in linear periodic systems with multiple coupling," *Journal of Sound and Vibration*, vol. 27, pp. 235–260, 1973.

21 F. Farzbod and M. J. Leamy, "Analysis of Bloch's method in structures with energy dissipation," *Journal of Vibration and Acoustics*, vol. 133, p. 051010, 2011.

22 M. Collet, M. Ouisse, M. Ruzzene, and M. N. Ichchou, "Floquet–Bloch decomposition for the computation of dispersion of two-dimensional periodic, damped mechanical systems," *International Journal of Solids and Structures*, vol. 48, pp. 2837–2848, 2011.

23 S. Mukherjee and E. H. Lee, "Dispersion relations and mode shapes for waves in laminated viscoelastic composites by finite difference methods," *Computers & Structures*, vol. 5, pp. 279–285, 1975.

24 R. Sprik and G. H. Wegdam, "Acoustic band gaps in composites of solids and viscous liquids," *Solid State Communications*, vol. 106, pp. 77–81, 1998.

25 M. I. Hussein, "Theory of damped Bloch waves in elastic media," *Physical Review B*, vol. 80, p. 212301, 2009.

26 M. I. Hussein and M. J. Frazier, "Band structure of phononic crystals with general damping," *Journal of Applied Physics*, vol. 108, p. 093506, 2010.

27 A. S. Phani and M. I. Hussein, "Analysis of damped Bloch waves by the Rayleigh perturbation method," *Journal of Vibration and Acoustics*, vol. 135, p. 041014, 2013.

28 J. D. Achenbach, *Wave Propagation in Elastic Solids*. North-Holland, 1999.

29 B. R. Mace and E. Manconi, "Modelling wave propagation in two-dimensional structures using finite element analysis," *Journal of Sound and Vibration*, vol. 318, pp. 884–902, 2008.

30 E. Manconi and B. R. Mace, "Estimation of the loss factor of viscoelastic laminated panels from finite element analysis," *Journal of Sound and Vibration*, vol. 329, pp. 3928–3939, 2010.

31 M. I. Hussein, M. J. Frazier, and M. H. Abedinnassab, "Chapter 1: Microdynamics of Phononic Materials," in *Handbook of Micromechanics and Nanomechanics* (S. Li and X.-L. Gao, eds.), Pan Stanford Publishing, 2013.

32 M. J. Frazier and M. I. Hussein, "Generalized Bloch's theorem for viscous metamaterials: Dispersion and effective properties based on frequencies and wavenumbers that are simultaneously complex," *arXiv:1601.00683 [cond-mat.mtrl-sci]*, 2016.

33 F. Bloch, "Über die quantenmechanik der elektronen in kristallgittern," *Zeitschrift für Physik*, vol. 52, no. 7–8, pp. 555–600, 1929.

34 F. Tisseur and K. Meerbergen, "The quadratic eigenvalue problem," *SIAM Review*, vol. 43, no. 2, pp. 235–286, 2001.

35 R. D. Mindlin, "Influence of rotatory inertia and shear on flexural motions of isotropic, elastic plates," *Journal of Applied Mechanics*, vol. 18, pp. 31–38, 1951.

36 E. Reissner, "The effect of transverse shear deformation on the bending of elastic plates," *Journal of Applied Mechanics*, vol. 12, pp. 68–77, 1945.

37 B. R. Mace and E. Manconi, "Modelling wave propagation in two-dimensional structures using finite element analysis," *Journal of Sound and Vibration*, vol. 318, no. 4, pp. 884–902, 2008.

38 M. I. Hussein, "Reduced Bloch mode expansion for periodic media band structure calculations," *Proceedings of the Royal Society A*, vol. 465, pp. 2825–2848, 2009.

39 M. Collet, M. Ouisse, M. Ruzzene, and M. Ichchou, "Floquet–Bloch decomposition for the computation of dispersion of two-dimensional periodic, damped mechanical systems," *International Journal of Solids and Structures*, vol. 48, no. 20, pp. 2837–2848, 2011.

5

Wave Propagation in Nonlinear Lattice Materials

Kevin L. Manktelow[1], Massimo Ruzzene[1] and Michael J. Leamy

George W. Woodruff School of Mechanical Engineering, Georgia Institute of Technology, Atlanta, Georgia, USA

5.1 Overview

The desire to achieve a more dynamic range of responses from periodic systems has inspired several investigations of nonlinear metamaterial properties and analysis methods [1–5]. The intimate connection between frequency response and amplitude in standalone nonlinear systems is well-known [6]. Still, understanding the complexities that arise in nonlinear dynamical systems remains an active area of research. For example, the interaction of two waves in a constitutively nonlinear material results in energy transfer between two waves at commensurate frequencies [7, 8]. In the same vein, finite-amplitude wave propagation almost always results in harmonic generation – a phenomenon that has been exploited for designing acoustic diodes [9].

Cubic nonlinearities that appear in force-displacement type relationships are a particularly important class of nonlinearity, often termed the *Duffing nonlinearity*. The Duffing nonlinearity is responsible for bi-stability, resonant amplitude-dependent frequency-shifting, and chaos as just a few examples [10]. Nonlinearities play an increasingly dominant role in the dynamics of small-scale applications such as resonators and gyroscopes [11–13]. Precise design at these scales relies on a complete understanding of how nonlinearities affect operation and performance. Individual Duffing oscillators have been purposed for enhancing energy-harvesting applications that utilize high-energy periodic orbits [14]. Moreover, broadening of the frequency response in the vicinity of linear natural frequencies enables energy harvesters to operate in wider frequency bands [15]. Forced and free vibration responses of coupled oscillators have received some attention, but less so than single oscillators. A present goal of acoustic metamaterial and phononic crystal research is to understand how nonlinearities influence material behaviour and how they may be used to enhance and control material properties.

Nonlinear phononic systems provide unique opportunities for tunable band gap engineering, through mechanisms such as wave–wave interactions [16], amplitude-dependent band structures [17–19], and extra harmonic generation. The effect of a cubic Duffing-like nonlinearity in force-displacement type relationships

---

1 KM and MR contributed equally to this work.

Dynamics of Lattice Materials, First Edition. Edited by A. Srikantha Phani and Mahmoud I. Hussein.

is of particular interest as it leads to amplitude-dependent dispersion [20–24], which has implications for device design. However, analysis methods and literature specifically addressing nonlinear phononic crystals and metamaterials is sparse.

A fundamental system of considerable interest in this area is the 1D monoatomic chain of nonlinear oscillators. The coupled-oscillator system finds application in, for example, anharmonic atomic lattice analysis, finite-difference modeling for continuous materials, and nonlinear modal analysis. Wave propagation in nonlinear Duffing oscillator chains was the subject of a recent review. The nonlinear dynamics and band gap behavior were investigated to determine approximate closed-form expressions for band gap shifting [2, 19]. Others have investigated wave propagation in similar systems and also in systems with locally-attached oscillators, which have shown chaotic responses [5, 25, 26]. Often, these analyses assume a weakly nonlinear constitutive force-displacement relation, which enables analysis through perturbation methods.

Continuous periodic systems have received less attention because of the inherent complexities associated with nonlinear partial-differential equations and geometrically complex material domains. A 1D bi-material system was investigated to determine the response of the system to second harmonic generation in and outside of the acoustic band gaps [27]. It was subsequently demonstrated to be a potential acoustic diode [9]. However, the study of complex media consisting of nonlinear continuous elements in periodic systems is a largely unexplored research area. As such, analysis methods specifically applicable to nonlinear phononic crystals and metamaterials are sparse.

With the motivation for studying nonlinear systems established, this chapter overviews a recently developed multiple-scale perturbation approach [16, 28–30] for assessing wave propagation in weakly nonlinear lattices, with specific focus on predicting amplitude-dependent band structure, wave–wave interactions, and device behavior. The analysis presented is broadly applicable to discrete nonlinear systems, and to continuous systems that have been discretized appropriately: via Galerkin methods, finite element methods, finite difference methods, and so on.

5.2   Weakly Nonlinear Dispersion Analysis

In linear systems, it is sufficient to consider a single unit cell to analyze Bloch wave propagation. However, for a nonlinear system, it is necessary to analyze a unit cell and partial neighbors so that nonlinear forces on the unit cell boundary $\partial\Omega$ are left unexposed in a free body diagram [31]. The fundamental reason for this is that Bloch's theorem does not apply in the nonlinear setting, and thus neighbor interactions cannot be removed via a propagation constant in the usual manner.

Consider 1D, 2D, or 3D systems as shown in Figure 5.1. The *global system* consists of a central unit cell (darker color) surrounded by identical neighboring unit cells and the associated internal forces $f_{(p,q,r)}$ on neighboring unit cells. All neighboring unit cells are included to avoid the appearance of internal forces for the central unit cell. Each unit cell can be indexed as convenient; here the central unit cell is indexed as zero. Suppose that the continuous domain has been discretized such that $\mathbf{q}_i$ denotes the degrees of freedom (DOFs) for the ith node. Furthermore, let $\mathbf{u}_{(p,q,r)} = [\mathbf{q}_1, \mathbf{q}_2, \dots \mathbf{q}_n]^\mathrm{T}$ denote the collection of DOFs within the unit cell indexed by $p, q,$ and r (only p for a 1D system,

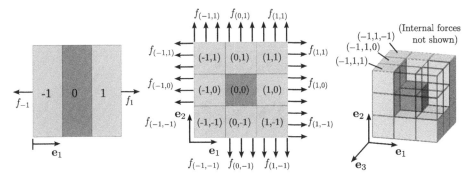

Figure 5.1 Central unit cell (darker color) surrounded by neighboring unit cells for 1D, 2D, and 3D periodic systems.

or p and q for a 2D system). Then, the total collection of generalized coordinates for the discretized system is given by

$$\mathbf{u} = [\mathbf{u}_{(-1,-1,-1)}, \mathbf{u}_{(-1,-1,0)}, \ldots, \mathbf{u}_{(0,0,0)}, \ldots, \mathbf{u}_{(0,1,1)}, \mathbf{u}_{(1,1,1)}]^{\mathrm{T}} \tag{5.1}$$

Given this ordering, the discretized equations of motion for the entire system are written in canonical form as

$$\mathbf{M\ddot{u}} + \mathbf{Ku} + \mathbf{f}^{NL}(\mathbf{u}) = \mathbf{f}^{ext} + \mathbf{f}^{int} \tag{5.2}$$

where $\mathbf{M}$ and $\mathbf{K}$ denote the mass and stiffness matrices for the central unit cell and its neighbors as shown in Figure 5.1, $\mathbf{f}^{NL}(\mathbf{u})$ denotes a nonlinear force vector, $\mathbf{f}^{ext}$ denotes a vector of externally applied forces, and $\mathbf{f}^{int}$ denotes the internal forces acting on the neighbor unit cells. The mass and stiffness matrices arise from the linear and positive definite differential operators in the associated wave equation. The nonlinear force vector $\mathbf{f}^{NL}$ accounts for all other terms resulting from the discretization of nonlinear operators.

Nonlinear terms arise from nonlinear force-deflection type relationships herein and so $\mathbf{f}^{NL}(\mathbf{u})$ depends only on $\mathbf{u}$. The following methods are, however, applicable to more complex nonlinear expressions, which may take the form $\mathbf{f}^{NL}(\mathbf{u}, \dot{\mathbf{u}}, \ddot{\mathbf{u}})$ [28]. Moreover, only systems that are weakly nonlinear are amenable to the perturbation analyses presented herein. Thus the nonlinear force vector is typically defined (or redefined) as $\mathbf{f}^{NL}(\mathbf{u}) \rightarrow \varepsilon \mathbf{f}^{NL}(\mathbf{u})$ in order to introduce a small perturbation parameter $|\varepsilon| < 1$, which controls where terms appear in an asymptotic perturbation expansion. Furthermore, free wave propagation is considered so that $\mathbf{f}^{ext} = \mathbf{0}$; then Eq. (5.2) becomes

$$\mathbf{M\ddot{u}} + \mathbf{Ku} + \varepsilon \mathbf{f}^{NL}(\mathbf{u}) = \mathbf{f}^{int} \tag{5.3}$$

The specific origin of ε varies from case to case. It is generally possible to redefine some nonlinear coefficient (say, Γ), which is known to be small such that $\Gamma = \varepsilon \hat{\Gamma}$. In other situations, a small parameter may arise out of dimensional analysis of the governing equations.

The dynamic equations (Eq. (5.3)) resulting from a discretization of the central unit cell and its neighbors constitute an *open set* of nonlinear difference equations whose solution is dependent upon all $\mathbf{u}_{(p,q,r)}$ surrounding the central unit cell. Assuming a lumped mass matrix, the open set of difference equations for the *central unit cell only* are given by

$$\text{1D system:} \quad \mathbf{M}_0 \ddot{\mathbf{u}}_0 + \sum_{p=-1}^{1} \mathbf{K}_p \mathbf{u}_p + \varepsilon \mathbf{f}_0^{NL} = \mathbf{0} \tag{5.4}$$

$$\text{2D system:} \quad \mathbf{M}_0 \ddot{\mathbf{u}}_{(0,0)} + \sum_{p=-1}^{1} \sum_{q=-1}^{1} \mathbf{K}_{(p,q)} \mathbf{u}_{(p,q)} + \varepsilon \mathbf{f}_{(0,0)}^{NL} = \mathbf{0} \tag{5.5}$$

$$\text{3D system:} \quad \mathbf{M}_0 \ddot{\mathbf{u}}_{(0,0,0)} + \sum_{p=-1}^{1} \sum_{q=-1}^{1} \sum_{r=-1}^{1} \mathbf{K}_{(p,q,r)} \mathbf{u}_{(p,q,r)} + \varepsilon \mathbf{f}_{(0,0,0)}^{NL} = \mathbf{0} \tag{5.6}$$

where $\mathbf{M}_0$ denotes a partition of total mass matrix corresponding to the central unit cell, $\mathbf{K}_{(p,q,r)}$ denote partitions of $\mathbf{K}$ responsible for linear restoring forces on the central unit cell as a result of $\mathbf{u}_{(p,q,r)}$, and $\mathbf{f}_{(p,q,r)}^{NL}$ denotes the partition of the nonlinear force vector. It has been noted that internal forces on the central unit cell are identically zero[2]. A few notation simplifications permit the open set of difference equations (for a general 3D system) to be rewritten as

$$\mathbf{M} \ddot{\mathbf{u}} + \sum_{(p,q,r)} \mathbf{K}_{(p,q,r)} \mathbf{u}_{(p,q,r)} + \varepsilon \mathbf{f}^{NL}(\mathbf{u}_{(p,q,r)}) = \mathbf{0} \tag{5.7}$$

where the simplified notation $\mathbf{f}_{(0,0,0)}^{NL} = \mathbf{f}^{NL}$ and $\sum_{p=-1}^{1} \sum_{q=-1}^{1} \sum_{r=-1}^{1} = \sum_{(p,q,r)}$ has been employed and zero subscripts have been dropped from $\mathbf{M}_0$ and $\mathbf{u}_0$. This form facilitates perturbation analysis.

Next, a multiple-scale analysis of Eq. (5.7) is sought. This postulates that the solution variables evolve on multiple time scales, formally stated as

$$\mathbf{u} = \mathbf{u}(t) = \mathbf{u}(T_0(t), T_1(t), \dots T_n(t)) \quad \forall \, n \geq 0 \in \mathbb{Z} \tag{5.8}$$

Individual time scales denoted by T_n are defined as $T_n = \varepsilon^n t$. The derivatives in Eq. (5.7) transform accordingly. Assuming that only time scales up to T_1 are relevant, a first total derivative $d()/dt$ is expressed using the chain rule as

$$\frac{d()}{dt} = \frac{dT_0}{dt} \frac{d()}{dT_0} + \frac{dT_1}{dt} \frac{d()}{dT_1} = D_0() + \varepsilon D_1() \tag{5.9}$$

where the specific time scale derivative operator notation $D_n = d()/dT_n$ has been introduced. The second time derivative follows accordingly as

$$\frac{d}{dt} \left(\frac{d()}{dt} \right) = D_0^2 + 2\varepsilon D_0 D_1 + \varepsilon^2 D_1^2 \tag{5.10}$$

where D_n^p denotes the pth derivative with respect to time scale T_n. Using Eq. (5.10) with Eq. (5.7) yields

$$\mathbf{M}(D_0^2 + 2\varepsilon D_0 D_1 + \varepsilon^2 D_1^2)(\mathbf{u}) + \sum_{p=-1}^{1} \sum_{q=-1}^{1} \sum_{r=-1}^{1} \mathbf{K}_{(p,q,r)} \mathbf{u}_{(p,q,r)} + \varepsilon \mathbf{f}^{NL} = \mathbf{0} \tag{5.11}$$

---

2 More specifically, the internal forces on the central unit cell are contained in the stiffness matrix such that $\mathbf{f}_0^{int} = -\mathbf{K}_{(p,q,r)} \mathbf{u}_{(p,q,r)}$.

Next, a standard asymptotic expansion for $\mathbf{u}_{(p,q,r)}$ is introduced, which allows the separation of individual solutions at each order of ε,

$$\mathbf{u}_{(p,q,r)}(T_0, T_1) = \mathbf{u}_{(p,q,r)}^{(0)}(T_0, T_1) + \varepsilon \mathbf{u}_{(p,q,r)}^{(1)}(T_0, T_1) + O(\varepsilon^2)$$
$$= \sum_i \varepsilon^i \mathbf{u}_{(p,q,r)}^{(i)}(T_0, T_1) \tag{5.12}$$

It is only necessary to retain terms up to $O(\varepsilon)$ to solve for a first perturbation correction term at $\mathbf{u}_{(p,q,r)}^{(0)}$. Substitution of Eq. (5.12) into Eq. (5.11) gives

$$\mathbf{M}(D_0^2 + 2\varepsilon D_0 D_1 + \varepsilon^2 D_1^2)(\mathbf{u}^{(0)} + \varepsilon \mathbf{u}^{(1)})$$
$$+ \sum_{p=-1}^{1} \sum_{q=-1}^{1} \sum_{r=-1}^{1} \mathbf{K}_{(p,q,r)}(\mathbf{u}_{(p,q,r)}^{(0)} + \varepsilon \mathbf{u}_{(p,q,r)}^{(1)}) + \varepsilon \mathbf{f}^{NL} = 0 \tag{5.13}$$

After some rearranging, this equation becomes

$$D_0^2 \mathbf{M} \mathbf{u}^{(0)} + \sum_{p=-1}^{1} \sum_{q=-1}^{1} \sum_{r=-1}^{1} \mathbf{K}_{(p,q,r)} \mathbf{u}_{(p,q,r)}^{(0)} + \varepsilon \left(D_0^2 \mathbf{M} \mathbf{u}^{(1)} + 2 D_0 D_1 \mathbf{M} \mathbf{u}^{(0)} \right.$$
$$\left. + \sum_{p=-1}^{1} \sum_{q=-1}^{1} \sum_{r=-1}^{1} \mathbf{K}_{(p,q,r)} \mathbf{u}_{(p,q,r)}^{(0)} + \mathbf{f}^{NL} \right) = 0 \tag{5.14}$$

where terms $O(\varepsilon^2)$ are neglected and the strict equality is no longer true. As ε is arbitrary, the coefficient multiplying each order of ε must vanish individually, yielding the ordered system of equations

$$O(\varepsilon^0): \quad D_0^2 \mathbf{M} \mathbf{u}^{(0)} + \sum_{p=-1}^{1} \sum_{q=-1}^{1} \sum_{r=-1}^{1} \mathbf{K}_{(p,q,r)} \mathbf{u}_{(p,q,r)}^{(0)} = 0 \tag{5.15}$$

$$O(\varepsilon^1): \quad D_0^2 \mathbf{M} \mathbf{u}^{(1)} + \sum_{p=-1}^{1} \sum_{q=-1}^{1} \sum_{r=-1}^{1} \mathbf{K}_{(p,q,r)} \mathbf{u}_{(p,q,r)}^{(1)} = -2 D_0 D_1 \mathbf{M} \mathbf{u}^{(0)} - \mathbf{f}^{NL}\left(\mathbf{u}_{(p,q,r)}^{(0)}\right) \tag{5.16}$$

Hence, the nonlinear system of open difference equations reduces to a cascading set of linear problems. The ordered equations reduce further when the specific form of the Bloch wave is specified. Equation (5.15) admits a Bloch wave solution. Moreover, a superposition of Bloch wave solutions is also a solution and may be used to assess nonlinear wave interactions

$$\mathbf{u}_{(p,q,r)}^{(0)} = \sum_i \frac{A_i(T_1)}{2} \boldsymbol{\phi}_i e^{i(\boldsymbol{\mu}_i \cdot \mathbf{r}_{pqr} - \omega_{0,i} T_0)} + c.c. \tag{5.17}$$

where individual Bloch wave solutions $\mathbf{u}^{(0)}$ have wave mode $\boldsymbol{\phi}_i$, frequency $\omega_{0,i}$, and slowly varying complex amplitude $A_i(T_1)$. Note that +c.c. is used herein to denote the complex conjugate of all preceding terms.

Substituting Eq. (5.17) into the right-hand side of the $O(\varepsilon^1)$ of Eq. (5.16) and evaluating the partial derivatives D_0 and D_1 yields

$$D_0^2 \mathbf{M} \mathbf{u}^{(1)} + \mathbf{K}(\boldsymbol{\mu}) \mathbf{u}^{(1)} = i \mathbf{M} \sum_i \omega_{0,i} A_i' \boldsymbol{\phi}_i e^{-i\omega_{0,i} T_0} - \mathbf{f}^{NL}(\mathbf{u}_{(p,q,r)}^{(0)}) + c.c. \tag{5.18}$$

where $\mathbf{K}(\mu_i)$ denotes the wavenumber-reduced stiffness matrix evaluated for the μ_ith wave vector and a prime denotes differentiation with respect to the argument. The right-hand side of Eq. (5.18) expands with the introduction of a polar form for amplitudes $A_i(T_1)$ such that

$$A_i(T_1) = \alpha_i(T_1)e^{-i\beta_i(T_1)} \tag{5.19}$$

The amplitudes $\alpha_i(T_1) \in \mathbb{R}$ and phases $\beta_i(T_1) \in \mathbb{R}$ vary slowly with T_1 such that they are constant in the T_0 time scale. The right-hand side of Eq. (5.18) evaluates to

$$(\ldots) = i\mathbf{M} \sum_i \omega_{0,i}(\alpha_i' - i\alpha_i\beta_i')\boldsymbol{\phi}_i e^{-i(\omega_{0,i}T_0+\beta_i(T_1))} - \mathbf{f}^{NL}\left(\mathbf{u}_{(p,q,r)}^{(0)}\right) + c.c. \tag{5.20}$$

We note that the linear kernel is the same as the linear kernel of the $O(\varepsilon^0)$ equation (Eq. (5.20)). Therefore, the Bloch wave mode matrix $\boldsymbol{\Phi}$ decouples the linear kernel. Introducing modal coordinates $\mathbf{u}^{(1)} = \boldsymbol{\Phi}\mathbf{z}^{(1)}(t)$ and pre-multiplying by $\boldsymbol{\phi}_j^H$, one obtains

$$D_0^2 m_{jj} z_j^{(1)} + k_{jj} z_j^{(1)} = \boldsymbol{\phi}_j^H i\mathbf{M} \sum_i \omega_{0,i}(\alpha_i' - i\alpha_i\beta_i')\boldsymbol{\phi}_i e^{-i(\omega_{0,i}T_0+\beta_i(T_1))} - \boldsymbol{\phi}_j^H \mathbf{f}^{NL} + c.c.$$

$$\tag{5.21}$$

Terms proportional to $\exp(\pm i\omega_{0,j}T_0)$ result in nonuniform perturbation expansions and must be removed. As the nonlinear force vector $\mathbf{f}^{NL}$ depends on $\mathbf{u}_{(p,q,r)}^{(0)}$, it must first be expanded to collect terms proportional to $\exp(i\omega_{0,j}T_0)$. The discrete frequency combinations found in $\mathbf{f}^{NL}$ are amenable to a multi-dimensional Fourier series expansion as (in the case of two frequencies denoted by i and j):

$$\mathbf{f}^{NL}(T_0) = \sum_i \sum_j \mathbf{c}_{ij} e^{i\omega_{0,i}T_0} e^{i\omega_{0,j}T_0} + c.c. \tag{5.22}$$

where the Fourier coefficients are generally a function of all amplitudes and wave vectors $\mathbf{c}_{ij}(\cup A_k, \cup \mu_k)$. This expansion specifically provides for the existence of higher generated harmonics as well as sum-and-difference frequency generations [32]. For simplicity, we denote the fundamental harmonic terms using a single index such that $\mathbf{c}_{i0} = \mathbf{c}_i$ and $\mathbf{c}_{0j} = \mathbf{c}_j$. The explicit dependence of $\mathbf{c}_i$ on the complex amplitude of each wave, as well as individual wavenumbers, has been noted. A special case arises when two frequencies $\omega_{0,m}$ and $\omega_{0,n}$ are commensurate[3] with one another such that these extraneous harmonics occur at one of the input frequencies $\omega_{0,i}$. Since cubic nonlinearities are considered herein, the coefficient vector $\mathbf{c}_i$ can be obtained symbolically (usually with the aid of symbolic manipulation software) or computationally with the aid of a Fourier transform.

Equations (5.21) and (5.22) combined facilitate the identification of secular terms

$$D_0^2 m_{jj} z_j^{(1)} + k_{jj} z_j^{(1)} = \boldsymbol{\phi}_j^H \sum_i \left[(i\mathbf{M}\omega_{0,i}\boldsymbol{\phi}_i(\alpha_i' - i\alpha_i\beta_i')e^{-i\beta_i(T_1)} - \mathbf{c}_i)e^{-i\omega_{0,i}T_0}\right] + c.c. \tag{5.23}$$

where other harmonic terms exist but do not contribute to secularity and so have been omitted from the above expression. Terms on the right-hand side of Eq. (5.23) with

---

3 Two frequencies ω_A and ω_B are commensurate when two integers n and m can be chosen that satisfy $n\omega_A + m\omega_B = 0$

exp $(i\omega_{0,i}T_0)$ time-dependence result in secular expansions, so we remove them by requiring that they vanish identically. Factoring out the $i = j$ term from the summation provides the required condition

$$i\omega_{0,j}m_{jj}(\alpha_j' - i\alpha_j\beta_j')e^{-i\beta_j} - \boldsymbol{\phi}_j^H\mathbf{c}_j = 0 \tag{5.24}$$

where the orthogonality properties of the mass matrix have been enforced.

When the frequencies $\omega_{0,i}$ are incommensurate it is convenient to multiply Eq.(5.24) by $e^{i\beta_j}$ to obtain for the jth equation

$$i\omega_{0,j}m_{jj}(\alpha_j' - i\alpha_j\beta_j') - \boldsymbol{\phi}_j^H\mathbf{c}_je^{i\beta_j} = 0 \tag{5.25}$$

Equation (5.25) requires that real and imaginary components vanish individually such that

$$\omega_{0,j}m_{jj}\alpha_j\beta_j' - \Re(\boldsymbol{\phi}^H\mathbf{c}_je^{i\beta_j}) = 0 \tag{5.26}$$

$$\omega_{0,j}m_{jj}\alpha_j' - \Im(\boldsymbol{\phi}^H\mathbf{c}_je^{i\beta_j}) = 0 \tag{5.27}$$

where $\Re(\cdot)$ and $\Im(\cdot)$ return real and imaginary components of the included expression, respectively. Equations (5.26) and (5.27) are termed *evolution equations* since, when solved, they describe the slow-time evolution of the amplitude $\alpha_j(T_1)$ and phase $\beta_j(T_1)$.

In general, Eqs. (5.26) and (5.27) constitute a set of $2N$ nonlinear differential equations when a total of N Bloch wave modes are considered. The evolution equations simplify considerably when the included frequencies are incommensurate. It happens that the coefficient $\mathbf{c}_j$ for the jth frequency always occurs, for cubic nonlinearities, in the fortuitous form $\mathbf{c}_j = \hat{\mathbf{c}}_j(\cup \alpha_k)\exp(-i\beta_j)$. Moreover, the scalar product $\boldsymbol{\phi}^H\mathbf{c}_j$ results in a purely real quantity since $\boldsymbol{\phi}_j$ has been derived from a Hermitian eigenvalue problem. Thus the evolution equations are simply

$$\omega_{0,j}m_{jj}\alpha_j\beta_j' - \boldsymbol{\phi}^H\hat{\mathbf{c}}_j = 0 \tag{5.28}$$

$$\omega_{0,j}m_{jj}\alpha_j' = 0 \tag{5.29}$$

Equation (5.29) implies that $\alpha_j \; \forall \, j$ is constant with respect to slow time T_1. Moreover, Eq. (5.28) can be solved and directly integrated to obtain $\beta_j(T_1)$, since $\hat{\mathbf{c}}_j$ is independent of β_j

$$\beta_j = \frac{\boldsymbol{\phi}^H\hat{\mathbf{c}}_j}{\omega_{0,j}\alpha_jm_{jj}}T_1 \tag{5.30}$$

Thus, it is often of interest to specify the slow-time behavior of $A(T_1)$ a priori, to reduce the complexity of the evolution equations. The nonlinear dispersion corrections, produced *for each Bloch wave* using the method of multiple scales for the case of incommensurate frequencies, read

$$\omega_j(\boldsymbol{\mu}) = \omega_{0,j}(\boldsymbol{\mu}) + \varepsilon\frac{\boldsymbol{\phi}^H\hat{\mathbf{c}}_j}{\omega_{0,j}\alpha_jm_{jj}} + O(\varepsilon^2) \tag{5.31}$$

Here again it is possible to manipulate the corrected frequency by introducing the modal stiffness according to $\omega_{0,j}^2m_{jj} = k_{jj}$ so that the updated frequency is given as

$$\omega_j(\boldsymbol{\mu}) = \omega_{0,j}(\boldsymbol{\mu})\left(1 + \varepsilon\frac{\omega_{0,j}\boldsymbol{\phi}^H\hat{\mathbf{c}}_j}{\alpha_jk_{jj}}\right) + O(\varepsilon^2) \tag{5.32}$$

In the more general case, the evolution equations (5.26) and (5.27) possess nontrivial solutions and must be solved as a coupled system of nonlinear differential equations for each $\alpha_j(T_1)$ and $\beta_j(T_1)$. Then, the instantaneous frequency correction yields frequency expansions in the form

$$\omega_j(\mu) = \omega_{0,j}(\mu) + \varepsilon\beta_j(T_1)/T_0 + O(\varepsilon^2) \tag{5.33}$$

such that the ith Bloch wave component is given by

$$\mathbf{u}_j^{(0)} = \frac{\alpha_i(T_1)}{2}\phi_j e^{-i[(\omega_{0,j}+\beta_j(\varepsilon t)/t)t]} + c.c. \tag{5.34}$$

This incommensurate case is generally more informative for assessing nonlinear dispersion.

5.3 Application to a 1D Monoatomic Chain

5.3.1 Overview

Acoustic devices may be subject to multi-harmonic excitations. In 1D linear materials, the only wave–wave interactions expected are constructive and destructive interference. In nonlinear materials, additional wave–wave interactions take place such that the strength of the interaction causes dispersion to become both amplitude and frequency dependent. The interaction of harmonic waves that are commensurate with frequency ratios close to 1:3 may be of particular concern for cubically nonlinear materials, where superharmonic generation may be present.

5.3.2 Model Description and Nonlinear Governing Equation

The monoatomic mass-spring chain with a cubic nonlinearity shown in Figure 5.2 is governed by the following equation of motion:

$$m\ddot{\tilde{u}}_p + k(2\tilde{u}_p - \tilde{u}_{p-1} - \tilde{u}_{p+1}) + \varepsilon\Gamma(\tilde{u}_p - \tilde{u}_{p-1})^3 + \varepsilon\Gamma(\tilde{u}_p - \tilde{u}_{p+1})^3 = 0, \quad \forall p \in \mathbb{Z} \tag{5.35}$$

where $\tilde{u}_p(t)$ is the displacement from equilibrium of the pth mass, ε is a small parameter, Γ characterizes the cubic nonlinearity, and m and k denote the mass and linear stiffness. We nondimensionalize the equation of motion Eq. (5.35) by first writing the equation of motion in canonical form, where $\omega_n^2 = k/m$

$$\ddot{\tilde{u}}_p + \omega_n^2(2\tilde{u}_p - \tilde{u}_{p-1} - \tilde{u}_{p+1}) + \frac{\varepsilon\Gamma}{m}(\tilde{u}_p - \tilde{u}_{p-1})^3 + \frac{\varepsilon\Gamma}{m}(\tilde{u}_p - \tilde{u}_{p+1})^3 = 0 \tag{5.36}$$

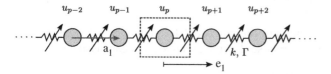

Figure 5.2 Monoatomic mass-spring chain with cubic stiffness and lattice vector $\mathbf{a}_1$.

We represent space and time by introducing characteristic length and time parameters (d_c and t_c) and nondimensional variables for space and time (u_p and t) such that $\tilde{u}_p \equiv d_c\, u_p$ and $\tilde{t} \equiv t_c\, t$, where characteristic length is $d_c = \sqrt{k/\Gamma}$ and the characteristic time is $t_c = 1/\omega_n$. The equation of motion is then re-written using nondimensional variables as

$$\ddot{u}_p + (2u_p - u_{p-1} - u_{p+1}) + \varepsilon(u_p - u_{p-1})^3 + \varepsilon(u_p - u_{p+1})^3 = 0, \quad \forall p \in \mathbb{Z} \quad (5.37)$$

Note that double dot corresponds to a second derivative with respect to nondimensional time t.

5.3.3 Single-wave Dispersion Analysis

We first perform a single-wave analysis presuming the presence of only one harmonic plane wave; later we perform a multi-wave analysis presuming the presence of two or more dominant harmonics such that wave–wave interactions may take place.

We begin with the equation of motion Eq. (5.37) and seek expressions for time-varying amplitude and phase. Correction terms up to $O(\varepsilon^2)$ are sought such that the only time scales of interest are $T_0 = t$ and $T_1 = \varepsilon t$. An asymptotic expansion for the dependent variable $u_p(t)$ is expressed in multiple time scales as

$$u_p(t) = \sum_n \varepsilon^n u_p^{(n)}(T_0, T_1, \dots T_n)$$

Keeping terms up to and including $O(\varepsilon^1)$ gives

$$u_p = u_p^{(0)}(T_0, T_1) + \varepsilon u_p^{(1)}(T_0, T_1) + O(\varepsilon^2) \quad (5.38)$$

Substituting Eq. (5.38) into Eq. (5.37), and collecting terms in orders of ε, gives the first two ordered equations, which may be written as

$$\varepsilon^0: \quad D_0^2 u_p^{(0)} + (2u_p^{(0)} - u_{p-1}^{(0)} - u_{p+1}^{(0)}) = 0 \quad (5.39a)$$

$$\varepsilon^1: \quad D_0^2 u_p^{(1)} + (2u_p^{(1)} - u_{p-1}^{(1)} - u_{p+1}^{(1)}) = -2D_0 D_1 u_p^{(0)} - f^{NL}(u_p^{(0)}, u_{p\pm1}^{(0)}) \quad (5.39b)$$

where

$$f^{NL} = (u_p^{(0)} - u_{p-1}^{(0)})^3 + (u_p^{(0)} - u_{p+1}^{(0)})^3 \quad (5.40)$$

Recall that D_0 and D_1 represent partial derivatives with respect to the time scales T_0 and T_1, respectively. In the presence of a single plane wave at frequency ω and wavenumber μ, the solution of Eq. (5.39a) is

$$u_p^{(0)}(T_0, T_1) = \frac{1}{2}A_0(T_1)e^{i(\mu p - \omega_0 T_0)} + \frac{1}{2}\overline{A_0}(T_1)e^{-i(\mu p - \omega_0 T_0)} \quad (5.41)$$

where $A_0(T_1)$ is the complex quantity that permits slow time evolution of the amplitude and phase and an over-bar indicates a complex conjugate. The distinction may be made explicit by using polar form such that $A_0(T_1) = \alpha(T_1)e^{-i\beta(T_1)}$, where both $\alpha(T_1)$ and $\beta(T_1)$ are real-valued functions. Substituting Eq. (5.41) into Eq. (5.39a) and simplifying gives the well-known linear dispersion relationship,

$$\omega_0 = \sqrt{2 - 2\cos\mu} \quad (5.42)$$

The linear kernel of the $O(\varepsilon^1)$ equation is identical to the linear kernel of the $O(\varepsilon^0)$ equation Eq. (5.39a), and thus the homogeneous solution has the same form as Eq. (5.41).

Therefore, any terms appearing on the right-hand side of Eq. (5.39b) with similar spatial and temporal forms result in a nonuniform expansion. Removal of these secular terms by setting them equal to zero results in a system of *evolution equations* for the functions $\alpha(T_1)$ and $\beta(T_1)$

$$\alpha' = 0 \tag{5.43a}$$

$$\beta' = \frac{3\alpha^2(\cos 2\mu - 4\cos\mu + 3)}{4\,\omega_0} = \frac{3}{8}\alpha^2\omega_0^3 \tag{5.43b}$$

where α' and β' denote the derivatives with respect to the slow time scale T_1. It is clear from Eq. (5.43a) that $\alpha(T_1) = \alpha_0$ and $\beta(T_1) = \beta_1 T_1 + \beta_0$, where α_0 and β_0 are arbitrary constants determined by imposing initial conditions on the plane wave given in Eq. (5.41). For the mass-spring chain under consideration, β_0 may be set to zero without loss of generality. Equation (5.41) may be expressed using trigonometric functions as

$$u_p^{(0)}(T_0, T_1) = \alpha_0 \cos(\mu p - (\omega_0 + \omega_1\varepsilon)T_0) \tag{5.44}$$

such that $\beta_1\varepsilon \equiv \omega_1\varepsilon$ may be regarded as a first-order frequency correction that results in a shift of the linear dispersion curve. Based on Eq. (5.44), the reconstituted dispersion relationship is then

$$\omega(\mu) = \sqrt{2 - 2\cos\mu} + \varepsilon\frac{3\alpha^2}{8}(2 - 2\cos\mu)^{3/2} + O(\varepsilon^2) \tag{5.45}$$

5.3.4 Multi-wave Dispersion Analysis

Due to linearity of the $O(\varepsilon^0)$ expression, a superposition of solutions in the form of Eq. (5.41) is also a solution such that $u_p^{(0)}$ can be replaced by $\sum_n u_{p,n}^{(0)}$. When two waves are present, we superimpose two traveling wave solutions, labeled A and B, such that

$$u_p^{(0)}(T_0, T_1) = u_{p,A}^{(0)}(T_0, T_1) + u_{p,B}^{(0)}(T_0, T_1) + O(\varepsilon^2) \tag{5.46}$$

where

$$u_{p,B}^{(0)}(T_0, T_1) = \frac{1}{2}B_0(T_1)e^{i(\mu_B p - \omega_{B0}T_0)} + c.c. = \frac{1}{2}\alpha_B(T_1)e^{i(\mu_B p - \omega_{B0}T_0 - \beta_B(T_1))} + c.c. \tag{5.47a}$$

$$u_{p,A}^{(0)}(T_0, T_1) = \frac{1}{2}A_0(T_1)e^{i(\mu_A p - \omega_{A0}T_0)} + c.c. = \frac{1}{2}\alpha_A(T_1)e^{i(\mu_A p - \omega_{A0}T_0 - \beta_A(T_1))} + c.c. \tag{5.47b}$$

and c.c. represents the complex conjugates of the preceding terms. Here, α_A, α_B, β_A, and β_B are functions for the amplitude and phase of each wave, which will be determined by imposing uniform asymptotic expansion conditions.

Substituting Eqs. (5.46), (5.47a), and (5.47b) into Eq. (5.39b) and expanding the nonlinear terms denoted by f^{NL} allows for identification of secular terms: those that are proportional to $e^{i\mu_A p}e^{i\omega_{A0}T_0}$ and $e^{i\mu_B p}e^{i\omega_{B0}T_0}$, labeled S_A and S_B, respectively. Due to the arbitrary naming convention of the A and B waves, the secular terms arising from each wave are identical. For the A wave, the secular terms are

$$S_A = e^{i\mu_A p}e^{-i\omega_{A0}T_0}\left[-iA_0'\omega_{A0} + \frac{3}{4}A_0|B_0|^2 e^{i(\mu_A - \mu_B)} + \frac{3}{4}A_0|B_0|^2 e^{i(\mu_B - \mu_A)} \right.$$
$$\left. + \frac{3}{8}A_0|A_0|^2 e^{i2\mu_A} + \frac{3}{8}A_0|A_0|^2 e^{-i2\mu_A} - \frac{3}{2}A_0|B_0|^2 e^{i\mu_A} - \frac{3}{2}A_0|B_0|^2 e^{-i\mu_A} \right.$$

$$- \frac{3}{2} A_0 |A_0|^2 e^{i\mu_A} - \frac{3}{2} A_0 |A_0|^2 e^{-i\mu_A} + \frac{3}{4} A_0 |B_0|^2 e^{i(\mu_B + \mu_A)} + \frac{3}{4} A_0 |B_0|^2 e^{-i(\mu_B + \mu_A)}$$

$$\left. - \frac{3}{2} A_0 |B_0|^2 e^{i\mu_B} - \frac{3}{2} A_0 |B_0|^2 e^{i\mu_B} + \frac{9}{4} A_0 |A_0|^2 + 3 A_0 |B_0|^2 \right] + c.c. \qquad (5.48)$$

Additional secular terms may arise if (μ_A, ω_{A0}) and (μ_B, ω_{B0}) are related such that their combination results in a secular term. When $\omega_{B0} \approx 3\omega_{A0}$ *and* $\mu_B \approx 3\mu_A$, these additional secular terms are

$$A_0^3 e^{i3\mu_A p} e^{-i3\omega_{A0}\tau_0} \left[-\frac{3}{8} e^{-i\mu_A} + \frac{1}{3} - \frac{1}{8} e^{i3\mu_A} + \frac{3}{8} e^{-i2\mu_A} - \frac{3}{8} e^{i\mu_A} - \frac{1}{8} e^{-i3\mu_A} + \frac{3}{8} e^{i2\mu_A} \right]$$

$$+ A_0^2 \overline{B_0} e^{i(\mu_B - 2\mu_A)p} e^{-i(\omega_{B0} - 2\omega_{A0})\tau_0} \left[-\frac{3}{8} e^{-i\mu_B} + \frac{3}{8} e^{2i\mu_A} - \frac{3}{8} e^{i\mu_B} - \frac{3}{4} e^{-i\mu_A} \right.$$

$$- \frac{3}{8} e^{i(\mu_B - 2\mu_A)} - \frac{3}{4} e^{i\mu_A} + \frac{3}{4} + \frac{3}{4} e^{i(\mu_B - \mu_A)}$$

$$\left. - \frac{3}{8} e^{-i(\mu_B + 2\mu_A)} + \frac{3}{4} e^{-i(\mu_B + \mu_A)} + \frac{3}{8} e^{-2i\mu_A} \right] + c.c. \qquad (5.49)$$

The equivalent secular terms for the B wave may be obtained by letting $\mu_A \rightarrow \mu_B$, $\omega_{A0} \rightarrow \omega_{B0}$, and $A_0 \leftrightarrow B_0$. In light of Eqs. (5.48) and (5.49), there exist two possibilities for secularity when two harmonic plane waves are present:

Case 1. $\{(\mu_B, \omega_{B0}) \in \mathbb{R}^2 : (\mu_B, \omega_{B0}) \neq a \cdot (\mu_A, \omega_{A0})$ where $a = 1/3$ or 3$\}$, such that wave–wave interactions are produced due to amplitude products (the most general case).

Case 2. $\{(\mu_B, \omega_{B0}) \in \mathbb{R}^2 : (\mu_B, \omega_{B0}) = a \cdot (\mu_A, \omega_{A0})$ where $a = 1/3$ or 3$\}$ such that wave–wave interactions are influenced by nonlinear frequency and wavenumber coupling, as in the long-wavelength limit.

Case 1 is the most general case in which there are two waves present, and nonlinear coupling occurs only due to the products of amplitude coefficients A_0 and B_0. Case 2 produces secular terms due to amplitude products *and* nonlinear coupling of both the frequency and wavenumber. The superharmonic ($a = 3$) and subharmonic ($a = 1/3$) cases are essentially the same due to the arbitrary choice in naming the A and B waves. Due to additional secular terms arising in the long-wavelength limit for the superharmonic (and subharmonic) case, a separate treatment is necessary.

Case 1. General Wave–Wave Interactions

We suppose there are two waves A and B such that there is no wave–wave interaction due to special combinations of frequency and wavenumber ($a = 1/3$ or 3). Then, two independent, complex secular terms arise, which lead to a nonuniform expansion at $O(\varepsilon^1)$, as in Eq. (5.48). Solvability conditions for a uniform expansion require these terms to vanish identically. Separating the real and imaginary components for each coefficient function and equating the resulting expressions to zero results in a set of four evolution equations

$$\alpha'_A = 0 \qquad\qquad\qquad \alpha'_B = 0$$

$$\beta'_A = \frac{3}{8} \omega^3_{A0} \alpha^2_A + \frac{3}{4} \omega_{A0} \omega^2_{B0} \alpha^2_B \qquad \beta'_B = \frac{3}{4} \omega^2_{A0} \omega_{B0} \alpha^2_A + \frac{3}{8} \omega^3_{B0} \alpha^2_B \qquad (5.50)$$

Equations (5.50) lead to some interesting observations regarding the dispersion relationship when two waves interact. As with the single-wave case, the amplitudes α_A

and α_B of both waves are constant at $O(\varepsilon^1)$. The separable equations β'_A and β'_B, when integrated, yield linear phase corrections that can be interpreted as frequency (and thus dispersion) corrections similar to Eq. (5.44). After integration, the expressions in Eq. (5.50) become

$$\alpha_A = \alpha_{A,0} \qquad\qquad\qquad \alpha_B = \alpha_{B,0}$$

$$\beta_A = \beta_{A,1}T_1 + \beta_{A,0} \qquad\qquad \beta_B = \beta_{B,1}T_1 + \beta_{B,0} \qquad (5.51)$$

where, as before, one of the constants of integration $\beta_{A,0}$ or $\beta_{B,0}$ may be set to zero without any loss of generality, while the remaining constant controls the phase relationship between the two waves (but has no effect on dispersion). The slopes of β_A and β_B, given by $\beta_{A,1} = \beta'_A$ and $\beta_{B,1} = \beta'_B$ in Eqs. (5.51), determine the magnitude of the dispersion correction for each wave.

It is convenient to express the relationship between ω_{A0} and ω_{B0} by a single frequency on the linear dispersion curve $\omega_0 \equiv \omega_{A0}$ and a frequency ratio r, so that both frequency corrections can be visualized on a single plot, and also so that $r = 3$ corresponds to Case 2 in the long-wavelength limit

$$r \equiv \frac{\omega_{B0}}{\omega_{A0}} \longrightarrow \omega_{B0} = r\omega_0 \qquad (5.52)$$

Furthermore, we assume, without loss of generality, that $\omega_{A0} < \omega_{B0}$, so that a single frequency ω_0 and a ratio $r > 1$ describe the dispersion corrections rather than two frequencies. The frequency-correction terms become

$$\omega_{A1} \equiv \beta_{A,1} = \frac{3}{8}\omega_0^3\alpha_A^2 + \frac{3r^2}{4}\omega_0^3\alpha_B^2 \qquad (5.53a)$$

$$\omega_{B1} \equiv \beta_{B,1} = \frac{3r}{4}\omega_0^3\alpha_A^2 + \frac{3r^3}{8}\omega_0^3\alpha_B^2 \qquad (5.53b)$$

The reconstituted dispersion relations, corrected to $O(\varepsilon^1)$, are given by

$$\omega_A = \sqrt{2 - 2\cos\mu_A} + \varepsilon\left(\frac{3}{8}\alpha_A^2 + \frac{3r^2}{4}\alpha_B^2\right)(2 - 2\cos\mu_A)^{3/2} + O(\varepsilon^2) \qquad (5.54a)$$

$$\omega_B = \sqrt{2 - 2\cos\mu_B} + \varepsilon\left(\frac{3r}{4}\alpha_A^2 + \frac{3r^3}{8}\alpha_B^2\right)(2 - 2\cos\mu_A)^{3/2} + O(\varepsilon^2) \qquad (5.54b)$$

so that each wave follows its own dispersion curve for a given frequency ratio r and given amplitudes α_A and α_B. Thus, when two waves nonlinearly interact they form two amplitude- and frequency-dependent dispersion branches. In the absence of nonlinear wave–wave interaction, only a single dispersion curve exists.

Figure 5.3 shows two possible dispersion relation corrections for both hardening and softening chains. These dispersion curves were plotted using $\alpha_A = \alpha_B = 4$, $r = 2.7$, and $\varepsilon = \pm 0.01$ to produce hardening and softening curves. Note that the dotted section of the ω_A curve corresponds to values that would cause the superharmonic $\omega_B = 3 \omega_{A0} + \varepsilon\omega_{B1}$ to be in the band gap. This section of the curve is not explored here, but has been included as a visual aid to make the trends more apparent. At large amplitudes the single-wave corrected dispersion curve in Eq. (5.45), labeled ω in Figure 5.3, underestimates the magnitude of the correction to both ω_A and ω_B. The failure is especially evident at the edge of the Brillouin zone.

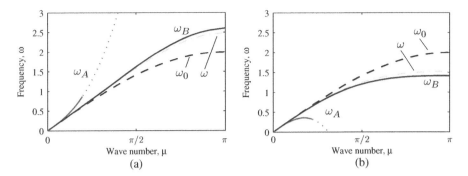

Figure 5.3 Multi-wave corrected dispersion curves compared with the linear curve ω_0 and the nonlinear single-wave corrected curve ω.

If the frequency-correction terms ω_{A1} and ω_{B1} in Eqs. (5.53a) and (5.53b) are interpreted as asymptotic frequency expansion terms as in the Lindstedt–Poincaré method; the ratios[4] of the $O(\varepsilon^1)$ correction terms to specified $O(\varepsilon^0)$ linear values $(\omega_{A0}, \omega_{B0})$ for each wave provide an estimate of the expansion uniformity and a normalized measure of the dispersion correction. These ratios are:

$$\rho_A = \left| \frac{\varepsilon \omega_{A1}}{\omega_{A0}} \right| \Rightarrow \left| \frac{\varepsilon \omega_{A1}}{\omega_0} \right| = \frac{[(3/8)\omega_0^3 \alpha_A^2 + (3r^2/4)\omega_0^3 \alpha_B^2]|\varepsilon|}{\omega_0} = \left[\frac{3}{8}\alpha_A^2 + \frac{3r^2}{4}\alpha_B^2 \right] |\varepsilon| \omega_0^2 \tag{5.55}$$

$$\rho_B = \left| \frac{\varepsilon \omega_{B1}}{\omega_{B0}} \right| \Rightarrow \left| \frac{\varepsilon \omega_{B1}}{r\omega_0} \right| = \frac{[(3r/4)\omega_0^3 \alpha_A^2 + (3r^3/8)\omega_0^3 \alpha_B^2]|\varepsilon|}{r\omega_0} = \left[\frac{3}{4}\alpha_A^2 + \frac{3r^2}{8}\alpha_B^2 \right] |\varepsilon| \omega_0^2 \tag{5.56}$$

Hence, α_A, α_B, r, and the quantity $|\varepsilon|\omega_0^2$ determine the uniformity and magnitude of the correction terms. Uniform expansions correspond to when the ratios in Eqs. (5.55) and (5.56) are much less than one. Furthermore, since $\rho_{A,B}$ represents a percentage of the linear frequency at specified point on the dispersion curve, we can calculate parameters to obtain a specific dispersion shift. Numerical simulations, presented in Section 5.3.5, have shown good agreement for $\rho_{A,B}$ less than about 0.2.

Case 2. Long-wavelength Limit Wave–Wave Interactions

We previously mentioned that in Case 2 additional wave–wave interactions may be produced due to nonlinear phase coupling when $\omega_{B0} \approx 3 \, \omega_{A0}$ *and* $\mu_B \approx 3\mu_A$. This corresponds to when the phase velocities of each wave are approximately equal, so that the waves travel together. The closeness of (μ_B, ω_{B0}) to $3(\mu_A, \omega_{A0})$ can be formally represented using detuning quantities $\overline{\sigma}_\omega$ and $\overline{\sigma}_\mu$, respectively, so that

$$\omega_{B0} = 3\omega_{A0} + \overline{\sigma}_\omega \text{ and } \mu_B = 3\mu_A + \overline{\sigma}_\mu \tag{5.57}$$

are exact relations. In the long-wavelength limit, $\overline{\sigma}_\omega = \overline{\sigma}_\mu = \overline{\sigma} \equiv \sigma\varepsilon$ to $O(\varepsilon^3)$ where the detuning parameter σ is defined to be a real, positive number of order $O(\varepsilon^0)$. To show this, let $\mu_A \to \varepsilon \hat{\mu}_A$ and $\hat{\mu}_A \equiv O(\varepsilon^0)$, so that μ_A is in the long-wavelength limit. Inverting

---

4 More specifically, the absolute value of the ratios in the event that $\varepsilon < 0$.

the linear dispersion relation $\omega_{B0}(\mu_B)$ gives

$$\mu_B(\omega_{B0}) = \cos^{-1}\left(1 - \frac{1}{2}\omega_{B0}^2\right)$$

Substituting $\omega_{B0} = 3\omega_{A0} + \sigma\varepsilon = 3\sqrt{2 - 2\cos(\varepsilon\hat{\mu}_A)} + \sigma\varepsilon$ in the previous expression and Taylor expanding $\mu_B(\hat{\mu}_A; \varepsilon, \sigma)$ in the small parameter ε gives

$$\mu_B(\hat{\mu}_A; \varepsilon, \sigma) = (3\hat{\mu}_A + \sigma)\varepsilon + O(\varepsilon^3)$$

Thus,

$$\mu_B \approx 3\mu_A + \overline{\sigma} \tag{5.58}$$

Hence, Case 2 may be realized for small μ_A (and consequently small ω_{A0}) that are at the long-wavelength limit. This one-to-one ratio of frequency to wavenumber is expected since the slope of the nondimensional linear dispersion relation is approximately 1 in the long-wavelength limit, where μ_A approaches zero.

One more implication of Eq. (5.58) arises when the substitutions (5.57) are made for ω_{B0} and μ_B in the $O(\varepsilon^0)$ solution (5.47b): an additional *long spatial scale* $J_1 \equiv \varepsilon p$ arises. Equation (5.47b) for Case 2, after substitutions, becomes

$$x_{p,B}^{(0)}(T_0, T_1, J_1) = \frac{B_0(T_1)}{2}e^{i(\mu_B p - \omega_{B0}T_0)} + c.c. = \frac{B_0(T_1)}{2}e^{i(3\mu_A + \overline{\sigma})p}e^{-i(3\omega_{A0} + \overline{\sigma})T_0} + c.c.$$

$$= \frac{B_0(T_1)}{2}e^{i(3\mu_A p + \sigma J_1)}e^{-i(3\omega_{A0}T_0 + \sigma T_1)} + c.c. \tag{5.59}$$

From this point forward the procedure follows the development in Section 5.3.4. However, secular terms are instead proportional to $\exp(i\mu_A p - i\omega_{A0}T_0)$, as well as the third harmonic $\exp(i3\mu_A p - i3\omega_{A0}T_0)$, so that the secular terms arise from both Eqs. (5.48) and (5.49). Separating the real and imaginary parts of the two solvability conditions yields a set of strongly coupled, nonlinear evolution equations with long spatial (J_1) and temporal dependence (T_1). It happens that the slow time variable T_1 exists in each evolution equation such that they may be written autonomously via substitution of a new variable $\gamma(T_1)$ and its slow time derivative

$$\gamma = -3\beta_A + \beta_B + \sigma T_1 - \sigma J_1 \tag{5.60a}$$

$$\gamma' = -3\beta_A' + \beta_B' + \sigma \tag{5.60b}$$

Using this fact, the evolution equations may be written in a fashion similar to Eqs. (5.50) as

$$\alpha_A' = \frac{3}{2}\alpha_A^2\alpha_B\xi\sin(\gamma) \tag{5.61a}$$

$$\alpha_B' = -\frac{1}{6}\alpha_A^3\xi\sin(\gamma) \tag{5.61b}$$

$$\beta_A' = -\frac{3}{2}\alpha_A\alpha_B\xi\cos(\gamma) + a_1\alpha_A^2 + a_2\alpha_B^2 \tag{5.61c}$$

$$\beta_B' = -\frac{1}{6}\frac{\alpha_A^3}{\alpha_B}\xi\cos(\gamma) + b_1\alpha_A^2 + b_2\alpha_B^2 - \sigma \tag{5.61d}$$

where the coefficients a_1, a_2, b_1, b_2, and ξ are most simply expressed as functions of $\omega_{A0} \equiv \omega_0$, given by

$$a_1 = \frac{3}{8}\omega_0^3 \qquad\qquad b_1 = \frac{1}{4}\omega_0^7 - \frac{3}{2}\omega_0^5 + \frac{9}{4}\omega_0^3$$

$$a_2 = \frac{3}{4}\omega_0^7 - \frac{9}{2}\omega_0^5 + \frac{27}{4}\omega_0^3 \qquad b_2 = \frac{1}{8}\omega_0^{11} - \frac{3}{2}\omega_0^9 + \frac{27}{4}\omega_0^7 - \frac{27}{2}\omega_0^5 + \frac{81}{8}\omega_0^3$$

$$\xi = -\frac{1}{4}\omega_0^5 + \frac{3}{4}\omega_0^3 \tag{5.62}$$

The large state space of Eqs. (5.61a) complicates the analysis. Some general observations can be made, however, about the character of the solution in certain limiting cases. The coefficients a_1, a_2, b_1, and b_2 occur in the general form $(a,b)_{1,2} = \sum_n c_n(-1)^{n+1}\omega_0^{2n+1}$, where $c_n > c_{n+1}$ by inspection. Thus for low frequencies in the long-wavelength limit, the $c_1\omega_0^3$ terms dominate. Comparison of Eqs. (5.62) with Eqs. (5.53a) and (5.53b) shows that these leading-order terms are identical for $r = 3$, and thus the primary difference in the resulting evolution equations of Cases 1 and 2 is the existence of the $\xi \sin(\gamma)$ and $\xi \cos(\gamma)$ terms in Eqs. (5.61a). These terms are responsible for coupling the α_A' and α_B' evolution equations with the β_A' and β_B' equations, and are *usually* negligible for small ω_0 (and hence small ξ).

Numerical integration of the evolution equations in Eq. (5.61a) provides additional insight into and verification of the previous assertions. The solutions of the differential equations depend on the initial values $\alpha_A(0) = \alpha_{A,0}$, $\alpha_B(0) = \alpha_{B,0}$, $\beta_A(0) = \beta_{A,0}$, and $\beta_B(0) = \beta_{B,0}$, and the parameters ω_0, σ, and ε. Some possible solutions for the evolution equations are illustrated in Figure 5.4 for $\alpha_{A,0} = \alpha_{B,0} = \alpha_0 = [1, 2, 5]$, $\beta_{A,0} = \beta_{B,0} = 0$, and $\omega_0 = 0.5$, $\sigma = 0$, and $\varepsilon = 0.01$.

The frequency was chosen to illustrate the oscillations clearly, but $\omega_0 = 0.5$ is at the edge of the long-wavelength limit. The amplitudes α_A and α_B tend to oscillate periodically about some fixed point, while the phase corrections β_A and β_B tend to oscillate about some linear functions. Therefore, in the long-wavelength limit the wave–wave interaction gives rise to amplitude and frequency modulation. Based on numerical observation, the modulations are negligible for small frequencies in the long-wavelength limit so that $\alpha_A' = 0$, $\alpha_B' = 0$, and β_A' and β_B' are both constant to $O(\varepsilon^1)$. Therefore, even

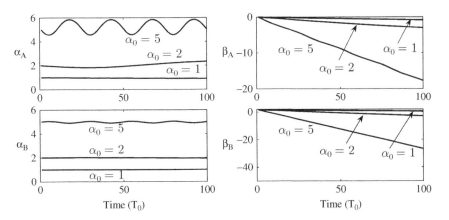

Figure 5.4 Typical solutions for α_A, α_B, β_A, and β_B.

when ω_{A0} and ω_{B0} are commensurate or nearly-commensurate, long-wavelength-limit wave–wave interactions may be negligible, and the corrected dispersion relations obtained for Case 1 provide a good (and potentially more useful) approximation.

5.3.5 Numerical Verification and Discussion

Numerical simulation of the mass-spring system depicted in Figure 5.2 and governed by Eq. (5.37) verifies the presented asymptotic approach. In deriving corrections to the linear dispersion curve, it was assumed that two interacting waves with $O(\varepsilon^0)$ amplitudes existed in a 1D nonlinear medium away from any forcing. By imposing plane-wave initial conditions and allowing a simulation to run for 100–200 nondimensional "seconds," a space–time matrix of displacements was produced. Points on the dispersion curve were located by performing a 2D fast Fourier transform (2D FFT) on the displacement matrix.

Figure 5.5 illustrates how the frequency ratio r may be considered as a design parameter in addition to the wave amplitudes for tunable acoustic metamaterials. We seek to obtain a 10% ($\rho_B = 0.1$) frequency shift for a B wave with amplitude $\alpha_A = 2$ and frequency $\omega_{B0}(\mu_B) = r\omega_0 = 1.8$. Then, the parameters α_A and r may be chosen freely to obtain $\rho_B = 0.1$ in Eq. (5.56), subject only to the restriction that $r > 1$ and $\alpha_A \in \mathbb{R}$ is order $O(\varepsilon^0)$. The parameter sets [$\alpha_A = 4.36$, $r = 3$] and [$\alpha_A = 8$, $r = 5.5$] were chosen such that these criteria were satisfied. The ratio $\rho_B = 0.1$ determines the dispersion curve of the B wave for *any* set of values α_A and r. However, the dispersion curve for the A wave varies, depending on the choice of parameters, according to Eq. (5.55). Several simulations were run and overlaid in Figure 5.5a. The zoomed view in Figure 5.5b plots the two different dispersion branches that the A wave follows for different choices of the parameters α_A and r.

The ratios ρ_A and ρ_B, introduced earlier to estimate the uniformity of the expansion, increase with ω_0^2. As the frequency $\omega_B = r\omega_0$ increases toward the cutoff frequency, the theoretical frequency corrections become less accurate. Numerical observation suggests that the propensity of the mass-spring chain to generate additional large-amplitude harmonics (comparable to the amplitude of the two injected waves) is responsible for this, owing to additional wave-wave interactions that are not accounted for herein.

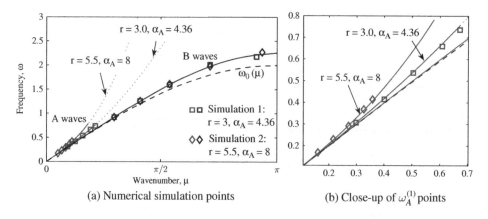

(a) Numerical simulation points   (b) Close-up of $\omega_A^{(1)}$ points

Figure 5.5 A 10% shift in frequency at $\omega_{B0} = 1.8$ can be achieved by injecting high-amplitude, low frequency waves, or low-amplitude, high-frequency waves.

5.4 Application to a 2D Monoatomic Lattice

5.4.1 Overview

Two-dimensional lattices and periodicity form the building blocks of many phononic crystals and metamaterials. Phononic crystal devices such as wave guides and resonators are formed from the strategic periodic arrangement of 2D unit cells [33–35]. Crystal lattice planes and graphene sheets also exhibit 2D lattice periodicity, where nonlinear restoring forces may arise from inter-atomic attraction and repulsion. For example, as early as 1973 dispersion relations and wave propagation in anharmonic atomic lattices were of interest to researchers [36]. The 2D anharmonic lattice considered herein is very similar to atomic systems and is one of the most fundamental and important systems for understanding and analyzing Bloch wave propagation in 2D geometry. The monoatomic lattice is also unique in that, in the long-wavelength limit, the lattice models wave propagation in membrane systems [37]. Membranes, like strings, exhibit cubic hardening nonlinearities as a result of stretch-induced tension, so studying the cubic nonlinearity in the monoatomic system is particularly appropriate.

Wave propagation in 2D systems exhibit directionality in addition to other dispersion behaviors exhibited by 1D systems. Directionality in 2D structures introduces conceptually new opportunities, such as wave beaming, spatial filtering, and imaging. Wave beaming in the 2D beam grillage was considered by Langley et. al. [38], while Ruzzene et. al. have considered it in 2D cellular structures [39]. The cubically nonlinear lattice and some variations were studied extensively for single-wave propagation by Narisetti et al. [40]. They showed that the nonlinear monoatomic lattice exhibits tunable dispersion under the influence of self-action frequency shifts. Tunable parameters included the amplitude of the fundamental Bloch wave and its wavenumber.

Nonlinear wave interaction in 2D offers a fundamentally different perspective for viewing tunability. The introduction of additional waves can predictably control or alter the behavior of the system, in addition to nonlinear self-action, whereby a wave self-adjusts its frequency according to local intensity. Wave interaction in 2D provides three new tunable parameters [5]: two wave vector components and an additional wave amplitude (usually termed the *control wave* or *pump wave*). The idea that nonlinear wave interactions can enhance traits in 2D periodic systems has been explored more recently in the photonic crystal community. Panoiu et. al. utilized the Kerr nonlinearity with a pump/control wave to enhance the "superprism" effect, whereby the direction of propagation in the photonic crystal is extremely sensitive to the wavelength and angle of incidence [41].

Next, multiple time-scale perturbation analysis is applied to determine dispersion frequency shifts for a Bloch wave solution containing two dominant harmonic components. The resulting frequency shift expressions are analyzed and three distinct propagation cases are identified: collinear propagation, orthogonal propagation, and oblique propagation. Numerical simulations validate the expected dispersion shifts. Finally, negative group velocity corrections induced through a control wave are explored as a viable means for achieving amplitude-tunable focus and beam steering that may, ultimately, lead to a phononic superprism effect.

---

5 Assuming the presence of two waves.

5.4.2 Model Description and Nonlinear Governing Equation

The monoatomic lattice equation of motion follows from Newton's second law, and is derived for a unit cell at location indices (p, q). Each mass adjoins four neighbors (top, bottom, left, and right) via springs with nonlinear force–displacement relationships, as shown in Figure 5.6.

Individual masses have only one degree of freedom in the out-of-plane direction $\mathbf{e}_3$. The out-of-plane displacements $u_{p,q}$ are described by the open set of nonlinear difference equations

$$m\ddot{u}_{p,q} + k_1(2u_{p,q} - u_{p+1,q} - u_{p-1,q}) + k_2(2u_{p,q} - u_{p,q+1} - u_{p,q-1}) + f^{NL} = 0 \quad (5.63)$$

where m denotes mass, k_1 and k_2 denote linear stiffnesses in the $\mathbf{e}_1$ and $\mathbf{e}_2$ directions, and f^{NL} results from nonlinear stiffness contributions. The nonlinear force term f^{NL} that arises from cubically nonlinear inter-atomic springs is

$$f^{NL} = \Gamma_1(u_{p,q} - u_{p+1,q})^3 + \Gamma_1(u_{p,q} - u_{p-1,q})^3 + \Gamma_2(u_{p,q} - u_{p,q+1})^3 + \Gamma_2(u_{p,q} - u_{p,q-1})^3$$
$$(5.64)$$

where Γ_1 and Γ_2 denote nonlinear stiffness coefficients along the $\mathbf{e}_1$ and $\mathbf{e}_2$ directions, respectively. Weak nonlinearity is enforced in the governing equation (Eq. (5.63)) by specifying nonlinear coefficients at $O(\varepsilon^1)$ for small parameter $|\varepsilon| < 1$

$$\Gamma_1 \equiv \varepsilon\hat{\Gamma}_1, \Gamma_2 \equiv \varepsilon\hat{\Gamma}_2 \quad (5.65)$$

and additionally specifying Γ_1 and $\Gamma_2 \approx O(\varepsilon^0)$. The resulting equation is amenable to perturbation analysis. Wave interactions are analyzed using the multiple-scale perturbation analysis procedure.

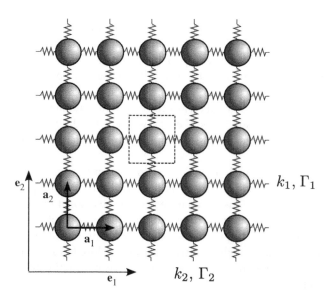

Figure 5.6 Monoatomic lattice configuration with lattice vectors $\mathbf{a}_1$ and $\mathbf{a}_2$. Dashed lines indicate boundaries for the unit cell.

5.4.3 Multiple-scale Perturbation Analysis

Here we introduce an asymptotic expansion for the displacement in addition to multiple time scales

$$u_{p,q}(T_i) = u_{p,q}^{(0)}(T_i) + \varepsilon u_{p,q}^{(1)}(T_i) + O(\varepsilon^2) \tag{5.66}$$

where slow time scales are defined according to $T_i \equiv \varepsilon^i t$. Derivatives transform according to Eq. (5.10), where D_i denotes derivatives with respect to the T_ith time scale. Ordered perturbation equations result from substitution of Eq. (5.66) and slow time derivatives D_i into Eq. (5.63)

$$O(\varepsilon^0): \quad mD_0^2 u_{p,q}^{(0)} + k_1(2u_{p,q}^{(0)} - u_{p+1,q}^{(0)} - u_{p-1,q}^{(0)}) + k_2(2u_{p,q}^{(0)} - u_{p,q+1}^{(0)} - u_{p,q-1}^{(0)}) \tag{5.67}$$
$$= 0$$

$$O(\varepsilon^1): \quad mD_0^2 u_{p,q}^{(1)} + k_1(2u_{p,q}^{(1)} - u_{p+1,q}^{(1)} - u_{p-1,q}^{(1)}) + k_2(2u_{p,q}^{(1)} - u_{p,q+1}^{(1)} - u_{p,q-1}^{(1)}) \tag{5.68}$$
$$= -2mD_0D_1 u_{p,q}^{(0)} - f^{NL}(u_{p,q}^{(0)}, u_{p\pm1,q\pm1}^{(0)})$$

where the nonlinear function f^{NL} depends only on the $O(\varepsilon^0)$ solution. The $O(\varepsilon^0)$ equation admits a Bloch wave solution

$$u_{p,q}^{(0)} = \frac{1}{2}A(T_1)e^{i(\mu \cdot r_{(p,q)} - \omega T_0)} + c.c. \tag{5.69}$$

for Bloch wavenumbers $\mu = [\mu_1, \mu_2]$ and position vector $r_{(p,q)} = [p,q]$. The complex amplitude $A(T_1)$ varies with slow time T_1 and is therefore considered constant at $O(\varepsilon^0)$. The zero-order equation, with application of the Bloch wave solution Eq. (5.69), reduces to

$$\left[-\omega^2 m + k_1(2 - e^{i\mu_1} - e^{-i\mu_1}) + k_2(2 - e^{i\mu_2} - e^{-i\mu_2})\right] u_{p,q}^{(0)} = 0 \tag{5.70}$$

Nontrivial solutions $u_{p,q}^{(0)} \neq 0$ exist only when the dispersion relation is satisfied. Setting the expression in brackets equal to zero produces the linear dispersion relationship

$$\omega(\mu) = \sqrt{\omega_{n1}^2(2 - 2\cos(\mu_1)) + \omega_{n2}^2(2 - 2\cos(\mu_2))} \tag{5.71}$$

where $\omega_{n1} \equiv \sqrt{k_1/m}$ and $\omega_{n2} \equiv \sqrt{k_2/m}$ denote characteristic frequencies. Next, corrections to this dispersion relationship due to weakly nonlinear wave interactions are considered.

As discussed earlier, dispersion shifts must be calculated for a specific frequency content and amplitude ordering. The input signal considered for wave interactions is composed of two Bloch waves at frequencies ω_{A0} and ω_{B0} such that both signals are present in the $O(\varepsilon^0)$ equation

$$u_{p,q}^{(0)} = \frac{1}{2}A(T_1)\exp\,(i\mu_A \cdot r - i\omega_{A0}T_0) + \frac{1}{2}B(T_1)\exp\,(i\mu_B \cdot r - i\omega_{B0}T_0) + c.c. \tag{5.72}$$

The wavenumbers $\mu_A = [\mu_{A1}, \mu_{A2}]$ and $\mu_B = [\mu_{B1}, \mu_{B2}]$ correspond to a primary wave at ω_{A0} and a control wave at ω_{B0} through the linear dispersion relationship Eq. (5.71), respectively. The frequencies ω_{A0} and ω_{B0} are incommensurate by definition. Indeed, the $O(\varepsilon^0)$ equation is identically satisfied for this multi-frequency Bloch wave solution.

A multiple time scales solution yields the resulting frequency corrections. The amplitude functions $A(T_1)$ and $B(T_1)$ take the form

$$A(T_1) = \alpha_A(T_1)e^{i\beta_A(T_1)} \quad \text{and} \quad B(T_1) = \alpha_B(T_1)e^{i\beta_B(T_1)} \tag{5.73}$$

for amplitudes α_A and α_B of $O(1)$. Solutions are sought where the amplitudes α_A and α_B will be assumed constant with respect to T_1 time scales. The time-varying phase terms β_A and β_B result in instantaneous frequency shifts.

We can take, without loss of generality, the indices $p = 0$ and $q = 0$ for the central unit cell. Terms appearing on the right of the $O(\varepsilon^1)$ equation with $\exp(i\omega_{A0}T_0)$ or $\exp(i\omega_{B0}T_0)$ time-dependence result in primary resonances, and therefore secular expansions. Collecting these terms and setting the coefficients S_A and S_B to zero, respectively, produces two complex equations. The first of the secular equations is

$$
\begin{aligned}
S_A = {} & \frac{3}{2}\Gamma_1\alpha_B{}^2\alpha_A e^{i\beta_A T_1}e^{i\mu_{B1}} + \frac{3}{2}\Gamma_1\alpha_B{}^2\alpha_A e^{i\mu_{A1}}e^{i\beta_A T_1} + \frac{3}{2}\frac{\Gamma_1\alpha_B{}^2\alpha_A e^{i\beta_A T_1}}{e^{i\mu_{A1}}} \\
& + \frac{3}{2}\frac{\Gamma_1\alpha_B{}^2\alpha_A e^{i\beta_A T_1}}{e^{i\mu_{B1}}} + m_1\alpha_A\beta_A\omega_A e^{i\beta_A T_1} - \frac{3}{4}\Gamma_2\alpha_B{}^2\alpha_A e^{i\mu_{A2}}e^{i\beta_A T_1}e^{i\mu_{B2}} \\
& - \frac{3}{4}\frac{\Gamma_2\alpha_B{}^2\alpha_A e^{i\mu_{A2}}e^{i\beta_A T_1}}{e^{i\mu_{B2}}} - \frac{3}{4}\frac{\Gamma_2\alpha_B{}^2\alpha_A e^{i\beta_A T_1}e^{i\mu_{B2}}}{e^{i\mu_{A2}}} - \frac{3}{4}\frac{\Gamma_2\alpha_B{}^2\alpha_A e^{i\beta_A T_1}}{e^{i\mu_{A2}}e^{i\mu_{B2}}} \\
& - \frac{3}{4}\Gamma_1\alpha_B{}^2\alpha_A e^{i\mu_{A1}}e^{i\beta_A T_1}e^{i\mu_{B1}} - \frac{3}{4}\frac{\Gamma_1\alpha_B{}^2\alpha_A e^{i\beta_A T_1}}{e^{i\mu_{A1}}e^{i\mu_{B1}}} - \frac{3}{4}\frac{\Gamma_1\alpha_B{}^2\alpha_A e^{i\beta_A T_1}e^{i\mu_{B1}}}{e^{i\mu_{A1}}} \\
& + \dots \\
& - 9/4\,\Gamma_2\alpha_A{}^3 e^{i\beta_A T_1} - 9/4\,\Gamma_1\alpha_A{}^3 e^{i\beta_A T_1} = 0
\end{aligned}
\tag{5.74}
$$

where some terms have been omitted for brevity; a similar equation results for ω_{B0} and has been omitted. The real and imaginary components of each of the two equations $S_A = 0$ and $S_B = 0$ must vanish independently such that

$$\mathfrak{R}(S_A) = \mathfrak{R}(S_B) = \mathfrak{I}(S_A) = \mathfrak{I}(S_B) = 0 \tag{5.75}$$

The imaginary equations reveal, as expected, that amplitudes α_A and α_B are constant with T_1 time scales. The other two evolution equations may be solved for the the instantaneous frequency corrections $\beta_A(T_1)$ and $\beta_B(T_1)$. The solution of the ordinary differential equations yields

$$
\begin{aligned}
\beta_A(T_1) = {} & \frac{3}{8}\frac{\hat{\Gamma}_1}{m\omega_{A0}}[\alpha_A^2 f(\mu_{A1})^2 + 2\alpha_B^2 f(\mu_{A1})f(\mu_{B1})]T_1 \\
& + \frac{3}{8}\frac{\hat{\Gamma}_2}{m\omega_{A0}}[\alpha_A^2 f(\mu_{A2})^2 + 2\alpha_B^2 f(\mu_{A2})f(\mu_{B2})]T_1,
\end{aligned}
\tag{5.76}
$$

and

$$
\begin{aligned}
\beta_B(T_1) = {} & \frac{3}{8}\frac{\hat{\Gamma}_1}{m\omega_{B0}}[\alpha_B^2 f(\mu_{B1})^2 + 2\alpha_A^2 f(\mu_{A1})f(\mu_{B1})]T_1 \\
& + \frac{3}{8}\frac{\hat{\Gamma}_2}{m\omega_{B0}}[\alpha_B^2 f(\mu_{B2})^2 + 2\alpha_A^2 f(\mu_{A2})f(\mu_{B2})]T_1,
\end{aligned}
\tag{5.77}
$$

where the function $f(\theta)$ is defined as

$$f(\theta) \equiv 2 - 2\cos(\theta). \tag{5.78}$$

The function $f(\theta)$ is recognized as the squared dispersion relation for the 1D mono-atomic chain[6].

The linearity of $\beta_A(T_1)$ and $\beta_B(T_1)$ on slow time scale T_1 permits the definitions $\omega_{A1} \equiv \beta_A/T_1$ and $\omega_{B1} \equiv \beta_B/T_1$, such that a Lindstedt–Poincaré type asymptotic series in ε describes frequency corrections to ω_A and ω_B as

$$\omega_A(\mu_A, \mu_B) = \omega_{A0}(\mu_A, \mu_B) + \varepsilon\omega_{A1}(\mu_A, \mu_B; \; \alpha_A, \alpha_B, \hat{\Gamma}_1, \hat{\Gamma}_2), \tag{5.79}$$

and

$$\omega_B(\mu_A, \mu_B) = \omega_{B0}(\mu_A, \mu_B) + \varepsilon\omega_{B1}(\mu_A, \mu_B; \; \alpha_A, \alpha_B, \hat{\Gamma}_1, \hat{\Gamma}_2), \tag{5.80}$$

where the frequency corrections are given explicitly as

$$\omega_{A1} = \frac{3}{8} \frac{\hat{\Gamma}_1}{m\sqrt{\omega_{n1}^2 f(\mu_{A1}) + \omega_{n2}^2 f(\mu_{A2})}} [\alpha_A^2 f(\mu_{A1})^2 + 2\alpha_B^2 f(\mu_{A1})f(\mu_{B1})]$$

$$+ \frac{3}{8} \frac{\hat{\Gamma}_2}{m\sqrt{\omega_{n1}^2 f(\mu_{A1}) + \omega_{n2}^2 f(\mu_{A2})}} [\alpha_A^2 f(\mu_{A2})^2 + 2\alpha_B^2 f(\mu_{A2})f(\mu_{B2})] \tag{5.81}$$

$$\omega_{B1} = \frac{3}{8} \frac{\hat{\Gamma}_1}{m\sqrt{\omega_{n1}^2 f(\mu_{B1}) + \omega_{n2}^2 f(\mu_{B2})}} [\alpha_B^2 f(\mu_{B1})^2 + 2\alpha_A^2 f(\mu_{A1})f(\mu_{B1})]$$

$$+ \frac{3}{8} \frac{\hat{\Gamma}_2}{m\sqrt{\omega_{n1}^2 f(\mu_{B1}) + \omega_{n2}^2 f(\mu_{B2})}} [\alpha_B^2 f(\mu_{B2})^2 + 2\alpha_A^2 f(\mu_{A2})f(\mu_{B2})] \tag{5.82}$$

5.4.4 Analysis of Predicted Dispersion Shifts

Equations (5.81) and (5.82) depend on the input wave vectors μ_A and μ_B in a complicated manner. The qualitative nature of these dispersion shifts relies heavily on the behavior of $f(\theta)$. Several key results are evident with the observations $f(0) = 0$ and $f(-\theta) = f(\theta)$:

1) **Collinear wave vectors:** Two waves that propagate collinearly along a lattice vector interact as if propagation were one-dimensional.
2) **Orthogonal wave vectors:** Two waves that propagate with orthogonal wave vectors aligned to the lattice vectors do not interact, and receive frequency shifts due only to self-action.
3) **Oblique wave vectors:** Two waves propagating with oblique wave vectors do interact according to the derived relationship.

Figure 5.7 depicts each of these scenarios schematically. Case 1 describes the situation where $\mu_A \cdot \mathbf{a}_2 = 0$ and $\mu_B \cdot \mathbf{a}_2 = 0$, or alternatively $\mu_A \cdot \mathbf{a}_1 = 0$ and $\mu_B \cdot \mathbf{a}_1 = 0$. This case reduces to the 1D monoatomic array analyzed in Section 5.3. A special case of this occurs during resonant wave propagation, where counter-propagating waves are described with $\mu_B = -\mu_A$. Case 2 describes orthogonal wave propagation only when wavenumbers are aligned to lattice vectors; in this case, dispersion corrections to $O(\varepsilon^1)$ do not indicate any wave interaction. Case 3 depicts the most general situation where arbitrary wave vectors interact. Each of these three cases is validated numerically.

---

6 With this definition, the dispersion relationship for the 2D monoatomic lattice may be written as
$\omega = \sqrt{\omega_{n1}^2 f(\mu_1) + \omega_{n2}^2 f(\mu_2)}$.

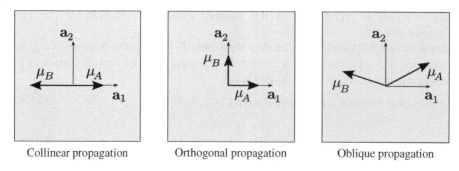

Figure 5.7 Three cases of wave–wave interaction in the monoatomic lattice.

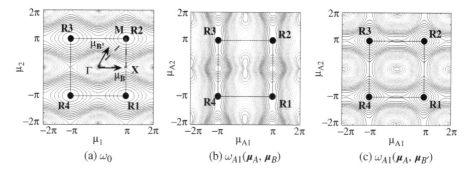

Figure 5.8 Brillouin zone symmetry is retained by dispersion shifts resulting from wave interactions. (a) Linear dispersion relationship with the FBZ identified by points R_i, $i = 1 \dots 4$. Wave vectors $\boldsymbol{\mu}_B = [3.0, 0.0]$ and $\boldsymbol{\mu}_{B'} = [1.0, 2.1]$ corresponding to horizontal and oblique control waves used for frequency corrections plotted in (b) and (c), respectively.

There are additional important aspects of the predicted dispersion shifts which are subtle: Brillouin zone symmetry is retained, and group velocity remains zero along the Brillouin zone boundaries. Figure 5.8 depicts Brillouin zone symmetry in the linear dispersion relation $\omega_0(\boldsymbol{\mu})$ and the frequency correction $\omega_{A1}(\boldsymbol{\mu}_A, \boldsymbol{\mu}_B)$ evaluated with horizontal and oblique control waves $\boldsymbol{\mu}_B$ and $\boldsymbol{\mu}_{B'}$, respectively. The linear dispersion relationship in Figure 5.8a depicts $\omega_0(\mu_1, \mu_2)$ and is symmetric with respect to both μ_1 and μ_2. Any reflection, or combination of reflections, of a wave vector $\boldsymbol{\mu}$ about axes μ_1 or μ_2 yields the same frequency. The same symmetry is retained by frequency corrections ω_{A1} and ω_{B1}. Figures 5.8b and 5.8c depict this symmetry for the ω_{A1} correction in Eq. (5.81) by evaluating the expression over the entire first Brillouin zone (FBZ), subject to a given control wave at wavenumber $\boldsymbol{\mu}_B$. It is easily verified that $\omega_{A1}(\boldsymbol{\mu}_A, \boldsymbol{\mu}_B)$ and $\omega_{B1}(\boldsymbol{\mu}_A, \boldsymbol{\mu}_B)$ are symmetric for all combinations of $\boldsymbol{\mu}_A$ and $\boldsymbol{\mu}_B$ since $f(-\theta) = f(\theta)$ and $f(\theta) \geq 0 \ \forall \ \theta \in \mathbb{R}$.

Less obvious is the fact that the group velocity of the corrected dispersion relationships ω_A and ω_B remains zero at the appropriate edges of the FBZ. Let $\mathbf{c}_{gA} = \mathbf{c}_{gA}^{(0)} + \mathbf{c}_{gA}^{(1)}$ and $\mathbf{c}_{gB} = \mathbf{c}_{gB}^{(0)} + \mathbf{c}_{gB}^{(1)}$, where $\mathbf{c}_{gA}^{(1)}$ and $\mathbf{c}_{gA}^{(1)}$ denote corrections to the linear group velocity from Eqs. (5.81) and (5.82), respectively. A group velocity calculation using the chain rule on the frequency correction of Eq. (5.81) expresses this relationship as

$$\mathbf{c}_{gA}^{(1)} = \nabla_\mu \omega_{A1} = \frac{\partial \omega_{A1}}{\partial f(\mu_{A1})} \frac{\partial f(\mu_{A1})}{\partial \mu_{A1}} \mathbf{b}_1 + \frac{\partial \omega_{A2}}{\partial f(\mu_{A2})} \frac{\partial f(\mu_{A2})}{\partial \mu_{A2}} \mathbf{b}_2 \tag{5.83}$$

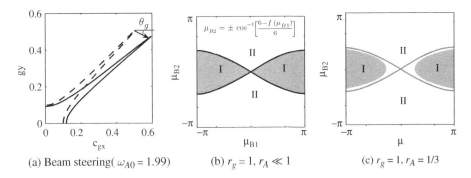

(a) Beam steering($\omega_{A0} = 1.99$) (b) $r_g = 1$, $r_A \ll 1$ (c) $r_g = 1$, $r_A = 1/3$

Figure 5.9 Negative group velocity corrections are a unique result of wave interactions. A linear wave beam (black, dashed) receives a negative group velocity correction (black arrow) to produce beam shifting (black, solid). Control waves that achieve $\theta_g < 0$ are found in region (I).

The function $f(\theta)$ has derivative $df/d\theta = 2\sin(\theta)$, so along the FBZ edges where $\mu_{1,2} = \pm\pi$ the group velocity evaluates to zero along boundary-normal directions.

The group velocity correction may be used to dynamically steer wave beams through the application of a control wave μ_B. The monoatomic lattice $k_1 = k_2$ exhibits a known singularity in the vicinity of $\mu_A = [\pi/2, \pi/2]$, whereby wave beams form along diagonals [40]. The group velocity correction is expressed by $c_{gA}^{(1)} = c_{gA}(\cos\theta_g \mathbf{b}_1 + \sin\theta_g \mathbf{b}_2)$, where c_{gA} denotes group velocity amplitude at an angle of θ_g. An interesting situation results when $\theta_g < 0$; group velocity corrections correspond to negative angles θ_g. Equation (5.83), evaluated in the vicinity of $\mu_A = [\pi/2, \pi/2]$, yields an expression describing the control wavenumbers, which produce negative angled corrections [7]:

$$\cos(\mu_{B2}) = \frac{1}{6}[6 - r_g r_A^2 + 7r_A^2 - 6r_g f(\mu_{B1})] \tag{5.84}$$

where $r_g \equiv \Gamma_1/\Gamma_2$ and $r_A \equiv \alpha_A/\alpha_B$ denote nonlinear stiffness and amplitude ratios, respectively. The solution of this equation yields two distinct solution regions: (I) $\theta_g < 0$ and (II) $\theta_g > 0$. Figure 5.9a depicts a wave beam of a linear system (black, dashed) and the corrected wave beam (black, solid) as a result of a negative group velocity. Figures 5.9b and 5.9c depict regions (I) and (II), where negative group velocity corrections exist for low and moderate amplitude ratios. This phenomenon is unique to nonlinear lattices subject to wave–wave interactions; negative group velocity corrections, as depicted in Figure 5.9, cannot be realized under nonlinear self-action corrections. This phenomenon offers a unique opportunity for beam steering and tunable focusing; example applications are presented in Section 5.4.6.

5.4.5 Numerical Simulation Validation Cases

Time-domain finite-element simulations confirm the accuracy of analytical perturbation calculations. Simulations are constructed using a large array of masses and springs with unit parameters $m = 1kg$ and $k_1 = k_2 = 1Nm^{-1}$, such that boundary reflections are not encountered. Bloch waves are introduced into the system by means of specifying

---

7 Equation (5.84) refers to the system described by $\omega_{n1} = \omega_{n2} = 1.0$.

initial conditions; numerical integration for a specified time interval of approximately 40 s results in space–time data for each mass. Each mass oscillates at a corrected frequency, depending on the initial amplitude, which is measurable from its time response signal.

Analysis Method

A nonlinear least-squares model provides an accurate method of quantifying the frequency shift over short time periods. This method also provides quantified uncertainty levels for the determined parameter values, although in the cases presented the uncertainty is negligible. The nonlinear least-squares curve fit method attempts to minimize the sum of squared errors between the trial function $\mathcal{F}(x)$ and a provided signal $u(t)$. This analysis method is first analyzed without the influence of wave interactions; that is, $\alpha_B = 0$. The left-hand part of Figure 5.10 depicts the wave field corresponding to $\mu_A = [1.8, 0]$ and amplitude $\alpha_A = 2$. The wave vector is aligned to the $\mathbf{a}_1$ lattice vector. The right-hand side shows the displacement field evaluated for $u_{p,q}(x, t)$ located centrally in the wave field such that boundary reflections have no influence for the times considered.

Figure 5.11 displays frequency shifts ω_{A1} as a function of the primary wave amplitude α_A. Frequency shifts obtained from numerical simulation (triangles) match the theoretical predictions almost exactly for amplitudes $\alpha_A < 2$. Simulation results for amplitudes $\alpha_A > 2$ diverge from theoretical perturbation calculations. This behavior is typical of asymptotic solutions where amplitude or frequency corrections exceed the $O(\varepsilon^0)$ values. Indeed, the frequency correction at $\alpha_A = 3$ results in a 7.2% difference to the linear frequency $\omega_{A0} = 1.567$. We note that the theoretical frequency corrections tend to overestimate the resulting frequency shift due to energy transfer from a primary wave to sub- and superharmonics. The dotted line in Figure 5.11 represents theoretical frequency corrections evaluated for the same probe location with an amplitude $\alpha_A^* = 0.95\alpha_A$. This curve better fits data at higher amplitudes (e.g. $\alpha_A = 3$), where additional harmonic generation causes energy leakage from the primary harmonic. Thus it is likely that frequency corrections will be less than predicted by a perturbation analysis when weakly nonlinear assumptions are no longer valid.

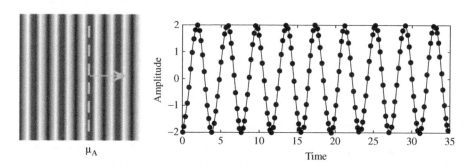

Figure 5.10 Initial wave field and corresponding displacement probe located centrally in the field ($p = 40, q = 40$). Markers denote a nonlinear least-squares fit while solid lines indicate the numerical simulation time signal.

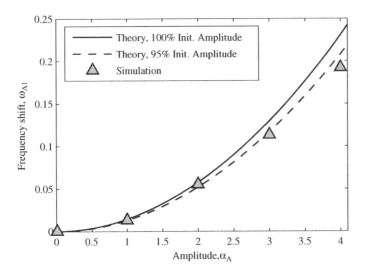

Figure 5.11 Numerical simulation results for frequency–amplitude relationship using a least-squares curve fitting method.

Orthogonal and Oblique Interaction

Numerical simulation of Case 2 (orthogonal interaction along a lattice vector) and Case 3 (general, oblique wave vectors) is presented next. Collinear propagation has been thoroughly addressed in Section 5.3. Recall that during the course of wave interactions, dynamic frequency shifts exist for both the primary wave ω_{A1} and the control wave ω_{B1}. In this section, emphasis is placed on validating ω_{A1}.

Figure 5.12 depicts initial wave fields for both orthogonal and oblique wave interactions. Each validation case considers a control wave of a constant amplitude $\alpha_B = 2.0$ while the amplitude of the primary wave α_A varies from 0 to 2 (as depicted in Figure 5.13). Primary and control Bloch wave fields correspond to randomly generated wave vectors $\mu_A = [1.811, 0.0]$ and $\mu_B = [0.0, 1.043]$ for orthogonal propagation and $\mu_A = [0.831, -2.528]$ and $\mu_B = [-1.391, 0.294]$ for oblique propagation. Initial wave fields are depicted on the left-hand side. The right-hand subplots depict the multifrequency numerical time-domain responses (solid), along with least-squares data fits of the form $f(t) = \tilde{B}\cos(\tilde{\omega}_A t + \tilde{\theta}_A) + \tilde{B}\sin(\tilde{\omega}_B t + \tilde{\theta}_B)$ (markers). Least-squares data fitting parameters are denoted by "~". The frequencies associated with the fit agree well with the time-domain responses.

Amplitude-dependent frequency shifts ω_{A1} are plotted in Figure 5.13. The solid lines denote theoretical frequency shifts, while markers indicate the frequencies extracted from simulation data. All simulation cases agree very well with perturbation theory. Figure 5.13a also overlays the result for no wave interaction, as presented in Figure 5.11. As expected, two waves propagating along a lattice vector experience no additional frequency shift from the presence of a control wave. In contrast, Figure 5.13b depicts a nonzero frequency shift even for $\alpha_A \approx 0$ owing to dynamic lattice anisotropy that is introduced by the presence of μ_B. This result, unique to wave interaction, has not been previously documented, to the author's knowledge.

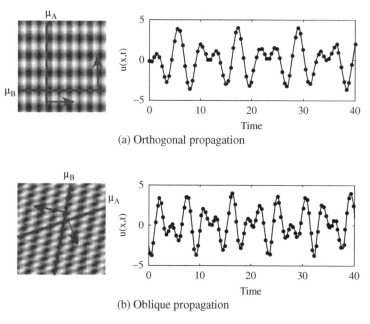

(a) Orthogonal propagation

(b) Oblique propagation

Figure 5.12 Initial wave field for orthogonal and oblique wave interaction (for $\alpha_A = 2.0$) and corresponding displacement probe located centrally at $p = 40, q = 40$. Symbols demarcate the time series corresponding to the nonlinear least-squares curve fit for frequency, phase, and amplitude.

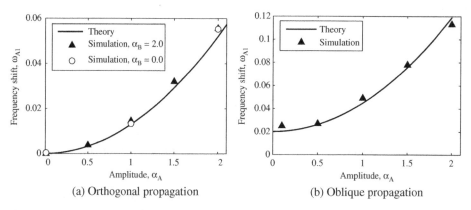

(a) Orthogonal propagation

(b) Oblique propagation

Figure 5.13 Numerical simulation results for orthogonal and oblique wave interactions. Theoretical results (line) are validated by numerical time-domain simulations (markers). Unlike orthogonal interactions (a), oblique interactions (b) result in nonzero frequency shifts for low-amplitude *A* waves.

The idea that lattice anisotropy can be dynamically introduced through nonlinear control-wave interaction is a powerful concept for nonlinear metamaterials and phononic devices. Dynamic anisotropy is explored in the next section, in relation to an amplitude-tunable focus device that employs wave dynamic wave beaming.

5.4.6 Application: Amplitude-tunable Focusing

The idea behind amplitude-tunable focusing was first explored in Section 5.4.4, where a group velocity analysis of dispersion frequencies exhibited possible negative group velocity corrections. Negative group velocity corrections offer the potential for tuning the direction of wave beams. Moreover, the constructive interference of two wave beams may be used to develop a metamaterial variant of the high-intensity focused ultrasound devices used in medical procedures to locally heat or destroy tissue [42].

Numerical validation of the conjectures in Section 5.4.4 are presented in Figure 5.14. A control wave with wavenumber $\mu_B = [3.0, 0.0]$ was injected into the nonlinear material $(r_g = 1)$. At time $t = 0$, a centrally located point source forms wave beams along the diagonals. Solid black lines depict the theoretical beam trajectories; dotted lines indicate low-amplitude (linear) trajectories for comparison. Root-mean-square displacements from the numerical simulations are overlaid on the theoretical beam trajectories. In order to visualize the primary wave field resulting from the point source, a notch filter was applied to the control-wave frequency for each frame by transforming spatial data into the wave vector domain.

Increasing the control-wave amplitude causes the beam to converge along the horizontal direction. At very high amplitudes of $\alpha_B = 4.0$, as shown in Figure 5.14c, the lattice response loses complete symmetry, which may suggest more efficient transfer of energy in the direction of the control-wave vector. Regardless of this, the numerical simulations agree strongly with the analytical group velocity calculations.

The same principle may be applied to an amplitude-tunable focus device (ATFD) as depicted schematically in Figure 5.15. The ATFD operates on the principle of constructive interference. Two identical sources located at the edge of a nonlinear metamaterial provide tunable wave beams. At low amplitudes, the focal point (FP) lies on the mid-plane (focal plane) at a distance of half the source separation [8]. The presence of a control wave produces dynamic lattice anisotropy, which adjusts the wave beam angle. By varying the intensity of the control wave, the FP moves toward or away from the sources. Numerical simulations of the ATFD at various amplitudes confirm

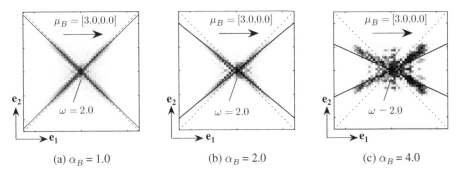

(a) $\alpha_B = 1.0$ (b) $\alpha_B = 2.0$ (c) $\alpha_B = 4.0$

Figure 5.14 Wave beam steering using a control wave with $\mu_B = [3.0, 0.0]$ and a point-source excitation at $\omega = 2.0$. Increasing levels of control-wave amplitude α_B, shown in (a)–(c), vary the beam angle. Solid lines indicate a theoretical beam path; dashed lines indicate the low-amplitude beam path for comparison.

---

8 This is because the wave beams form at 45° angles in the system considered.

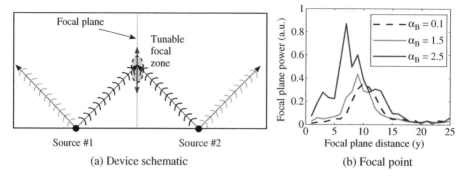

(a) Device schematic (b) Focal point

Figure 5.15 (a) Schematic of a tunable focus device that utilizes constructive interference on a central focal plane. (b) Power distributions calculated along the focal plane from numerical simulation results reveal a sharpening of the focal point (see also Figure 5.16) and tunable distance.

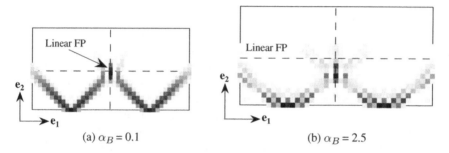

(a) $\alpha_B = 0.1$ (b) $\alpha_B = 2.5$

Figure 5.16 Time-domain simulation results for a tunable focus device. In the presence of a control-wave field ($\mu_B = [3.0, 0.0]$), dynamic anisotropy introduced into the lattice alters the focal point (FP) distance and sharpness. Dashed lines indicate the focal plane.

the expected behavior, as depicted in Figures 5.15b and 5.16. Figure 5.15b depicts the power distribution (in arbitrary units, a.u.) evaluated along the focal plane for three control-wave amplitudes. An unexpected and positive consequence of the focusing mechanism is a distinct peak sharpening relative to the low-amplitude/linear scenario. The low-amplitude case (black dashed) exhibits a broader FP than do the cases with greater control-wave amplitude. Figure 5.16 depicts a distinct change in the FP location. Indeed, the FP has shifted toward the sources, from $y = 10.0$ to approximately $y = 6.0$, a 40% location change.

Summary

The effect of finite-amplitude wave propagation in nonlinear periodic structures was analyzed within the framework of dispersion band structures. Nonlinearities in stan-dalone nonlinear systems such as the classical Duffing oscillator are responsible for a variety of dynamic effects not found in linear systems: frequency conversion, harmonic generation, chaos, and amplitude-dependence, for example. Engineered materials com-posed of periodically repeating nonlinear elements exhibit nonlinear amplitude-tunable wave propagation characteristics. Several analysis methods and techniques were presented as strategies for analyzing a variety of such systems.

Perturbation-based analysis methods applied to discretized unit cells lead to nonlinear dispersion-relation corrections. These first-order corrected dispersion relations describe amplitude-tunable qualities that are enhanced with nonlinearity, such as pass and stop bands, group velocity, and wave beaming. A multiple time-scales perturbation analysis was presented, this approach providing the additional generality needed for analyzing nonlinear wave interactions that result from multiharmonic excitation. Analysis of discrete-parameter atomic systems in 1D and 2D connected with cubically nonlinear spring elements demonstrated that the interaction of two waves results in different amplitude- and frequency-dependent dispersion branches for each wave. This contrasts with nonlinear dispersion relations derived from monochromatic excitation signals, where only a single amplitude-dependent branch is present. A theoretical development utilizing multiple time scales results in a set of evolution equations that are validated by numerical simulation. For the specific case where the wavenumber and frequency ratios are *both* close to 1:3 as in the long-wavelength limit, the evolution equations suggest that small amplitude- and frequency-modulations may be present. In the case of 2D systems, it was shown that lattice anisotropy could be introduced by injecting a control plane wave. The resulting lattice anistropy can be used to control wave directionality for application in tunable-focus devices.

Acknowledgements

The authors of this chapter would like to acknowledge financial support from the US National Science Foundation under awards CMMI 0926776 and CMMI 1332862. All techniques and results described in this chapter were developed by the authors while supported by these two grants.

References

1 O. R. Asfar and A. H. Nayfeh, "The application of the method of multiple scales to wave propagation in periodic structures," *SIAM Review*, vol. 25, no. 4, pp. 455–480, 1983.

2 G. Chakraborty and A. K. Mallik, "Dynamics of a weakly non-linear periodic chain," *International Journal of Nonlinear Mechanics*, vol. 36, no. 2, pp. 375–389, 2001.

3 J.-H. Jang, C. K. U. Ullal, T. Gorishnyy, V. V. Tsukruk, and E. L. Thomas, "Mechanically tunable three-dimensional elastomeric network/air structures via interference lithography," *Nano Letters*, vol. 6, no. 4, pp. 740–743, 2006.

4 A. Marathe and A. Chatterjee, "Wave attenuation in nonlinear periodic structures using harmonic balance and multiple scales," *Journal of Sound and Vibration*, vol. 289, no. 4–5, pp. 871–888, 2006.

5 V. Rothos and A. Vakakis, "Dynamic interactions of traveling waves propagating in a linear chain with a local essentially nonlinear attachment," *Wave Motion*, vol. 46, no. 3, pp. 174–188, 2009.

6 Nayfeh and Mook, *Nonlinear Oscillations*. Wiley, 1995.

7 J. J. Rushchitsky and C. Cattani, "Evolution equations for plane cubically nonlinear elastic waves," *International Applied Mechanics*, vol. 40, no. 1, pp. 70–76, 2004.

8 J. J. Rushchitsky and E. V. Saveleva, "On the interaction of cubically nonlinear transverse plane waves in an elastic material," *International Applied Mechanics*, vol. 42, no. 6, pp. 661–668, 2006.

9 B. Liang, B. Yuan, and J. Cheng, "Acoustic diode: rectification of acoustic energy flux in one-dimensional systems," *Physical Review Letters*, vol. 103, no. 10, p. 104301, 2009.

10 I. Kovacic and M. J. Brennan, *The Duffing Equation: Nonlinear Oscillators and their Behaviour*. Wiley, 2011.

11 W. O. Davis and A. P. Pisano, "Nonlinear mechanics of suspension beams for a micromachined gyroscope," in *International Conference on Modeling and Simulation of Microsystems*, 2001.

12 V. Kaajakari, T. Mattila, A. Oja, and H. Seppä, "Nonlinear limits for single-crystal silicon microresonators," *Journal of Microelectromechanical Systems*, vol. 13, no. 5, 2004.

13 V. Kaajakari, T. Mattila, A. Lipsanen, and A. Oja, "Nonlinear mechanical effects in silicon longitudinal mode beam resonators," *Sensors and Actuators A: Physical*, vol. 120, no. 1, pp. 64–70, 2005.

14 A. Erturk and D. J. Inman, *Piezoelectric Energy Harvesting*. Wiley, 2011.

15 N. Elvin and A. Erturk, *Advances in Energy Harvesting Methods*. Springer, 2013.

16 K. L. Manktelow, M. J. Leamy, and M. Ruzzene, "Multiple scales analysis of wave-wave interactions in a cubically nonlinear monoatomic chain," *Nonlinear Dynamics*, vol. 63, pp. 193–203, 2011.

17 C. Daraio, V. F. Nesterenko, E. B. Herbold, and S. Jin, "Strongly nonlinear waves in a chain of teflon beads," *Physical Review E*, vol. 72, no. 1, p. 016603, 2005.

18 C. Daraio, V. Nesterenko, E. Herbold, and S. Jin, "Tunability of solitary wave properties in one-dimensional strongly nonlinear phononic crystals," *Physical Review E*, vol. 73, no. 2, p. 026610, 2006.

19 R. K. Narisetti, M. J. Leamy, and M. Ruzzene, "A perturbation approach for predicting wave propagation in one-dimensional nonlinear periodic structures," *ASME Journal of Vibration and Acoustics*, vol. 132, no. 3, p. 031001, 2010.

20 P. P. Banerjee, *Nonlinear Optics: Theory, Numerical Modeling and Applications*. Marcel Dekker, 2004.

21 S. D. Gupta, "Progress in optics," in *Nonlinear Optics of Stratified Media* (E. Wolf, ed.), vol. 38, pp. 1–84, Elsevier, 1998.

22 J. W. Haus, B. Y. Soon, M. Scalora, C. Sibilia, and I. V. Melnikov, "Coupled-mode equations for Kerr media with periodically modulated linear and nonlinear coefficients," *Journal of the Optical Society of America B*, vol. 19, no. 9, p. 2282, 2002.

23 S. Inoue and Y. Aoyagi, "Design and fabrication of two-dimensional photonic crystals with predetermined nonlinear optical properties," *Physical Review Letters*, vol. 94, no. 10, p. 103904, 2005.

24 I. S. Maksymov, L. F. Marsal, and J. Pallarès, "Band structures in nonlinear photonic crystal slabs," *Optical and Quantum Electronics*, vol. 37, no. 1, pp. 161–169, 2005.

25 A. F. Vakakis and M. E. King, "Resonant oscillations of a weakly coupled, nonlinear layered system," *Acta Mechanica*, vol. 128, no. 1, pp. 59–80, 1998.

26 A. F. Vakakis and M. E. King, "Nonlinear wave transmission in a monocoupled elastic periodic system," *Journal of the Acoustical Society of America*, vol. 98, no. 3, pp. 1534–1546, 1995.

27 Y. Yun, G. Miao, P. Zhang, K. Huang, and R. Wei, "Nonlinear acoustic wave propagating in one-dimensional layered system," *Physics Letters A*, vol. 343, no. 5, pp. 351–358, 2005.

28 K. Manktelow, M. J. Leamy, and M. Ruzzene, "Comparison of asymptotic and transfer matrix approaches for evaluating intensity-dependent dispersion in nonlinear photonic and phononic crystals," *Wave Motion*, vol. 50, pp. 494–508, 2013.

29 K. Manktelow, R. K. Narisetti, M. J. Leamy, and M. Ruzzene, "Finite-element based perturbation analysis of wave propagation in nonlinear periodic structures," *Mechanical Systems and Signal Processing*, vol. 39, p. 32–46, 2012.

30 K. L. Manktelow, M. J. Leamy, and M. Ruzzene, "Analysis and experimental estimation of nonlinear dispersion in a periodic string," *Journal of Vibration and Acoustics*, vol. 136, no. 3, p. 031016, 2014.

31 R. K. Narisetti, *Wave propagation in nonlinear periodic structures*. PhD thesis, Georgia Institute of Technology, 2010.

32 R. W. Boyd, *Nonlinear Optics*. Academic Press, 1992.

33 R. H. Olsson III, I. F. El-Kady, M. F. Su, M. R. Tuck, and J. G. Fleming, "Microfabricated VHF acoustic crystals and waveguides," *Sensors and Actuators A*, pp. 87–93, 2008.

34 A. Khelif, B. Djafari-Rouhani, J. Vasseur, and P. Deymier, "Transmission and dispersion relations of perfect and defect-containing waveguide structures in phononic band gap materials," *Physical Review B*, vol. 68, no. 2, p. 024302, 2003.

35 F. Casadei, T. Delpero, A. Bergamini, P. Ermanni, and M. Ruzzene, "Piezoelectric resonator arrays for tunable acoustic waveguides and metamaterials," *Journal of Applied Physics*, vol. 112, no. 6, pp. 064902–064902, 2012.

36 A. Askar, "Dispersion relation and wave solution for anharmonic lattices and Korteweg de Vries continua," *Proceedings of the Royal Society of London. A. Mathematical and Physical Sciences*, vol. 334, no. 1596, pp. 83–94, 1973.

37 P. M. Morse and K. U. Ingard, *Theoretical Acoustics*. McGraw Hill, 1987.

38 R. Langley, N. Bardell, and H. Ruivo, "The response of two-dimensional periodic structures to harmonic point loading: a theoretical and experimental study of a beam grillage," *Journal of Sound and Vibration*, vol. 207, no. 4, pp. 521–535, 1997.

39 M. Ruzzene, F. Scarpa, and F. Soranna, "Wave beaming effects in two-dimensional cellular structures," *Smart Materials and Structures*, vol. 12, no. 3, p. 363, 2003.

40 R. K. Narisetti, M. Ruzzene, and M. J. Leamy, "A perturbation approach for analyzing dispersion and group velocities in two-dimensional nonlinear periodic lattices," *Journal of Vibration and Acoustics*, vol. 133, p. 061020, 2011.

41 N. Panoiu, M. Bahl, R. Osgood Jr, *et al.*, "Optically tunable superprism effect in nonlinear photonic crystals," *Optics Letters*, vol. 28, no. 24, pp. 2503–2505, 2003.

42 J. Kennedy, "High-intensity focused ultrasound in the treatment of solid tumours," *Nature Reviews Cancer*, vol. 5, no. 4, pp. 321–327, 2005.

6

Stability of Lattice Materials

Filippo Casadei[1], Pai Wang[1] and Katia Bertoldi

School of Engineering and Applied Sciences, Harvard University, Cambridge, Massachusetts, USA

6.1 Introduction

Lattice structures may significantly change their architecture in response to an applied deformation. When subjected to excessive deformation, they may eventually become unstable and, beyond the instability threshold, rapid and dramatic changes of the structural geometry occur. Traditionally, mechanical instabilities have been viewed as failure modes, and a number of numerical methods, including the finite-difference method [1–4], the finite-element method [5, 6], and the Bloch-wave method [7–12] have been developed and used to determine the conditions of the onset of bifurcation.

However, mechanical instabilities are not always deleterious. For elastic materials the geometric reorganization occurring at instability is both reversible and repeatable and it occurs over a narrow range of the applied load, providing opportunities for the design of tunable/adaptive structures—ones that can change their properties in response to variations in their environment—with applications in sensors, microfluidics, bioengineering, robotics, acoustics, and photonics [13–18].

In particular, instabilities in periodic porous structures comprising square and triangular arrays of circular holes have been found to lead to the transformation of the pores into ordered arrays of high-aspect ratio (almost closed) ellipses [19–21]. These have been demonstrated to be important for the design of phononic switches [15, 22], color displays [23], and materials with unusual properties such as large negative Poisson's ratio [24, 25].

Here, we focus on a square lattice of elastic beams. First, we investigate the stability of the structure both under uniaxial and equibiaxial loading conditions and show that in the latter case a short-wavelength instability is critical, inducing an homogeneous pattern transformation. Then, we demonstrate that under equibiaxial loading the abrupt changes introduced into the architecture by instability can be exploited to tune the propagation of elastic waves. In particular, the pattern transformation induced by

---

1 F.C. and P.W. contributed equally to this work.

Dynamics of Lattice Materials, First Edition. Edited by A. Srikantha Phani and Mahmoud I. Hussein.
© 2017 John Wiley & Sons Ltd. Published 2017 by John Wiley & Sons Ltd.

buckling is found to open a wide band gap, paving the road to the design of phononic switches.

6.2 Geometry, Material, and Loading Conditions

In this study we consider a periodic square lattice of identical Timoshenko beams with length L and bending stiffness EI (see Figure 6.1). We assume the cross section of each beam to be square with edge d and choose $L/d = 10$. The material is modeled as elastic, with Young modulus E, and a large displacement formulation is used [26]. For all the analyses, finite-element models are constructed using 2D Timoshenko beam elements (element type B21 in ABAQUS) and the accuracy of the mesh is ascertained through a mesh refinement study.

Recognizing that the finite-sized specimens are necessarily influenced by boundary conditions, the deformation of both finite (see Figure 6.1a) and infinite (see Figure 6.1b) periodic arrays is investigated. Furthermore, in this study we consider both uniaxial and equibiaxial loading conditions.

Finite-sized Periodic Structures To simulate equibiaxial loading conditions for the finite-sized periodic specimen, the bottom and left edges are fixed in the vertical and horizontal directions, respectively, whereas the top and right edges are uniformly compressed in the vertical and horizontal directions. Similarly, to simulate uniaxial loading conditions, the bottom edge is fixed in the vertical direction (the bottom-left corner is also fixed in the horizontal direction to prevent rigid-body motion), whereas the top edge is uniformly compressed in the vertical direction.

Infinite Periodic Structures The stability of the infinite periodic structure is investigated using a unit cell consisting of four half beams (see Figure 6.1b). However, guided by the stability analysis, we use enlarged unit cells comprising 2×2 and 1×5 original unit cells (see Figure 6.1b) for the postbuckling and wave propagation analysis under equibiaxial and uniaxial loading conditions, respectively.

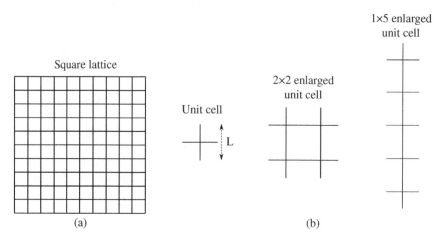

Figure 6.1 The period square lattice considered: (a) schematic of a finite-sized square lattice; (b) schematic of the unit cell and 2×2 and 1×5 enlarged unit cells.

To subject the unit cell to a macroscopic deformation gradient $\bar{\mathbf{f}}$, periodic boundary conditions are imposed on all cell boundaries [27, 28]

$$u_\alpha^{A_i} - u_\alpha^{B_i} = (\bar{F}_{\alpha\beta} - \delta_{\alpha\beta})(X_\beta^{A_i} - X_\beta^{B_i}), \qquad \theta^{A_i} = \theta^{B_i}, \quad \alpha = 1, 2, \quad \zeta = 1, 2, ..., N \quad (6.1)$$

where $\delta_{\alpha\beta}$ is the Kronecker delta, $u_\alpha^{A_i}$ and $u_\alpha^{B_i}$ ($\alpha = 1, 2$) are displacements of points periodically located on the boundary of the unit cell and θ^{A_i} and θ^{A_i} are the corresponding rotations. Moreover, N denotes the number of pairs of nodes periodically located on the boundary of the unit cell. For the square lattice considered in this study $N = 2$ when the unit cell is considered, while $N = 4$ and $N = 6$ for the enlarged 2×2 and 1×5 unit cells, respectively. Note that the components of $\bar{F}$ can be conveniently prescribed within the finite-element framework using a set of virtual nodes. The corresponding macroscopic first Piola–Kirchhoff stress $\bar{P}$ is then obtained through virtual work considerations [27, 28].

In this study, the following macroscopic loading conditions are considered:

- Equibiaxial compression, so that the macroscopic deformation gradient $\bar{F}$ is given by

$$\bar{F} = (1 + \varepsilon) \quad \hat{e}_1 \otimes \hat{e}_1 + (1 + \varepsilon) \quad \hat{e}_2 \otimes \hat{e}_2 + \hat{e}_3 \otimes \hat{e}_3 \qquad (6.2)$$

  where ε denotes the applied strain. In our analysis a compressive strain up to $\varepsilon = -4\%$ is applied to investigate the effect of the nonlinear deformation on the propagation of small-amplitude elastic waves.
- Uniaxial compression in the x_2 direction, so that the macroscopic deformation gradient $\bar{F}$ is given by

$$\bar{F} = \lambda_{11} \quad \hat{e}_1 \otimes \hat{e}_1 + (1 + \varepsilon) \quad \hat{e}_2 \otimes \hat{e}_2 + \hat{e}_3 \otimes \hat{e}_3 \qquad (6.3)$$

  where ε denotes the applied strain in x_2 direction, and λ_{11} is determined from $\sigma_{11} = 0$. A compressive engineering strain up to $\varepsilon = -4\%$ is applied to investigate the effect of the nonlinear deformation on the propagation of small-amplitude elastic waves.

6.3 Stability of Finite-sized Specimens

The stability of periodic finite-sized specimens is first examined using eigenvalue analyses. A linear perturbation procedure is used, and is accomplished within the commercial finite element code ABAQUS/Standard using the *BUCKLE module.

Figure 6.2a shows the evolution of the critical strain ε_{cr} as a function of the size of the structure for both uniaxial and equibiaxial loading conditions. For both loading cases, ε_{cr} is found to first decrease as a function of the structure size and then to plateau to $\varepsilon_{cr}^{uni} = -0.459\%$, and $\varepsilon_{cr}^{equi} = -0.826\%$, when there are more than ten unit cells in both directions.

In Figure 6.2b we report the first eigenmode for equibiaxial loading. Interestingly, for this loading condition the critical mode is characterized by an homogeneous pattern throughout the specimen (with the exception of the rows neighboring the boundaries). Focusing on the central part of the sample, we find that each beam buckles into a half sinusoid and simultaneously preserves the angles with all its neighbors at the joints. By contrast, under uniaxial loading a long-wavelength mode is found to be associated with the lowest critical strain, as shown in Figure 6.2c.

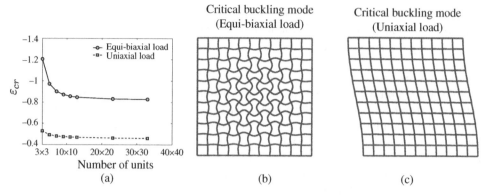

Figure 6.2 Stability of finite-sized specimens: (a) critical strains for equibiaxial and uniaxial loading as function of the sample size; (b) critical buckling mode under equibiaxial loading obtained for a finite-sized sample of 10×10 unit cells; (c) critical buckling mode under unibiaxial loading obtained for a finite-sized sample of 10×10 unit cells.

6.4 Stability of Infinite Periodic Specimens

Upon application of deformation, an infinite periodic structure can suddenly change its periodicity due to mechanical instability. Such instability could be either microscopic (with wavelength that is of the order of the size of the microstructure) or macroscopic (with much larger wavelength than the size of the microstructure) [9, 28, 29].

6.4.1 Microscopic Instability

Microscopic (local) buckling modes are characterized by wavelengths that are of the order of the size of the microstructure and may alter the initial periodicity of the solid. The simplest, but computationally expensive path for investigating them is to construct enlarged unit cells of various sizes and to use a linear perturbation procedure to calculate their critical strains and corresponding modes. The critical strain of the infinite periodic structure is then defined as the minimum of the critical strains on all possible enlarged unit cells [12].

Interestingly, the computational cost of such analysis can be significantly reduced by considering a single primitive unit cell and applying Bloch-periodic boundary conditions [30]:

$$\mathbf{u}(\mathbf{x} + \mathbf{r}) = \mathbf{u}(\mathbf{x})e^{i\mathbf{k}\cdot\mathbf{r}} \tag{6.4}$$

where $\mathbf{k}$ denotes the wave vector and

$$\mathbf{r} = r_{a1}\mathbf{a}_1 + r_{a2}\mathbf{a}_2 \tag{6.5}$$

r_{a1} and r_{a2} being arbitrary integers and $\mathbf{a}_1$ and $\mathbf{a}_2$ denoting the lattice vectors spanning the single unit cell. In fact, the response of an enlarged cell comprising $m_1 \times m_2$ unit cells can be investigated using Eq. (6.4) with [28, 31]:

$$\mathbf{r} = m_1\mathbf{a}_1 + m_2\mathbf{a}_2 \quad \text{and}$$
$$\mathbf{k} = \frac{1}{m_1}\mathbf{b}_1 + \frac{1}{m_2}\mathbf{b}_2 \tag{6.6}$$

b_1 and b_2 being the reciprocal primitive vectors defined as

$$b_1 = 2\pi \frac{a_2 \times \hat{e}_3}{\|a_1 \times a_2\|}, \quad b_2 = 2\pi \frac{\hat{e}_3 \times a_1}{\|a_1 \times a_2\|} \tag{6.7}$$

where $\hat{e}_3 = (a_1 \times a_2)/\|a_1 \times a_2\|$.

For this particular choice of r and k, it is easy to see that Eq. (6.4) reduces to:

$$u(x + r) = u(x), \tag{6.8}$$

indicating that a single primitive unit cell can be used to investigate the response of an enlarged unit cell spanned by the lattice vectors $m_1 a_1$ and $m_2 a_2$, when boundary conditions specified by Eqs. (6.4) and (6.6) are applied.

Therefore, the stability of the enlarged unit cell comprising $m_1 \times m_2$ unit cells can be investigated within the finite-element framework by detecting when the tangent stiffness matrix of the corresponding single unit cell subjected to Bloch-type boundary conditions defined by Eqs. (6.4) and (6.6) becomes singular along the loading path (when $\det[K] = [0]$). Alternatively, it can be also investigated by detecting the applied load at which the smallest eigenfrequency associated to a nontrivial eigenmode of the single unit cell subjected to Bloch-type boundary conditions is zero [32].

Finally, the onset of instability for the infinite periodic structure is defined as the minimum critical strain on all the considered enlarged unit cells defined by $m_1 a_1$ and $m_2 a_2$. Here, we investigated the stability of 25 enlarged primitive units by choosing $m_1 = 1, \ldots, 5$ and $m_2 = 1, \ldots, 5$ in Eq. (6.6). To detect the onset of instability for each enlarged unit cell, we perform eigenfrequency analysis along the loading path at increasing values of applied deformation and detect the smallest load for which an eigenfrequency associated with a nontrivial eigenmode becomes zero[2]. The critical strain of the infinite periodic structure is then defined as the minimum of such loads on all 25 considered enlarged unit cells.

To work with the complex-valued relations of the Bloch-periodic conditions (Eq. (6.4)) in a commercial software such as ABAQUS/Standard, we split all fields into real and imaginary parts. In this way, the equilibrium equations are divided into two sets of uncoupled equations for the real and imaginary parts [28, 33]. Thus the problem is solved using two identical finite-element meshes for the unit cell, one for the real part and one for the imaginary part, which are coupled by Bloch-periodic displacement boundary conditions:

$$\mathrm{Real}(u_i^B) = \mathrm{Real}(u_i^A) \cos[k \cdot r_{A_i B_i}] - \mathrm{Imag}(u_i^A) \sin[k \cdot r_{A_i B_i}]$$
$$\mathrm{Imag}(u_i^B) = \mathrm{Real}(u_i^A) \sin[k \cdot r_{A_i B_i}] + \mathrm{Imag}(u_i^A) \cos[k \cdot r_{A_i B_i}] \tag{6.9}$$

where $r_{A_i B_i} = x_i^B - x_i^A$ denotes the distance in the current/deformed configuration between the two nodes A_i and B_i periodically located on the boundary. Note that Eqs. (6.9) can be easily implemented within ABAQUS/Standard using a user-subroutine MPC.

The results obtained for a square lattice subjected to equibiaxial and uniaxial compression are shown in Figure 6.3a and b, respectively. In the undeformed configuration

---

2 Note that this procedure is significantly simpler than calculating $\det[K] = [0]$, since it does not require the stiffness matrix calculated by ABAQUS to be exported and its determinant evaluated using a numerical package.

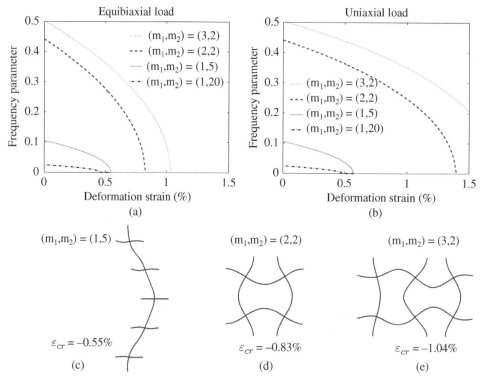

Figure 6.3 Evolution of the frequency parameter as a function of the applied strain for: (a) biaxial and (b) uniaxial loading conditions. Mode shapes associated with: (c) $(m_1, m_2) = (1, 5)$, (d) $(m_1, m_2) = (2, 2)$ and (e) $(m_1, m_2) = (3, 2)$ obtained at their corresponding critical load under equibiaxial compression.

$(\varepsilon = 0)$, all eigenfrequencies ω associated with the considered wave vectors $\mathbf{k}$ are positive. However, as ε increases, the eigenfrequencies associated with each wave vector $\mathbf{k}$ gradually decrease and eventually become negative. The critical strain parameter ε_{cr} associated with each wave vector can be easily extracted from the plot, since it corresponds to the intersection point between each curve and the horizontal line $\omega = 0$.

Similar trends are observed for the case of equibiaxial and uniaxial loading. The lowest critical strain for the 25 considered enlarged unit cells is $\varepsilon_{cr}^{equi} = -0.55\%$ and $\varepsilon_{cr}^{uni} = -0.58\%$ and is associated with $(m_1, m_2) = (1, 5)$, yielding the mode shown in Figure 6.3c. Moreover, slightly lower values of critical strain are found when the number of unit cells in the vertical direction is increased (as shown for the case $(m_1, m_2) = (1, 20)$ in Figure 6.3a and b), associated with eigenmodes similar to that corresponding to $(m_1, m_2) = (1, 5)$.

Focusing on such long-wavelength modes (very similar to the critical mode reported in Figure 6.2c for the finite-sized sample under uniaxial compression), we first note that they are compatible with the boundary conditions experienced by the finite-sized specimen during uniaxial compression, so that we expect them to be observed during the loading of real (finite-sized) samples. By contrast, the boundary conditions applied during an equibiaxial compression test to finite-sized specimens are not compatible with such modes, so that we do not expect them to be triggered during equibiaxial a loading.

To make sure the critical mode is compatible with the boundary conditions experienced by finite-sized samples under equibiaxial compression, we then fix the displacement of the central node of the unit cell. In this case we find that the critical strain $\varepsilon_{cr}^{equi} = -0.83\%$ is that associated with $(m_1, m_2) = (2, 2)$, resulting in a pattern similar to that found for the corresponding finite-sized sample (see Figure 6.2b,d).

As shown above, macroscopic (or long-wavelength) instabilities can be detected by considering $k_i \to 0$ in Eq. (6.6) when performing the microscopic instability analysis. Alternatively, it has been rigorously shown that macroscopic instabilities can also be detected from the loss of strong ellipticity of the overall response of the material [9]. Specifically for the cellular structure considered in this study, macroscopic instability may occur whenever the condition [34]:

$$(\mathbf{m} \otimes \mathbf{N}) : \overline{\mathbb{L}} : (\mathbf{m} \otimes \mathbf{N}) > 0, \quad \overline{\mathbb{L}}_{ijkL}\, m_i\, N_j\, m_k N_L > 0 \tag{6.10}$$

is first violated along the loading path, $\mathbf{m}$ and $\mathbf{N}$ denoting unit vectors defined in the current and the initial configurations, respectively. Note that the homogenized mixed elasticity tensor $\overline{\mathbb{L}}$ relates the macroscopic deformation gradient increment $\dot{\overline{\mathbf{F}}}$ to the macroscopic first Piola–Kirchhoff stress increment $\dot{\overline{\mathbf{P}}}$ as:

$$\dot{\overline{\mathbf{P}}} = \overline{\mathbb{L}} : \dot{\overline{\mathbf{F}}}, \quad \dot{\overline{\mathbf{P}}}_{ij} = \overline{\mathbb{L}}_{ijkL}\dot{\overline{\mathbf{F}}}_{kL} \tag{6.11}$$

In this study, 2D finite-element simulations on a single primitive unit cell with spatially periodic boundary conditions (Eq. (6.1)) are performed to monitor the loss of ellipticity of the homogenized tangent modulus $\overline{\mathbb{L}}$. After determining the principal solution, the components of $\overline{\mathbb{L}}$ are identified by subjecting the unit cells to four independent linear perturbations of the macroscopic deformation gradient $\dot{\overline{\mathbf{F}}}$, calculating the corresponding averaged stress components $\dot{\overline{\mathbf{P}}}$ and comparing to Eq. (6.11). Then, the loss of ellipticity condition is examined by checking the positive definite condition of Eq. (6.10) with $\mathbf{m}$ and $\mathbf{N}$ separately explored at every $\pi/360$ radian increment.

For the structure considered in this study we find that the critical strain associated with macroscopic instability is $\varepsilon_{cr} = -5\%$ for both uniaxial and equibiaxial loading conditions, showing nice agreement with the results obtained from the microscopic instability analysis. Again, we note that the long-wavelength mode is compatible with the boundary conditions applied to the finite-sized sample when loaded uniaxially, so that we expect it to be triggered along the loading path. By contrast, given the incompatibility of such a mode with the boundary conditions applied to the finite-sized sample for equibiaxial compression, we do not expect it to emerge when the samples are deformed equibiaxially.

6.5 Post-buckling Analysis

Load-displacement analyses for both the finite-sized and infinite periodic structures are then performed with ABAQUS/Standard to capture the post-transformation behavior. After determining the pattern transformation (the lowest eigenmode) from the eigen analysis, an imperfection in the form of the most critical eigenmode is introduced into the mesh[3]. Note that to study the postbuckling behavior of the infinite period structure,

---

3 Note that we have also considered imperfections in the form of linear combination of the first three modes. However, we found these to lead to the same post-transformation behavior.

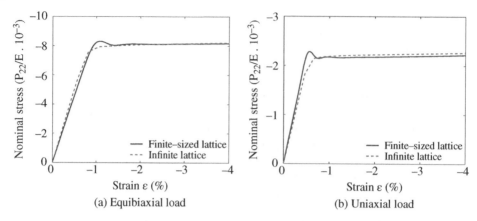

Figure 6.4 Macroscopic stress–strain curves for the square lattice under (a) equibiaxial and (b) uniaxial loading conditions. The departure from linearity indicates the onset of instability.

we use enlarged unit cells with size dictated by the new periodicity introduced by buckling.

In particular, we consider an enlarged unit cell with $(m_1, m_2) = (2, 2)$ for equibiaxial loading, while the long-wavelength behavior expected for uniaxial loading is approximately described using an enlarged unit cell with $(m_1, m_2) = (1, 5)$. The macroscopic stress–strain relationships for equibiaxial and uniaxial compression are shown in Figure 6.4a,b, respectively. For both loading cases, the lattice exhibits an initial linear elastic behavior with a sudden departure from linearity to a plateau stress. This behavior is the result of a sudden instability that alters the initial architecture of the lattice, as shown in Figures 6.5 and 6.6.

Specifically, snapshots of deformed configurations at different levels of strain are shown in Figures 6.5 and 6.6 for equibiaxial and unixial loading conditions, respectively. As predicted by the stability analysis, when the lattice is compressed equibiaxially an homogeneous pattern transformation is induced by buckling. Here, we emphasize the uniformity and robustness of the transformation where the change occurs essentially uniformly throughout the structure. In other words, the instability does not localize deformation in a row or diagonal band, but instead results in a homogeneous pattern transformation throughout the structure. The transformed structure is then accentuated with continuing deformation, and the entire process is reversible and repeatable. Finally, our post-buckling simulations confirm that under uniaxial compression a long-wavelength mode is triggered during loading.

6.6 Effect of Buckling and Large Deformation on the Propagation Of Elastic Waves

Next, we investigate the effect of the applied deformation on the propagation of small-amplitude elastic waves, superimposing incremental strains to a given state of finite deformation and using the finite-element method to compute the band structure [28, 35].

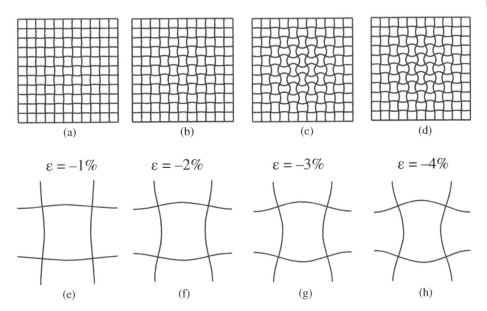

Figure 6.5 Deformation field obtained for different values of the global equibiaxial strain: (a)–(d) finite-sized sample; (e)–(h) unit cell of the infinite periodic system.

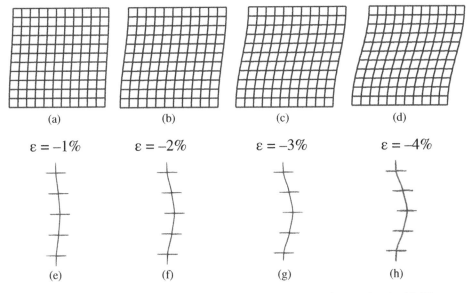

Figure 6.6 Deformation filed obtained for different values of the global uniaxial strain: (a)–(d) finite-sized sample; (e)–(h) unit cell of the infinite periodic system.

Phononic band gaps are identified by computing the eigenfrequencies $\omega(\mathbf{k})$ for wave vectors $\mathbf{k}$ spanning the perimeter of the irreducible Brillouin zone. In particular, phononic band gaps (frequency ranges for which the propagation of waves is barred) are identified by regions within which no $\omega(\mathbf{k})$ exist. For the simulations presented

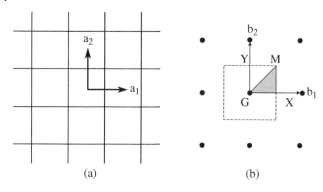

Figure 6.7 Square periodic lattice: (a) lattice vectors; (b) first Brillouin Zone.

in this chapter, 20 uniformly spaced points on each edge of the irreducible Brillouin zone are used (Figure 6.7). Also, results are presented in terms of the nondimensional frequency parameter $\Omega = \omega/\overline{\omega}$, defined as the ratio between the frequency of wave propagation ω and the first natural frequency $\overline{\omega}$ of a beam of length L with clamped boundaries (that is, $\overline{\omega} = (4.73004)^2 \sqrt{EI/(\rho AL^4)}$)

The dynamic behavior of the system is further characterized by computing the frequency response of a finite-sized sample comprising $10{\times}10$ cells. This study is conducted in ABAQUS by performing a *STEADY STATE DYNAMICS analysis after deforming the lattice to the desired level of strain, either uniaxially or equibiaxially. The frequency response function is computed by exciting the central node of the structure and by recording the horizontal component (U1) of the displacement field at the top-right-hand corner.

The results of this study are shown in Figures 6.8 and 6.9 for equibiaxial and uniaxial compression, respectively. We start by noting that the undeformed lattice does not feature any band gaps. This is in line with the behavior previously observed for a square undeformed lattice of beams [36]. However, for the case of equibiaxial compression the dispersion relations of the system reveal that when $\varepsilon_{equi} < -2.8\%$ a complete band gap forms in the vicinity of $\Omega = 1$ and this widens for increasing levels of deformation. For example, Figure 6.8e shows that a relatively wide band gap (shaded region) extends between $\Omega_{low} = 0.91$ and $\Omega_{up} = 1.37$ for $\varepsilon = -4\%$. The unit cell predictions are corroborated by the frequency response analyses shown in Figure 6.8b,d,f. The results, in fact, clearly indicate that waves are free to propagate in the lattice for small values of the applied strain, and that a strong attenuation region forms corresponding to the band gap frequency range when $\varepsilon = -4\%$. As already noted for the case of periodic and porous hyper-elastic structures [35], this behavior is triggered by the pattern transformation induced by instability. Interestingly, the propagation of elastic waves in a stress-free lattice structure with the deformed, buckled-like geometry has been recently investigated [37], and a similar dynamic response has been observed. The important role played by geometric nonlinearities is further confirmed by the results shown in Figure 6.9 for uniaxial loading. In this case, the critical instability mode is characterized by a global deformation pattern with a characteristic wavelength of the order of the structure size. As shown in Figure 6.6, this deformation mechanism does not

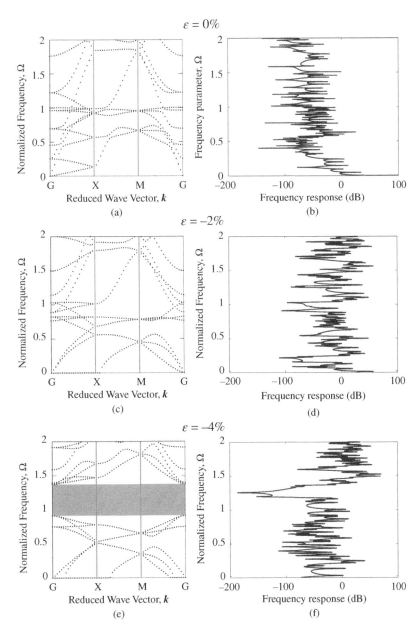

Figure 6.8 Comparison between the dispersion relations (a,c,e) and frequency response functions (b,d,f) of the system obtained for different values of the applied equibiaxial strain.

significantly alter the local periodicity of the system. As such, both the dispersion relations and the frequency response results shown in Figure 6.9 reveal only moderate changes in the dynamic behavior of the lattice as a function of the applied deformation, and that no band gap appears for the considered strain levels.

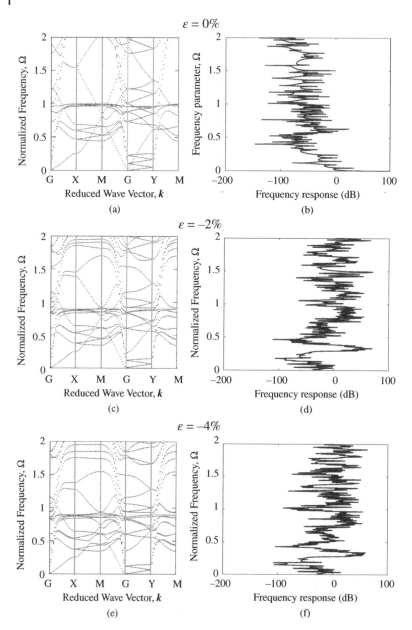

Figure 6.9 Comparison between the dispersion relations (a,c,e) and frequency response functions (b,d,f) of the system obtained for different values of the applied uniaxial strain.

6.7 Conclusions

To summarize, we used numerical simulations to study the stability of a square lattice of elastic beams. Our results indicate that microscopic (local) instabilities are critical when the structure is compressed equibiaxially, while macroscopic (global) instability is first encountered along the loading path in the case of uniaxial compression. In contrast to the case of porous structures, for which the critical modes are found to be quite insensitive to the boundary conditions applied to the finite-sized samples, we find that for lattice structures the boundary conditions applied to the finite-sized samples play an important role in determining the mode triggered during loading.

We also investigated the effect of deformation on the propagation of small-amplitude elastic waves and find that the band gaps of the lattice are significantly affected by the pattern change induced by microscopic (local) modes. By contrast, macroscopic (global) modes are found to have very limited effects on the band gaps of the structure.

References

1 S. E. Forman and J. W. Hutchinson, "Buckling of reticulated shell structures," *International Journal of Solids and Structures*, vol. 6, no. 7, pp. 909–932, 1970.

2 J. Renton, "Buckling of long, regular trusses," *International Journal of Solids and Structures*, vol. 9, no. 12, pp. 1489–1500, 1973.

3 T. Wah, "The buckling of gridworks," *Journal of the Mechanics and Physics of Solids*, vol. 13, no. 1, pp. 1–16, 1965.

4 T. Wah and L. R. Calcote, *Structural Analysis by Finite Difference Calculus*. Van Nostrand Reinhold, 1970.

5 A. K. Noor, M. S. Anderson, and W. H. Greene, "Continuum models for beam-and platelike lattice structures," *AIAA Journal*, vol. 16, no. 12, pp. 1219–1228, 1978.

6 S. Papka and S. Kyriakides, "In-plane biaxial crushing of honeycombs: Part II: Analysis," *International Journal of Solids and Structures*, vol. 36, no. 29, pp. 4397–4423, 1999.

7 M. S. Anderson, "Buckling of periodic lattice structures," *AIAA Journal*, vol. 19, no. 6, pp. 782–788, 1981.

8 M. Anderson and F. Williams, "Natural vibration and buckling of general periodic lattice structures," *AIAA Journal*, vol. 24, no. 1, pp. 163–169, 1986.

9 G. Geymonat, S. Muller, and N. Triantafyllidis, "Homogenization of nonlinearly elastic-materials, microscopic bifurcation and macroscopic loss of rank-one convexity," *Archive for Rational Mechanics and Analysis*, vol. 122, no. 3, pp. 231–290, 1993.

10 M. Schraad and N. Triantafyllidis, "Scale effects in media with periodic and nearly periodic microstructures. II. Failure mechanisms," *Journal of Applied Mechanics-Transactions of the ASME*, vol. 64, pp. 762–771, 1997.

11 R. Hutchinson and N. Fleck, "The structural performance of the periodic truss," *Journal of the Mechanics and Physics of Solids*, vol. 54, no. 4, 2006.

12 K. Bertoldi, M. C. Boyce, S. Deschanel, S. M. Prange, and T. Mullin, "Mechanics of deformation-triggered pattern transformations and superelastic behavior in periodic elastomeric structures," *Journal of the Mechanics and Physics of Solids*, vol. 56, no. 8, pp. 2642–2668, 2008.

13 T. S. Horozov, B. P. Binks, R. Aveyard, and J. H. Clint, "Effect of particle hydrophobicity on the formation and collapse of fumed silica particle monolayers at the oil–water interface," *Colloids and Surfaces A: Physicochemical and Engineering Aspects*, vol. 282, pp. 377–386, 2006.

14 E. P. Chan, E. J. Smith, R. C. Hayward, and A. J. Crosby, "Surface wrinkles for smart adhesion," *Advanced Materials*, vol. 20, no. 4, pp. 711–716, 2008.

15 J.-H. Jang, C. Y. Koh, K. Bertoldi, M. C. Boyce, and E. L. Thomas, "Combining pattern instability and shape-memory hysteresis for phononic switching," *Nano Letters*, vol. 9, no. 5, pp. 2113–2119, 2009.

16 S. Yang, K. Khare, and P.-C. Lin, "Harnessing surface wrinkle patterns in soft matter," *Advanced Functional Materials*, vol. 20, no. 16, pp. 2550–2564, 2010.

17 J. Kim, J. A. Hanna, M. Byun, C. D. Santangelo, and R. C. Hayward, "Designing responsive buckled surfaces by halftone gel lithography," *Science*, vol. 335, no. 6073, pp. 1201–1205, 2012.

18 J. Shim, C. Perdigou, E. R. Chen, K. Bertoldi, and P. M. Reis, "Buckling-induced encapsulation of structured elastic shells under pressure," *Proceedings of the National Academy of Sciences of the United States of America*, vol. 109, no. 16, pp. 5978–5983, 2012.

19 T. Mullin, S. Deschanel, K. Bertoldi, and M. Boyce, "Pattern transformation triggered by deformation," *Physical Review Letters*, vol. 99, no. 8, p. 084301, 2007.

20 Y. Zhang, E. A. Matsumoto, A. Peter, P.-C. Lin, R. D. Kamien, and S. Yang, "One-step nanoscale assembly of complex structures via harnessing of an elastic instability," *Nano Letters*, vol. 8, no. 4, pp. 1192–1196, 2008.

21 S. Singamaneni, K. Bertoldi, S. Chang, J.-H. Jang, S. L. Young, E. L. Thomas, M. C. Boyce, and V. V. Tsukruk, "Bifurcated mechanical behavior of deformed periodic porous solids," *Advanced Functional Materials*, vol. 19, no. 9, pp. 1426–1436, 2009.

22 L. Wang and K. Bertoldi, "Mechanically tunable phononic band gaps in three-dimensional periodic elastomeric structures," *International Journal of Solids and Structures*, vol. 49, no. 19, pp. 2881–2885, 2012.

23 J. Li, J. Shim, J. Deng, J. T. Overvelde, X. Zhu, K. Bertoldi, and S. Yang, "Switching periodic membranes via pattern transformation and shape memory effect," *Soft Matter*, vol. 8, no. 40, pp. 10322–10328, 2012.

24 J. Overvelde, S. Shan, and K. Bertoldi, "Compaction through buckling in 2D periodic, soft and porous structures: Effect of pore shape," *Advanced Materials*, vol. 24, no. 17, pp. 2337–2342, 2012.

25 K. Bertoldi, P. M. Reis, S. Willshaw, and T. Mullin, "Negative Poisson's ratio behavior induced by an elastic instability," *Advanced Materials*, vol. 22, no. 3, p. 361, 2010.

26 Hibbitt, Karlsson, and Sorensen, *ABAQUS/Standard User's Manual*, vol. 1. Hibbitt and Karlsson and Sorensen, 2001.

27 M. Danielsson, D. M. Parks, and M. C. Boyce, "Three-dimensional micromechanical modeling of voided polymeric materials," *Journal of the Mechanics and Physics of Solids*, vol. 50, no. 2, pp. 351–379, 2002.

28 K. Bertoldi and M. C. Boyce, "Wave propagation and instabilities in monolithic and periodically structured elastomeric materials undergoing large deformations," *Physical Review B*, vol. 78, no. 18, 2008.

29 N. Triantafyllidis and B. N. Maker, "On the comparison between microscopic and macroscopic instability mechanisms in a class of fiber-reinforced composites," *Journal of Applied Mechanics-Transactions of the ASME*, vol. 52, no. 4, pp. 794–800, 1985.

30 C. Kittel, *Introduction to Solid State Physics*. Wiley, 8th ed., 2005.

31 N. Triantafyllidis, M. Nestorovic, and M. Schraad, "Failure surfaces for finitely strained two-phase periodic solids under general in-plane loading," *Journal of Applied Mechanics*, vol. 73, pp. 505–515, 2006.

32 K.-J. Bathe, *Finite Element Procedures*. Prentice Hall, 1996.

33 M. Aberg and P. Gudmundson, "The usage of standard finite element codes for computation of dispersion relations in materials with periodic microstructure," *Journal of the Acoustical Society of America*, vol. 102, no. 4, pp. 2007–2013, 1997.

34 J. E. Marsden and T. J. R. Hughes, *Mathematical Foundations of Elasticity*. Prentice-Hall, 1983.

35 P. Wang, J. M. Shim, and K. Bertoldi, "Effects of geometric and material nonlinearities on tunable band gaps and low-frequency directionality of phononic crystals," *Physical Review B*, vol. 88, no. 1, 2013.

36 A. S. Phani, J. Woodhouse, and N. Fleck, "Wave propagation in two-dimensional periodic lattices," *The Journal of the Acoustical Society of America*, vol. 119, no. 4, pp. 1995–2005, 2006.

37 Y. Liebold-Ribeiro and C. Körner, "Phononic band gaps in periodic cellular materials," *Advanced Engineering Materials*, vol. 16, no. 3, pp. 328–334, 2014.

7

Impact and Blast Response of Lattice Materials

Matthew Smith[1], Wesley J. Cantwell[2] and Zhongwei Guan[3]

[1] *AMRC with Boeing, Rolls-Royce Factory of the Future, University of Sheffield, Sheffield, UK*
[2] *Aerospace Research and Innovation Center, Khalifa University of Science, Technology and Research, Abu Dhabi, UAE*
[3] *School of Engineering, University of Liverpool, Liverpool, UK*

7.1 Introduction

The design of energy-absorbing and blast-resistant structures requires the use of efficient energy-absorbing structures. In actual blast-loading conditions a large amount of kinetic energy is input to the protective structure. This sacrificial layer is responsible for the dissipation of the energy as well as ensuring the forces transmitted to the main structure are kept below acceptable levels [1]. Recent attention has focused on the need to design and construct lightweight structures that offer protection to both civil and military personnel. Foam-based materials, such as highly ductile metallic foams, have been implemented in the design and manufacture of blast-resistant sandwich structures [2, 3]. However, many of the first generation of core materials offered highly irregular cell structures, making safe design both difficult and highly conservative. Lattice structures have been identified as possible energy-absorbing core structures, which can offer greater strength-to-weight and stiffness-to-weight ratios than those offered by traditional foams [3–4].

Several studies have been carried out to assess the quasi-static response of lattice structures subject to compressive loads [10–37]. The aim of the work presented here is to demonstrate the impact and blast resistance of lightweight lattice architectures for use in energy-absorbing applications. Initial attention focuses on investigating the impact and blast response of lattice blocks and is subsequently extended to consider the blast response of larger lattice-sandwich structures.

7.2 Literature Review

This section discusses some of the previous work on the impact, blast, shock, and indentation responses of cellular structures.

7.2.1 Dynamic Response of Cellular Structures

Deformation of cellular materials at high strain rates may result in an increase in the mechanical properties and energy absorption of the structure. The effect of strain rate

Dynamics of Lattice Materials, First Edition. Edited by A. Srikantha Phani and Mahmoud I. Hussein.
© 2017 John Wiley & Sons Ltd. Published 2017 by John Wiley & Sons Ltd.

has been shown to depend on the material, geometry, and the method of manufacture [18]. The energy absorption of batch-cast aluminium foam (Alporas), at dynamic strain rates, has been shown to increase by over 50% compared to the quasi-static rate. In contrast, aluminium foam formed by a powder metallurgical technique (Alulight), did not show any strain-rate effect [19].

Yungwirth et al. [20] measured the impact response of pyramidal lattice core sandwich plates made from 304 stainless steel and 6061-T6 aluminium alloy subjected to a spherical steel projectile whose impact velocity ranged from 250–1300 ms^{-1}. They concluded that sandwich plates have the potential to outperform monolithic plates in resisting both blast and ballistic loads. The impact response of 6061-T6 aluminium corrugated-core sandwich panels was also studied [21]. Here, the mechanisms of projectile penetration of the sandwich panels were investigated. It was found that low-momentum impacts are laterally deflected by interactions with the inclined webs of the empty core. Complete perforation was initiated by shear-off within the impacted front face sheet, followed by stretching, bending, and tensile fracture of the core webs and finally shear-off within the back face sheet.

Evans [22] indicated that metallic sandwich panels with unfilled cellular cores could exhibit superior bending stiffness and strength compared to monolithic plates of the same alloy and mass per unit area. When such sandwich structures are subjected to ballistic impacts they offer potentially useful multifunctionality. Many unfilled-core topologies have been explored for structural load support applications including honeycombs [23], prismatic corrugations [24], and truss structures (including some with hollow trusses) [7, 25]. With the optimized distribution of panel mass between the front face sheet, core, and back face sheet, the back-face deflection caused by impulsive load can be reduced due to the high bending-resistance of sandwich panels [26–38]. The benefit of such distribution can be enhanced if fluid–structure interaction effects are exploited to minimize impulse reflection [29]. When sandwich structures are impulsively loaded by high-velocity sand [30] or air [31], small reductions (of the order of 10%) in impulse may be obtained. In general, the higher bending-resistance of sandwich structures can further reduce back-side deflections in comparison to the equivalent solid plates. Zhang et al. [32] studied the dynamic response of sandwich structures with aluminium alloy pyramidal-truss cores under impact loading, aiming to maximize the specific bending stiffness and strength, as well as to obtain excellent energy-absorption capability.

Lee et al. [33, 34] investigated the response of stainless steel pyramidal-truss structures under quasi-static and dynamic compressive loads. Quasi-static tests were performed using a miniature loading stage while a Kolsky bar apparatus was used to investigate intermediate strain rates (263–550 s^{-1}). High strain rates (7257–9875 s^{-1}) were examined using a light gas gun. Compared to the quasi-static rate, an increase of approximately 50% in the peak stress was observed at intermediate rates. This was attributed to a micro-inertia effect. The increase in peak stress at high strain rates was more pronounced, with increases of between 130 and 190% observed. The deformation of the structure was different at high strain rates and the inertia associated with the bending and buckling of the struts played a more significant role. The inertia effect dominated the initial response of the truss core because of two effects:

- the propagation of a plastic wave along the truss members, which delayed buckling of the member

- buckling-induced lateral motion.

Vaughn et al. [35] discusses the inertia effect in more detail. The effect of strain rate on a range of 'first generation' stainless steel lattice structures built with the selective laser melting (SLM) process has shown an increase in the yield stress of 20% from quasi-static at strain rates of around 1×10^3 s^{-1} [13]. The collapse mechanisms in the quasi-static and the dynamic tests were compared and found to be similar for each type of unit-cell structure. In the study, a MCP Realiser I SLM system was used to build lattice structures. This produced lattice microstruts with elliptical cross-sections. Mines et al. [36] investigated the drop-weight impact response of sandwich panels with SLM lattice cores. It was found that the current Ti_6Al_4V BCC microlattice cores are competitive with aluminium honeycomb, but that there is scope to improve their impact performance in Ti_6Al_4V microstruts by quantifying micro-inertia and material strain-rate effects in the core and by adapting the microlattice structure architecture.

7.2.2 Shock- and Blast-loading Responses of Cellular Structures

Radford et al. [37] have shown that metal foam projectiles can be used to generate dynamic pressure–time histories representative of shock loading in water and air. McShane et al. [38] used this technique to measure the dynamic response of monolithic and sandwich plates with steel pyramidal or square-honeycomb cores. The resistance to shock loading was quantified by the permanent transverse deflection at the mid-span of the plates. It was observed that under dynamic loading, the sandwich panels outperformed the monolithic panels of equal mass and that the honeycomb panels had a higher shock resistance than the pyramidal core plates. McShane et al. [39] investigated the underwater blast response of free-standing sandwich panels with a square honeycomb core and a corrugated core. It was found that core topology dictates the final degree of core compression: the stronger the core, the smaller the core compression. Mori et al. [40, 41] used a conical shock tube (with the diameter increasing towards the target end) to generate underwater blast pulses to load edge-clamped sandwich panels with metallic lattice cores. It was found that the back face of the sandwich plates deflected less than monolithic plates of equal mass. They indicated that the sandwich plate acquires a smaller impulse than a monolithic plate of equal mass, due to fluid-structure interaction effects. The shock response of metallic pyramidal lattice core sandwich panels to high-intensity impulsive loading in air was also investigated [42]. Here, a small-scale explosive loading was applied to sandwich panels with low-relative-density pyramidal lattice cores to study the large-scale bending and fracture responses of a model sandwich panel system in which the core has little stretch resistance.

Hanssen et al. [43] investigated the close-range blast response of aluminium foam panels when used as a sacrificial cladding structure. It was shown that such structures are effective under blast loading, although counterintuitively, the results showed that the energy and impulse transferred to the ballistic pendulum was increased by adding a face sheet to the foam panel. This is thought to be due to complex face sheet deformation into a concave shape and it is this which controls the energy transfer rather than the deformation of the core material.

Langdon et al. [44] investigated the effect of core density and cover-plate thickness of Cymat aluminium foam cladding panels under blast loading. Core relative densities of 10, 15, and 20% and steel-cover plate thicknesses of 2 and 4 mm were tested. The cover-plate thickness had a significant effect on the panel response, with the 2-mm face-plates showing variable levels of crush across the section, with significant permanent deformation. The 4-mm face-plates were more rigid, causing the core to crush uniformly. The effect of bonding the face-sheets to the foam core was also examined and was found to increase the level of fracture in the core.

The design of sandwich panels for blast protection is discussed in detail by Ashby [45]. The author notes that it is beneficial to attach a heavy face-plate to the front of the energy absorber, because the blast impulse imparts a momentum to the face-plate, accelerating it to a certain velocity with an associated kinetic energy. Heavier face-plates result in a lower velocity and hence a lower kinetic energy for the absorber to dissipate.

Cladding and sandwich specimens were tested by Karagiozova et al. [46, 47] and it was concluded that a quasi-static approach could not be used to accurately estimate the absorbed energy due to a strong dynamic effect. The energy dissipated by the foam compaction depended on the speed of compaction through different strain histories. Sandwich-type plates with polystyrene foam and aluminium honeycomb cores were also tested under blast loads. The permanent deflection of the back plate was measured and could be determined by the velocity-attenuation properties of the core. For panels of comparable mass, the honeycomb cores outperformed the polystyrene cores. Further research was undertaken on aluminium honeycomb sandwich panels subjected to air blast and impact loadings [32–48].

7.2.3 Dynamic Indentation Performance of Cellular Structures

Studies of the indentation performance of cellular metals are mainly concerned with the structural response, load-carrying capacity, failure modes, and energy absorption of the panels, due to indentation damage [53]. A review of the indentation performance of some cellular metals, typically used in the aerospace industry, can be found in the literature [54].

The indentation performance of SLM-built lattice structures has also been assessed [18, 55]. Static penetration tests, performed on stainless steel lattice cores and sandwich panels, have shown that the SLM structures are comparable to Alporas aluminium foam and that the performance can be further improved by changing the parent material or by optimizing the unit-cell topology. Drop-weight impact tests were also conducted on the sandwich panels and damage mechanisms similar to those seen in the quasi-static tests were observed.

The impact performance of titanium lattice core sandwich panels was compared to aluminium honeycomb based sandwich panels by Hasan et al. [56]. The deformation mechanisms within the cores were assessed by CT scans of the specimens. The impact resistance of each panel was comparable at high-impact energies but the titanium lattice core showed a more localized damage area than the aluminium honeycomb. This was said to be an advantage as it is a requirement, outlined by aircraft manufacturers, for damage areas to be of similar dimensions to the impactor. Localized damage in a panel or component is beneficial as less of the structure must be replaced after damage occurs.

7.3 Manufacturing Process

This section briefly describes the manufacturing process used to build stainless steel microlattice structures using the SLM process.

7.3.1 The Selective Laser Melting Technique

Selective laser melting (SLM) is an additive manufacturing technology that uses a high-powered ytterbium fiber laser to fuse fine metallic powders, forming functional three-dimensional parts [57].

The MCP Realizer II, shown in Figure 7.1, is a commercial SLM workstation with a 200-W continuous wave ytterbium fiber laser (SPI, UK) with an operating wavelength of 1068–3095 nm. The build envelope of the machine is 250 × 250 × 240 mm and is capable of building from 5 to 20 cm^3 of dense steel per hour. The scanning system features a dual-axis mirror-positioning system and a galvanometer optical scanner, which directs the laser beam in the x- and y-axes. The variable focusing optic is a Sill 300-mm focal length f-theta lens, which produces a beam spot size of 90 μm across the build envelope [58].

Two lattice unit-cell topologies were used to develop the lattice structures investigated in this study: a body-centered cubic (BCC) repeating unit and a similar structure with vertical columns, termed a BCC-Z unit cell. These are shown in Figure 7.2. The BCC unit cell can be described as a point at the center of a cube from which eight struts radiate out to the corners of the cube. The images in Figure 7.2 differ from the description above in order to clarify that there is one complete vertical strut per unit cell in the BCC-Z unit cell.

The MCP Realizer II machine was used to create various 316L stainless steel lattice cube structures, shown in Figure 7.3, and larger panels in order to investigate the

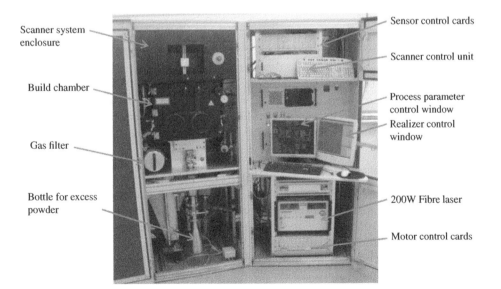

Figure 7.1 The MCP Realizer II machine.

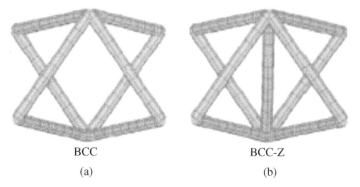

BCC   BCC-Z

(a)   (b)

Figure 7.2 Microlattice geometries used in this study: (a) BCC and (b) BCC-Z.

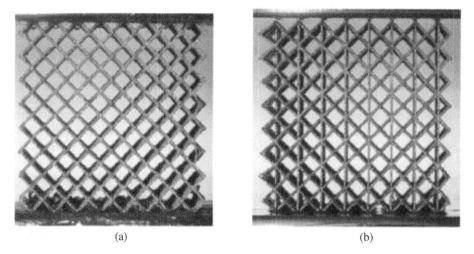

(a)   (b)

Figure 7.3 316L stainless steel lattice blocks with (a) BCC unit cells and (b) BCC-Z unit cells.

dynamic impact and blast response of these structures. Details of the lattice structures used in blast tests are given in Table 7.1, while details of the lattices used in the drop-weight impact tests are given in Table 7.4. Further details of the manufacturing process are given in the relevant papers [12, 13].

7.3.2 Sandwich Panel Manufacture

A compression molding process was used to create sandwich panels with microlattice cores. The lattice cores had overall dimensions of $100 \times 100 \times 20$ mm and a unit-cell size of 2.5 mm (lattice D). In each specimen, the top and bottom skins were made with four plies of carbon fiber reinforced plastic (CFRP). The lay-up was then placed in a hot press where the processing pressure was maintained below the yield stress of the core and the temperature was maintained at 120°C for 90 min. The combination of heat and pressure allowed the resin to flow over the core material and the microstruts at the surface and then to penetrate the skins, resulting in excellent adhesion, as shown by Shen [18]. The

Table 7.1 Details of the manufactured lattice structures used in the blast tests.

Lattice ID	Unit cell type	Unit cell volume (mm³)	Relative density, ρ^* (%)	Average strut diameter approx. (mm)	Lattice block dimensions approx. (mm)
A	BCC	1.25 ³	13.6	0.2	20 × 20 × 18.75
B	BCC	1.5 ³	6.86	0.2	20 × 20 × 19.5
C	BCC	2 ³	3.85	0.2	21 × 20 × 19.5
D	BCC	2.5 ³	2.95	0.2	20 × 20 × 20
E	BCC-Z	1.25 ³	14.5	0.2	20 × 20 × 18.75
F	BCC-Z	1.5 ³	8.03	0.2	20 × 20 × 19.5
G	BCC-Z	2 ³	4.77	0.2	21 × 20 × 19.5
H	BCC-Z	2.5 ³	2.84	0.2	20 × 20 × 20

specimens were allowed to cool in the hot press before being removed. The skin was subsequently measured and found to be approximately 1 mm thick.

CFRP was used as the skin material to make the sandwich panels. EP121-C15–33 is a fiber-reinforced thermosetting pre-impregnated material suitable for use in aircraft parts. It has a plain-weave woven structure consisting of 3k HTA carbon fabric, impregnated with epoxy resin EP121. It can be cured at temperatures between 120 and 160°C and at pressures of at least 0.07 MPa and is suitable for hot-pressing or vacuum bag processing [59]. Key properties of the CFRP are presented in Table 7.2 and a typical manufactured sandwich panel is shown in Figure 7.4.

7.4 Dynamic and Blast Loading of Lattice Materials

This section briefly describes the experimental procedures used to investigate the dynamic and blast response of the lattice structures.

7.4.1 Experimental Method – Drop-hammer Impact Tests

To investigate the mechanical properties of the lattice structures at dynamic rates of strain, an instrumented drop-hammer impact tower, shown in Figure 7.5, was used to conduct dynamic compression tests on a range of lattice structures. An instrumented carriage, with a flat impactor, was released from heights of up to 1.5 m in order to crush the specimens. The variation of load with time was measured using a Kistler

Table 7.2 Properties of the CFRP used in the sandwich panels.

Prepreg material	Skin lay-up	Areal density (g/m²)	Tensile modulus (GPa)	Tensile strength (MPa)	Cure temperature/time (°C/min)
Plain weave carbon fiber/epoxy matrix	4-ply 0/90 – nominal thickness 1 mm	410 ± 15	58 (8 ply)	850 (8 ply)	155/35

Figure 7.4 A 100 × 100 × 20-mm sandwich panel with BCC lattice core.

piezo-electric load cell, located between the carriage and the impactor, and a dedicated PC. The movement of a target point on the impactor during the test was recorded using a 10 000 frame-per-second high-speed camera (MotionPro X4, Integrated Design Tools, Inc.) and recording software. The software package 'ProAnalyst' was used to convert the recorded motion of the target point into a plot of velocity against time, which was subsequently integrated to obtain a plot of displacement against time. A total of nine drop-hammer compression tests were conducted on BCC and BCC-Z lattice structures with various relative densities. The compression plate and impactor were greased before the tests to reduce the effect of friction. The lattice structures were loaded in the build direction; in other words, the vertical pillars of the BCC-Z lattice were parallel to the loading direction.

7.4.2 Experimental Method – Blast Tests on Lattice Cubes

Blast tests were conducted on a ballistic pendulum, shown in Figure 7.6, at the Blast Impact and Survivability Research Unit (BISRU) at the University of Cape Town. The blast loads were applied to the lattice structures via a steel plate, referred to as a 'striker'. Two guide rails were used to ensure that the lattice structures were loaded uniformly. A schematic of the pendulum test set-up is shown in Figure 7.7. Three cylindrical steel plate strikers were used and their masses and dimensions are given in Table 7.3. Damage to the strikers occurred during testing, due to the severity of the loading conditions, so several strikers had to be used. The masses of the strikers varied by a maximum of 4.6 g (2.3%), which was considered sufficiently small to permit comparisons between each test.

A locating jig was used to ensure that the lattices were placed centrally on the rig and the guide rails were greased to reduce friction between the rails and the striker. The blast-impulse load was generated by detonating a disc of PE4 explosive at a standoff distance of 13 mm. The standoff distance was achieved by placing the explosive on a polystyrene pad with a thickness of 13 mm and attaching the pad to the striker with an adhesive. The standoff distance was used to reduce damage to the face of the

Figure 7.5 Drop-hammer impact tower test rig.

striker under the action of the applied impulse. After detonation, the polystyrene pad disintegrated and therefore had no influence on the loading of the striker. Changing the thickness of the PE4 disc varied the mass of explosive, and therefore the impulse applied to the striker and lattice. The amplitude of the blast pendulum swing was recorded and used to calculate the applied impulse using a similar method to that described by Theobald et al. [14]. A total of 29 blast tests were conducted on various BCC and BCC-Z lattice structures.

7.4.3 Experimental Method – Blast Tests on Composite-lattice Sandwich Structures

A series of blast tests was also performed on lattice core sandwich structures with CFRP skins. These tests were conducted in order to evaluate how the sandwich structures performed under blast-loading conditions. The tests were also performed on the ballistic

Figure 7.6 Ballistic pendulum used for the blast tests.

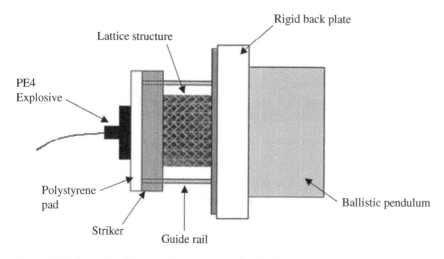

Figure 7.7 Schematic of the experimental set-up for the blast tests.

pendulum and the experimental test set-up for the tests on the sandwich structures is shown in Figure 7.8.

A square blast-tube arrangement, similar to that employed by Theobald [14], was used to generate a uniform load on the sandwich structures. In this set-up, the back plate of the sandwich panel is fully supported and the front plate is free to move as the sandwich core is deformed. The sandwich panels were attached to the solid bake plate and

Table 7.3 Properties of the strikers used in the blast tests.

Striker ID	Mass (g)	Thickness approx. (mm)	Diameter approx. (mm)	Details
S1	199.7	20	40	PTFE bushings used
S2	197.6	20	40	No bushings – heat treated
S3	196.3	20	40	Brass bushings used

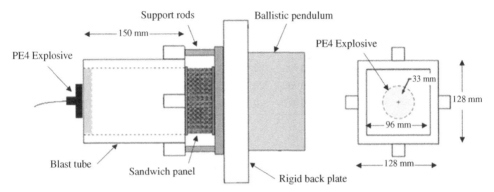

Figure 7.8 Schematic of the blast tube used in the blast tests on the sandwich panel.

positioned so that there were no gaps between the test panel and the blast tube. Detonating circular discs of PE4 explosive with diameters of 20–33 mm generated the blast loads. The explosive was placed on a 96 × 96-mm polystyrene pad, which was located in the other end of the blast tube, giving a standoff distance of 150 mm. The initial and final thicknesses of the sandwich panels were measured at 10-mm intervals across the panel using a BATY CL1 plunge gauge.

7.5 Results and Discussion

7.5.1 Drop-hammer Impact Tests

Dynamic compression tests were conducted on a drop-hammer impact tower to characterize the effect of higher strain rates on the material properties of the lattice structures. Details of the lattice structures tested and the test conditions are given in Table 7.4.

Figure 7.9 shows typical engineering stress–strain curves following drop-weight impact tests on the lattice structures. The stress–strain curves are similar in form to those tested under quasi-static loading, where there is a steep rise in the elastic region followed by a stress plateau, which continues until the onset of densification, associated with another abrupt rise in stress. The stress–strain trace for the drop-hammer compression test shows small oscillations compared to the smooth stress–strain traces observed in quasi-static testing. This effect is caused by vibrations in the drop-hammer carriage as it strikes the lattice structure and the supporting rails of the impact rig.

Table 7.4 Summary of the lattice specimens used in the drop-hammer tests.

Lattice ID	Unit cell Type	Mass (g)	Height (mm)	Width (mm)	Depth (mm)	Relative density (%)	Test speed (m/s)	Strain rate (s⁻¹)	Plateau stress (at 25% strain) (MPa)	Impact energy (J)
A5-D	BCC	6.1	14.1	20	20.35	13.29	3.71	263.1	6.37	15.5
A6-D	BCC	6.6	15	20	20.4	13.48	4.43	295.3	7.26	22.2
A7-D	BCC	6.7	15.23	19.95	20.32	13.56	5.43	356.5	8.24	33.2
B4-D	BCC	4.7	14.6	20.1	20.4	9.81	3.71	254.1	4.02	15.5
D5-D	BCC	1.7	15.75	20.2	20.3	3.29	2.80	177.9	0.66	4.3
D6-D	BCC	1.8	15.27	20.3	20.41	3.56	2.80	183.5	0.53	4.3
H5-D	BCC-Z	1.9	14.9	19.8	20.4	3.95	2.80	188.0	1.452	4.3
H6-D	BCC-Z	1.7	14.95	19.9	20.15	3.54	3.43	229.5	1.42	6.4
H7-D	BCC-Z	1.7	14.91	19.85	20.36	3.53	3.43	230.1	1.32	6.4

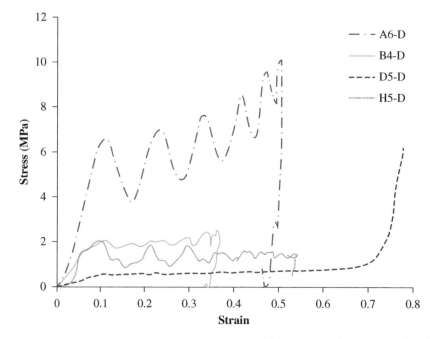

Figure 7.9 Stress–strain curves for the lattice structures following compression tests on the drop-hammer impact tower.

Figure 7.10 shows the engineering stress–strain curves following a drop-hammer test on the BCC structures of lattice D, along with compression test data from tests on similar structures carried out on an Instron 4024 test machine at increasing rates of strain [12, 60]. This graph shows the influence of loading velocity on the compressive response

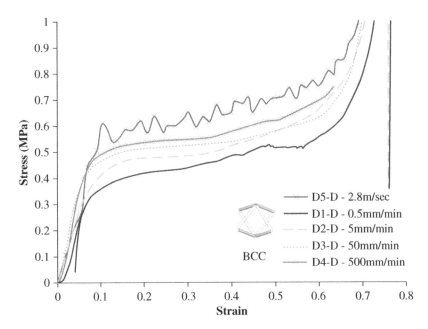

Figure 7.10 Stress–strain curves for BCC (lattice D) structures following compression tests on the drop-hammer impact tower.

of the lattice structure. As the loading velocity is increased, the yield strength and plateau stress of the lattices also increase.

The strain rate associated with each drop-weight impact test ranged from 177.9–356.5 s^{-1} and the effect of strain rate on the plateau stress of the lattice structures is shown in Figure 7.11. Plateau stress is the stress at which the structures collapse plastically at a relatively constant level. It was measured at a strain of 25% for all of the lattices tested. The graphs show that all of the BCC and BCC-Z structures tested exhibit strain-rate sensitivity. The data for the BCC and BCC-Z geometries suggest that the plateau stress of this lattice increases steadily with strain rate, with the values at strain rates in the region of 300 s^{-1}, being between 20 and 35% higher than the quasi-static value. Lattice A shows a high degree of scatter in the recorded plateau stress, which suggests that the specimens were of varying quality.

Similar dynamic tests, performed by McKown et al. [13] on BCC and BCC-Z stainless steel lattice structures with a unit cell size of 2.5 mm, also exhibited similar strain-rate sensitivity. Tests were performed at strain rates between 8×10^{-3} and 150 s^{-1} and the yield stress (taken at a 5% strain offset) was observed to increase by 25% over the range of test conditions considered. Previous work on plain 316L stainless steel specimens under tensile loading [15] has shown that the material is moderately strain-rate sensitive.

7.5.2 Blast Tests on the Lattice Structures

A range of blast tests were conducted on lattice structures with various relative densities and unit-cell topologies and the results are presented in Table 7.5. The mass of PE4 given in Table 7.5 refers to the mass of the explosive disc applied to the striker, plus the mass

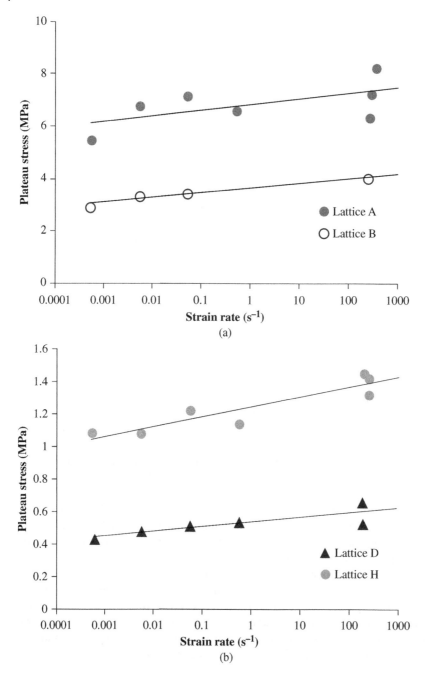

Figure 7.11 Variation of plateau stress with increasing strain rate for (a) lattice structures A and B and (b) lattice structures D and H.

Table 7.5 Summary of the blast conditions adopted during the lattice tests.

Unit cell type	Lattice ID	Mass of explosive (g)	Impulse (Ns)	Initial striker velocity, $\overline{V}_0$ (ms^{-1})	Average nominal strain rate (s^{-1})	Percentage crush (%)
BCC	A1	3 + 1	4.6	23.4	1252.3	49.9
	A2	5 + 1	5.2	26.7	1427.1	59.5
	A3	6 + 1	6.2	32.0	1708.9	64.3
	A4	2 + 1	3.1	15.7	839.2	37.4
BCC	B1	3 + 1	4.3	21.4	1099.4	70.3
	B2	1 + 1	1.6	8.0	409.6	40.5
	B3	2 + 1	3.1	15.8	796.1	69.6
	B4	1.5 + 1	2.9	15.0	760.5	65.7
	B5	1 + 0.5	1.7	8.9	428.4	50.6
BCC	C1	0 + 1	0.8	4.1	211.0	16.2
	C2	1 + 0.5	1.7	8.9	452.8	74.3
	C3	1 + 1	2.3	11.8	597.5	78.2
BCC-Z	E1	3 + 1	4.1	20.6	1102.1	16.6
	E2	5 + 1	5.3	27.0	1441.8	25.9
	E3	6 + 1	5.9	30.3	1622.2	33.3
	E4	4 + 1	5.1	26.5	1419.4	26.7
BCC-Z	F1	3 + 1	4.2	21.2	1079.6	47.9
	F2	2 + 1	3.3	16.6	840.5	33.3
	F3	5 + 1	5.3	26.9	1365.4	64.8
	F4	1 + 1	2.2	11.0	557.2	13.4
	F5	3 + 1	4.2	21.4	1087.0	47.7
	F6	5 + 1	5.2	26.2	1344.9	65.0
	F7	6 + 1	5.8	29.6	1504.1	71.7
BCC-Z	G1	3 + 1	4.1	21.1	1076.0	83.2
	G2	1 + 1	1.7	8.8	448.7	41.3
	G3	2 + 1	3.3	16.8	857.1	76.7
	G4	1.5 + 1	2.5	12.8	655.6	66.7
	G5	1 + 0.5	1.9	9.9	509.8	42.0
BCC-Z	H1	1.5 + 1	2.9	15.3	763.6	85.8

of the detonator. The number of tests on low-density (lattices C, D and H) structures was limited, due to the minimum quantity of PE4 explosive that could be detonated.

The initial velocity of the striker was calculated by dividing the measured applied impulse by the mass of the striker used in the test. The average nominal strain rates given in Table 7.5 were calculated using the initial velocity of the striker and the height of the lattice.

Images of typical blast-loaded lattices are shown on the left in Figure 7.12. The images show the uniformity of crush in the samples. The percentage crush was determined from the average of the four corner heights of the deformed lattice relative to the original height of the lattice.

The percentage crush against the applied impulse of the structures is presented in Figure 7.13, and the relationship is shown to be linear up to the point of densification, where the curve reaches a plateau, since the lattices cannot deform easily. The graphs also suggest that there is a threshold impulse for the onset of crushing to occur.

The strain rate associated with each blast test ranged between 210 and 1710 s^{-1}. These values exceed those observed in the drop-weight impact tests, which were in the region of 178–357 s^{-1}. Increasing the strain rate was shown, in dynamic compression tests, to result in an increase in the yield strength of the 316L steel in the lattice structures. An increase of 30% in the plateau stress of the lattice structures was observed in the dynamic tests [60]. Extrapolating the dynamic tests to the strain rates predicted in the blast tests suggests an increase in the yield strength of approximately 35%.

7.5.3 Blast Tests on the Sandwich Panels

Blast tests were carried out on sandwich panels based on a BCC lattice core and CFRP skins. Photographs of the sandwich panels after blast testing are shown in Figure 7.14 and the residual thicknesses of the panels are given in Table 7.6. Due to the presence of delamination and partial delamination in specimens SP1 and SP2 respectively, the residual thickness could not be accurately measured. The post-blast thicknesses of specimens SP1 and SP2 were estimated to be 17 mm, representing a level of 85% crush. This value corresponds to the point at which the lattice structure reaches full densification under compression testing. The front plate of the sandwich panel rebounds forwards when the core reaches full densification and detaches from the core at the skin–core interface [12].

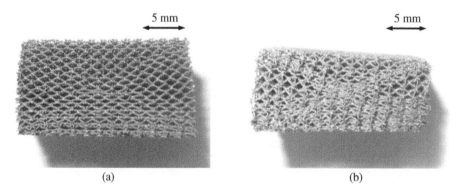

Figure 7.12 Images of blast loaded specimen: (a) lattice B2 (BCC); (b) lattice F1 (BCC-Z).

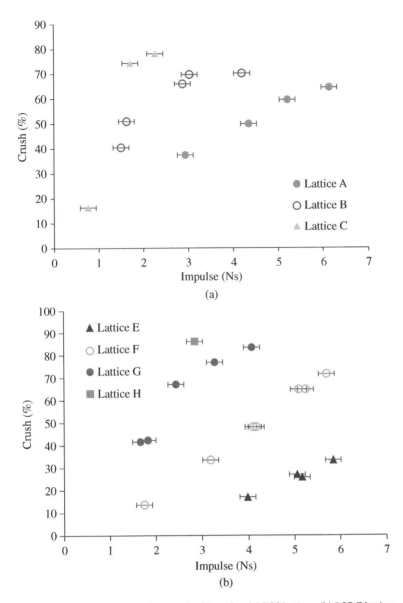

Figure 7.13 Percentage crush vs applied impulse: (a) BCC lattices; (b) BCC-Z lattices.

Figure 7.15 shows a contour plot of panel thickness after the blast test on panel SP3. This plot shows that the permanent deformation of the panel is increased at the four corner regions. This could be due to the fact that the impulse created by the blast may not be uniform or that the panel edges are weaker due either to damage of the outermost unit cells or a lack of constraint compared to the unit cells in the center of the panel. This phenomenon is not yet fully understood and further research is needed to fully elucidate these effects.

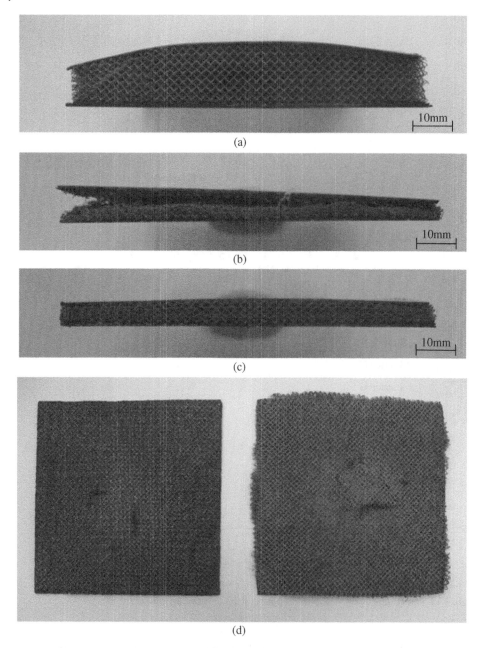

Figure 7.14 Sandwich panels based on the BCC unit cell: (a) specimen SP4 (impulse $= 4$ Ns); (b) specimen SP2 (impulse $= 7$ Ns); (c) specimen SP3 (impulse $= 5.2$ Ns); (d) specimen SP1 (impulse $= 14$ Ns).

Table 7.6 Summary of the blast conditions during the sandwich panel tests.

Sandwich panel ID	Panel dimensions (mm)	Mass of explosive (g)	Impulse (ns)	Crush level (%)	Comments
SP1	101.5 × 100 × 21.2	8+1	14.0	~85	Front plate detached
SP2	101.5 × 99.8 × 21.2	3+1	7.0	~85	Partial delamination of front plate
SP3	103 × 101.8 × 20.5	2+1	5.2	63.2	
SP4	100 × 99.8 × 20.9	1+1	4.0	13.9	

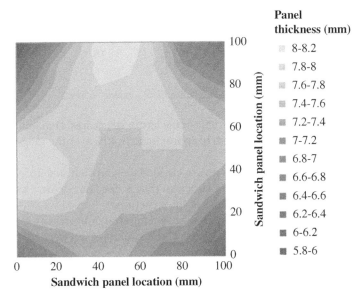

Figure 7.15 Contour plot showing the residual thickness of sandwich panel SP3 (impulse = 5.2 Ns).

Figure 7.16 shows how the average degree of crush in the sandwich panel varies with applied impulse. The graph shows that the sandwich panels exhibit similar trends to the lattice structures under blast loading; there is a threshold impulse to initiate crush, followed by a steep rise until the threshold for densification, at which point the curve reaches a plateau.

Concluding Remarks

This chapter has investigated the dynamic impact and blast response of lattice structures based on both BCC and BCC-Z unit cells, both as lattice cubes and as cores within sandwich panels. The study has shown that the behavior of the lattices is predictable and that the collapse mechanisms, for a particular geometry, remain unchanged with increasing strain rate.

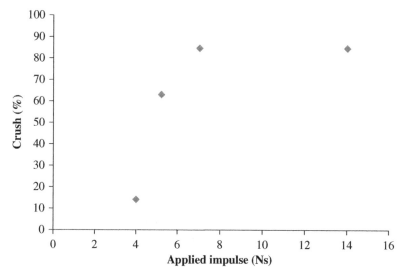

Figure 7.16 Percent crush versus applied impulse for the BCC lattice CFRP sandwich panels.

Strain-rate sensitivity is an important characteristic of the lattice structures. Dynamic testing has shown an increase in the plateau stress of approximately 30% over the range of strain rates investigated.

The addition of a face sheet to the lattices, to create a cladded or sandwich structure, has been explored. The face sheet constrains the lattice so that it does not deform as easily, thereby improving the mechanical properties of the structure.

Improvements in the SLM manufacturing technique, producing straighter struts with fewer imperfections, could also lead to an enhancement of the structures' properties.

Acknowledgements

The authors are grateful to Professors Genevieve Langdon and Gerald Nurick from BISRU, Cape Town for their help in conducting the blast tests on their ballistic pendulum.

References

1 M.D. Theobald, G.N. Nurick, Numerical investigation of the response of sandwich-type panels using thin-walled tubes subject to blast loads. *International Journal of Impact Engineering*, Vol. 34: pp. 134–156, (2007).

2 A.G. Hanssen, L. Enstock, and M. Langseth, Close-range blast loading of aluminium foam panels. *International Journal of Impact Engineering*, Vol. 27(6): pp. 593–318, (2002).

3 M.D. Theobald, G.S. Langdon, G.N. Nurick, et al, Large inelastic response of unbonded metallic foam and honeycomb core sandwich panels to blast loading. *Composite Structures*, Vol. 92: pp. 2465–2475, (2010).

4 F.W. Zok, H.J. Rathbun, Z. Wei, and A.G. Evans, Design of metallic textile core sandwich panels. *International Journal of Solids and Structures*, Vol. 40(21): pp. 5707–3722, (2003).

5 S. Chiras, D.R. Mumm, A.G. Evans, N. Wicks, J.W. Hutchinson, K. Dharmasena, H.N.G. Wadley, and S. Fichter, The structural performance of near-optimized truss core panels. *International Journal of Solids and Structures*, Vol. 39(15): pp. 4093–3115, (2002).

6 J. Wang, A.G. Evans, K. Dharmasena and H.N.G. Wadley, On the performance of truss panels with Kagomé cores. *International Journal of Solids and Structures*; Vol. 40(25): pp. 6981–6988, (2003).

7 V.S. Deshpande, N.A. Fleck, and M.F. Ashby, Effective properties of the octet-truss lattice material. *Journal of the Mechanics and Physics of Solids*, Vol. 49(8): pp. 1747–3769, (2001).

8 V.S. Deshpande, and N.A. Fleck, Collapse of truss core sandwich beams in 3-point bending. *International Journal of Solids and Structures*, Vol. 38(36–37): pp. 6275–3305, (2001).

9 V.S. Deshpande, M.F. Ashby, and N.A. Fleck, Foam topology: bending versus stretching dominated architectures. *Acta Materialia*, Vol. 49(6): pp. 1035–3040, (2001).

10 G.W. Kooistra, V.S. Deshpande, H.N.G. Wadley, Compressive behavior of age hard-enable tetrahedral lattice truss structures made from aluminium. *Acta Materialia* Vol. 52: pp. 4229–4237, (2004).

11 D.T. Queheillalt, Y. Murty and H.N.G. Wadley, Mechanical properties of an extruded pyramidal lattice truss sandwich structure. *Scripta Materialia*, Vol. 58: pp. 76–79, (2008).

12 M. Smith, W.J. Cantwell, Z. Guan, S. Tsopanos, M.D. Theobald, G.N. Nurick and G.S. Langdon, The quasi-static and blast response of steel lattice structures, *Journal of Sandwich Structures and Materials*, Vol. 13(4), pp. 479–301, (2010).

13 S. McKown, Y. Shen, W.K. Brooks, C.J. Sutcliffe, W.J. Cantwell, G.S. Langdon, G.N. Nurick, and M.D. Theobald, The quasi-static and blast loading response of lattice structures. *International Journal of Impact Engineering*, Vol. 35(8): pp. 795–310, (2008).

14 O. Rehme, *Cellular Design for Laser Freeform Fabrication*, Cuvillier Verlag, 2009.

15 R. Gümrük, R.A.W. Mines, S. Karadeniz, Static mechanical behaviours of stainless steel micro lattice structures under different loading conditions, *Materials Science & Engineering*, Vol. A586: pp. 392–406, (2013)

16 R. Gümrük, R.A.W. Mines, Compressive behaviour of stainless steel micro-lattice structures. *International Journal of Mechanical Sciences*, Vol. 68: pp. 125–339, (2013).

17 L. St-Pierre, N.A. Fleck, V.S. Deshpande. The predicted compressive strength of a pyramidal lattice made from case hardened steel tubes. *International Journal of Solids and Structures*, Vol. 51: pp. 41–52, (2014).

18 Y. Shen, High performance sandwich structures based on novel metal cores. PhD thesis, University of Liverpool, (2009).

19 T. Mukai, H. Kanahashi, T. Miyoshi, Experimental study of energy absorption in closed-cell aluminium foam under dynamic loading, *Scripta Metallurgica*. Vol. 40, p. 921, (1999).

20 C.J. Yungwirth, H.N.G. Wadley, J.H. O'Connor, A.J. Zakraysek and V.S. Deshpande, Impact response of sandwich plates with a pyramidal lattice core. *International Journal of Impact Engineering*, Vol. 35: pp. 920–936, (2008).

21 H.N.G. Wadley, K.P. Dharmasena, M.R. O'Masta, J.J. Wetzel., Impact response of aluminum corrugated core sandwich panels. *International Journal of Impact Engineering*, Vol. 62: pp. 114–128, (2013).

22 A.G. Evans, J.W. Hutchinson and M.F. Ashby. Multifunctionality of cellular metal systems. *Progress in Materials Science*, Vol. 43: pp. 171–321, (1999).

23 K.P. Dharmasena, H.N.G. Wadley, Z. Xue, J.W. Hutchinson. Mechanical response of metallic honeycomb sandwich panel structures to high intensity dynamic loading. *International Journal of Impact Engineering* Vol. 35: pp. 1063–3074, (2008).

24 J.W. Hutchinson, Z. Xue. Metal sandwich plates optimized for pressure impulses. *International Journal of Mechanical Sciences*, Vol. 47: pp. 545–359, (2005).

25 S.M. Pingle, N.A. Fleck, V.S. Deshpande, H.N.G. Wadley. Collapse mechanism maps for a hollow pyramidal lattice. *Proceedings of The Royal Society A*, Vol. 47: pp. 985–3011, (2011).

26 V.S. Deshpande, N.A. Fleck. One-dimensional response of sandwich plates to underwater shock loading. *Journal of the Mechanics and Physics of Solids*, Vol. 53: pp. 2347–3383, (2005).

27 N. Wicks, J.W. Hutchinson. Optimal truss plates. *International Journal of Solids and Structures*, Vol. 38: pp. 5165–3183, (2001).

28 Y. Liang, A.V. Spuskanyuk, S.E. Flores, D.R. Hayhurst, J.W. Hutchinson, R.M. McMeeking, et-al. The response of metallic sandwich panels to water blast. *Journal of Applied Mechanics*, Vol. 74: pp. 81–39, (2007).

29 J.W. Hutchinson. Energy and momentum transfer in air shocks. *Journal of Applied Mechanics*, Vol. 76: p. 051307-3, (2009).

30 J.J. Rimoli, B. Talamini, J.J. Wetzel, K.P. Dharmasena, R. Radovitzky, H.N.G. Wadley. Wet-sand impulse loading of metallic plates and corrugated core sandwich panels. *International Journal of Impact Engineering*, Vol. 38: pp. 837–348, (2011).

31 N. Kambouchev, L. Noels, R. Radovitzky. Fluide structure interaction effects in the loading of free-standing plates by uniform shocks. *Journal of Applied Mechanics*, Vol. 74: pp. 1042–3045, (2007).

32 G. Zhang, B. Wang, L. Ma, J. Xiong, L. Wu. Response of sandwich structures with pyramidal truss cores under the compression and impact loading. *Composite Structures*, Vol. 100: pp. 451–463, (2013).

33 S. Lee, F. Barthelat, N. Moldovan, H.D. Espinosa, H.N.G. Wadley, Deformation rate effects on failure modes of open-cell Al foams and textile cellular materials. *International Journal of Solids and Structures*, Vol. 43: pp. 53–73, (2006).

34 S. Lee, F. Barthelat, J.W. Hutchinson, H.D. Espinosa, Dynamic failure of metallic pyramidal truss core materials – Experiments and modeling, *International Journal of Plasticity*, Vol. 22: pp. 2118–2145, (2006).

35 D. Vaughn, M. Canning and J.W. Hutchinson, Coupled plastic wave propagation and column buckling. *Journal of Applied Mechanics*, 72, (1), pp. 1–8, (2005).

36 R.A.W. Mines, S. Tsopanos, Y. Shen, R. Hasan, S.T. McKown. Drop weight impact behaviour of sandwich panels with metallic micro lattice cores. *International Journal of Impact Engineering*, Vol. 60: 120–332, (2013).

37 D.D. Radford, V.S. Deshpande, N.A. Fleck, The use of metal foam projectiles to simulate shock loading on a structure, *International Journal of Impact Engineering*, Vol. 31, pp. 1152–1171, (2005).

38 G.J. McShane, D.D. Radford, V.S. Deshpande, N.A. Fleck, The response of clamped sandwich plates with lattice cores subjected to shock loading, *European Journal of Mechanics A/Solids*, Vol. 25: pp. 215–229, (2006).

39 G.J. McShane*, V.S. Deshpande, N.A. Fleck, Underwater blast response of free-standing sandwich plates with metallic lattice cores. *International Journal of Impact Engineering*, Vol. 37: pp. 1138–3149, (2010).

40 L.F. Mori, S. Lee, Z.Y. Xue, A. Vaziri, D.T. Queheillalt, K.P. Dharmasena, et al. Deformation and fracture modes of sandwich structures subjected to underwater impulsive loads. *Journal of the Mechanics of Materials and Structures*, Vol. 2: pp. 1981–3006, (2007).

41 L.F. Mori, D.T. Queheillalt, H.N.G. Wadley, H.D. Espinosa. Deformation and failure modes of I-core sandwich structures subjected to underwater impulsive loads. *Experimental Mechanics*, Vol. 49: pp. 257–375, (2009).

42 K.P. Dharmasena, H.N.G. Wadley, K. Williams, Z. Xue, J.W. Hutchinson. Response of metallic pyramidal lattice core sandwich panels to high intensity impulsive loading in air. *International Journal of Impact Engineering*, Vol. 38: pp. 275–389, (2011).

43 A.G. Hanssen, L. Enstock, and M. Langseth, Close-range blast loading of aluminium foam panels. *International Journal of Impact Engineering*, Vol. 27(6): pp. 593–318. (2002).

44 G.S. Langdon, D. Karagiozova, M.D. Theobald, G.N. Nurick, G. Lu, and R.P. Merrett, Fracture of aluminium foam core sacrificial cladding subjected to air-blast loading. *International Journal of Impact Engineering*, Vol. 37(6): pp. 638–351, (2010).

45 M.F. Ashby, A.G. Evans, N.A. Fleck, L.J. Gibson, J.W. Hutchinson and H.N.G Wadley. *Metal Foams: a Design Guide*. Butterworth-Heinemann, 2000.

46 D. Karagiozova, G.S. Langdon, and G.N. Nurick, Blast attenuation in Cymat foam core sacrificial claddings. *International Journal of Mechanical Sciences*, Vol. 52(5): pp. 758–376, (2010).

47 D. Karagiozova, G.N. Nurick, G.S. Langdon, S.C.K. Yuen, Y. Chi, and S. Bartle, Response of flexible sandwich-type panels to blast loading. *Composites Science and Technology*, Vol. 69(6): pp. 754–363, (2009).

48 M.D. Theobald, G.S. Langdon, G.N. Nurick, et al. Large inelastic response of unbonded metallic foam and honeycomb core sandwich panels to blast loading. *Composite Structures*, Vol. 92: pp. 2465–2475, (2010).

49 Y. Chi, G.S. Langdon, G.N. Nurick. The influence of core height and face plate thickness on the response of honeycomb sandwich panels subjected to blast loading. *Materials and Design*, Vol. 31: pp. 1887–1899, (2010).

50 B. Hou, A. Ono, S. Abdennadher, S. Pattofatto, Y.L. Li, H. Zhao, Impact behavior of honeycombs under combined shear-compression. Part I: Experiments. *International Journal of Solids and Structures*, Vol. 48: pp. 687–697, (2011).

51 B. Hou, S. Pattofatto, Y.L. Li, et al. Impact behavior of honeycombs under combined shear-compression. Part II: Analysis. *International Journal of Solids and Structures*, Vol. 48: pp. 698–705, (2011).

52 X. Li, P. Zhang, Z. Wang, G. Wu, L. Zhao. Dynamic behavior of aluminum honeycomb sandwich panels under air blast: Experiment and numerical analysis. *Composite Structures*, Vol. 108: pp. 1001–1008, (2014).

53 F. Zhu, G. Lu, D. Ruan, Z. Wang, Plastic deformation, failure and energy absorption of sandwich structures with metallic cellular cores, *International Journal of Protective Structures*, Vol. 1(4), pp. 507–341, (2010).

54 J. Tomblin, T. Lacy, B. Smith, S. Hooper, A. Vizzini and S. Lee, Review of damage tolerance for composite sandwich airframe structures. Federal Aviation Administration report No. DOT/FAA/AR-99/49, (1999).

55 R.A.W. Mines, S. McKown, S. Tsopanos, E. Shen, W. Cantwell, W. Brooks, C. Sutcliffe, Local effects during indentation of fully supported sandwich panels with micro lattice cores, *Applied Mechanics and Materials*, Vol. 13–34: pp. 85–30, (2008).

56 R. Hasan, R. Mines, E. Shen, S. Tsopanos, W. Cantwell, W. Brooks and C. Sutcliffe, Comparison of the drop weight impact performance of sandwich panels with aluminium honeycomb and titanium alloy micro lattice cores, *Applied Mechanics and Materials*, Vol. 24–35, pp. 413–318, (2010).

57 Renishaw website http://www.renishaw.com/mtt-group/us/selective-laser-melting .html, accessed 2011.

58 S. Tsopanos, Micro heat exchangers by selective laser melting, PhD thesis, University of Liverpool, 2009.

59 EP121-C15–33 data sheet, Gurit (Zullwill) AG. Switzerland.

60 M. Smith, The compressive response of novel lattice structures subjected to static and dynamic loading. PhD thesis, University of Liverpool, 2012.

8

Pentamode Lattice Structures

Andrew N. Norris

Mechanical and Aerospace Engineering, Rutgers University, Piscataway, New Jersey, USA

8.1   Introduction

Pentamode materials (PMs) were defined and introduced by Milton and Cherkaev in 1995 [1] as limiting cases of elastic solids, with five easy modes of deformation, hence "penta"; see also Milton's *Theory of Composites* [2], page 666. An inviscid compressible fluid, such as water, serves as a useful reference material for PMs since it has a single bulk modulus but zero shear rigidity. PMs can therefore be thought of as generalizations of liquids but without the ability to flow. However, unlike liquids, for which the stress is isotropic, PMs can display anisotropy.

Interest in PMs has increased after the observation that they have the potential to realize transformation acoustics (TA) [3]. Like its counterpart, transformation electro-magnetics, TA describes how wave equations are altered under spatial transformations mapping one region to another; cloaking maps a large volume to a small one in order to reduce the scattering cross-section, but other transformations are of interest. Unlike the more complex equations of electromagnetics, for which the transformed equations are unique, the central question in TA is simply how the Laplacian operator changes under a mapping from one coordinate system to another. The transformed acoustic wave equation turns out to have a nonunique [3] physical interpretation:

- as the equations for a fluid with anisotropic inertia (density) [4, 5]
- as the equation for a PM with anisotropic inertia [3].

The reason for the nonuniqueness can be associated with the ability to define the displacement field in the transformed domain, which introduces a degree of freedom characterized by a divergence-free, second-order symmetric tensor [3, 6]. This tensor gives rise to the variety of physically equivalent transformed equations of motion. Under a wide range of transformations, including cylindrical and spherical ones, the tensor can be chosen to make the inertia isotropic (that is, normal) fixing a unique PM. This was the author's original motivation for studying PMs.

Application of TA to sound in water using PMs requires the ability to make "anisotropic" water. If such a material is to be made it must possess a small amount of rigidity, otherwise it would be unstable under arbitrary loading. At the same time the material must be able to mimic water in its isotropic state. It is therefore desired to

Dynamics of Lattice Materials, First Edition. Edited by A. Srikantha Phani and Mahmoud I. Hussein.
© 2017 John Wiley & Sons Ltd. Published 2017 by John Wiley & Sons Ltd.

have a generic type of material that can meet the PM or near-PM requirements and that can also be designed to have the mass density dictated by TA. The material should also possess subwavelength microstructure so that it behaves as the PM, with the desired stiffness and density in a quasi-static effective sense. In the continuum limit, the dynamics of PMs are relatively simple because of the single wave property. This leads to a scalar wave equation in the homogenized limit that has acoustic-like solutions (see Eq. (8.15)). Therefore, in the long-wavelength limit PM lattices have simple scalar wave dynamics, whether isotropic or not. This is one of the key properties of PMs.

PMs can be realized from specific microstructures with tetrahedral-like unit cells [1]. Figure 8.1 shows the basic unit cell behind the periodic lattice structure, which is considered in detail below. The crucial feature of such structures is the ability to support a single type of stress and strain, which can be achieved by confining the force transmission to small contact zones connecting axial force members. Quasi-static elastic moduli can be estimated using standard homogenization procedures [7] or from the low-frequency Bloch–Floquet phase speeds [8–10]. A wide range in PM anisotropy and density is obtainable by varying the microstructural design of the unit cell [7, 8]. As an example, simulations of waves in a slab of PM defined by TA with hexagonal unit-cell microstructure [11] exhibit a two-dimensional (2D) acoustical illusion, as described elsewhere by the author [6]. Numerical calculations based on fabricated three-dimensional (3D) microstructures with small contact zones suggested that bulk-to-shear modulus ratios of the order of 10^3 should be realistically achievable [12, 13]. Values in this range have subsequently been reported from experimental measurements of 3D printed polymer-based samples [14]. By combining small contact zones with large inertial masses, a wide range of elastic properties and densities are achievable [15]. The phonon band structures of 3D PMs show, among other effects, that passbands exist in which only the single quasi-compressional wave exists, indicative of a one-wave material [16]. 3D anisotropic PM simulations [10] demonstrate the possibility of quasi-longitudinal waves along different axes with wave speeds varying by a factor of ten. The wave propagation properties of microlattice structures similar to PM structures have been explored using numerical methods [17], the authors noting that the elastic moduli and the dispersion properties can be independently modified with mass-like elements. Further numerical studies of PM properties based on designs inspired by Bravais lattices have been published by Méjica and Lantada [9]. A PM-based "unfeelability" cloak, which removes and hides imposed elastic strain fields was recently proposed

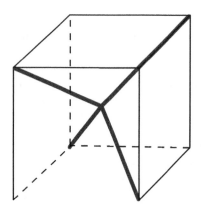

Figure 8.1 The fundamental cell for a diamond-like lattice of thin beams. The effective elasticity is cubic, with principal axes shown by the dashed lines. The effective properties approximate pentamodal when the bending compliance of the rod is much greater than the axial compliance, which is achieved by long slender beams and/or junctions with large bending compliance.

[18]. The opposite extreme of pentamode behavior, in which Poisson's ratio $v \approx -1$ and the material displays a unimode response, is achievable with "dilational" elastic materials [19]. A review of recent work on acoustic and electromagnetic metamaterials, including developments in PMs, can be found the paper by Kadic et al. [20].

In order for a PM lattice structure to achieve the approximate wave properties of water yet retain some rigidity, Figure 8.1 suggests that one needs to use materials with far larger inherent stiffness. Metals provide the necessary reservoir of both stiffness and mass. Thus an aluminum honeycomb 2D PM structure, called metal water [22], approximates the acoustical properties of water by providing the correct density and bulk modulus in the quasi-static limit, with small shear (see Figure 8.2). At higher frequencies, metal water also displays Floquet branches with negative group velocity, which bring the possibility of negative refractive indices [21, 23]. A 3D pentamode structure requires a regular diamond lattice of slender beams. Figure 8.3 shows the dispersion curves for a lattice of thin steel rods designed to make the PM have the wave speed of water. Note the relatively slow shear waves corresponding to small but finite rigidity, and the existence of a passband in which only the pseudo-acoustic wave propagates.

The focus of this chapter is on theoretical modeling of the quasi-static properties of PM lattice structures. Consider, for instance, the regular diamond lattice of slender beams in Figure 8.1. Assuming that the beams are circularly symmetric in cross-section and occupy volume fraction ϕ, then classical axial and flexural beam theory yields

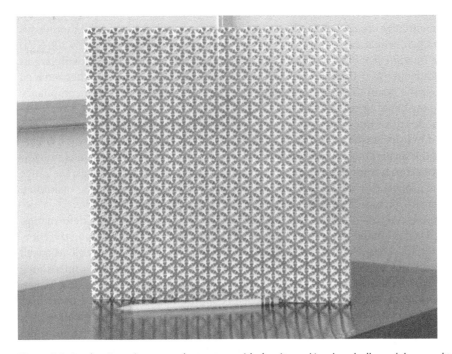

Figure 8.2 An aluminum honeycomb structure with density and in-plane bulk modulus equal to that of water. The thin struts have large bending compliance, which minimizes the shear modulus ($< 5\%$) relative to the bulk modulus. The "islands" of metal provide matched density, with little extra stiffness. Phononic properties of a similar metal water structures are given in Hladky–Hennion et al. [21]. Photograph courtesy of J. Cipolla.

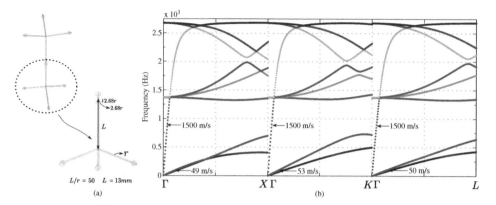

Figure 8.3 Diamond-like lattice of thin steel rods designed to have PM wave speed equal to that in water, 1500 m/s: (a) unit cell; (b) dispersion curves, from a simulation by A.J. Nagy.

effective moduli of cubic elasticity, with bulk modulus K $(= (C_{11} + 2C_{12})/3)$ and shear moduli μ_1 $(= C_{44})$, μ_2 $(= (C_{11} - C_{12})/2)$ given by (see Eq. (8.29) and [24]):

$$K = \frac{\phi}{9}E, \quad \mu_1 = \frac{9M}{2N + 4M}K, \quad \mu_2 = \frac{3M}{2N}K \tag{8.1}$$

where M and N are the axial and bending compliances, with $M/N = 4\phi/(\pi\sqrt{3})$ for rods of uniform circular cross-section. The ratio of the shear moduli to the bulk modulus is therefore $O(\phi)$, which can be made arbitrarily small. For instance, steel ($E = 200$ GPa) at volume fraction $\phi = 0.045$ yields an effective bulk modulus of $K = 1$ GPa with largest shear modulus $\mu_1/K = 0.149$. The shear-to-bulk modulus ratio can be further reduced by tapering the beam cross-section along its length to decrease the axial compliance relative to the bending compliance.

Pentamode lattice microstructures, such as the tetrahedral-like unit cell of Figure 8.1, are related to low-density materials, such as foams, in which the low density is a consequence of the small filling fraction of the solid phase. Christensen [25] provides a review of theoretical models for the mechanical properties of low-density materials; the range of material properties – including stiffness, strength and fracture toughness – exhibited by low-density microarchitectured materials is reviewed by Fleck et al. [26].

The response of low-density lattice structures depends on whether the deformation under load is dominated by stretching or bending. This in turn depends upon the coordination number, Z, the number of nearest neighboring joints in the unit cell. Maxwell [27] derived the necessary although not sufficient condition for a d-dimensional ($d = 2, 3$) space frame of b struts and j pin joints to be just rigid: $b - 3 = (j - 3)d$. For an infinite periodic structure ($b \approx jZ/2$), Maxwell's condition becomes $Z = 2d$. Structures with $Z = 2d$, known as isostatic lattices, are at the threshold of mechanical stability [28]. A closer examination of the issue, taking into account the degrees of freedom in the applied strain field, $d(d + 1)/2$, leads to the conclusion that the necessary and sufficient condition for rigidity of frameworks with coordination number Z is $Z \geq d(d + 1)$ [29]. The octet truss lattice ($Z = 12$) is an example of a 3D lattice that satisfies the rigidity condition [30]. 3D frameworks with $Z < 12$ admit soft modes. Zero-frequency modes, "floppy" modes, which occur for $Z < 2d$, correspond to collapse mechanisms, a topic that has also been examined for truss-like 2D lattices [31]. We note that closed-cell foams are an interesting alternative to the beam-lattice model of PM microstructures; Spadoni et al. [32] used theoretical and numerical models to study a Kelvin (body-centered cubic) foam, a face-centered cubic foam, and a Weaire–Phelan structure, demonstrating that such foams can display highly anisotropic pentamodal behavior.

The fundamental properties of continuous PMs are reviewed in Section 8.2, starting with the universal PM stress–strain relation $C = K \, S \otimes S$; see Eq. (8.2). Applications to lattices are given in Section 8.3, which describes how the PM stress–strain relation can be determined using simple beam theory for lattices with coordination number $d + 1$, where d is the spatial dimension. The quasi-static elastic moduli for isotropic and anisotropic PMs for specific lattice microstructures follow from the explicit relation of Eq. (8.19), which is perhaps our main result. Details of the model are given in Section 8.4.

8.2 Pentamode Materials

8.2.1 General Properties

Let C_{ijkl} be the elements of the elastic stiffness C in an orthonormal basis in three dimensions, with the symmetries $C_{ijkl} = C_{jikl}$ and $C_{ijkl} = C_{klij}$. Elements of stress σ and strain ϵ are σ_{ij}, ϵ_{ij}, related by $\sigma = C\epsilon$ or, equivalently, $\sigma_{ij} = C_{ijkl}\epsilon_{kl}$. The at-most 21 elements can be expressed in terms of the Voigt notation via $C_{ijkl} \rightarrow C_{IJ} = C_{JI}, I, J \in \overline{16}$. The Voigt form for the elasticity, characterized by a 6×6 matrix $[C_{IJ}]$, emphasizes the fact that the elastic stiffness C defines a positive definite mapping of a six-vector (strain) to a six-vector (stress). Hence, the elasticity C is characterized by six positive eigenvalues, the Kelvin moduli [33]. *A PM is the special case of elasticity with five zero eigenvalues.* A PM familiar to all is the inviscid acoustic fluid, for which $C = K_0 \, I \otimes I \Leftrightarrow C_{ijkl} = K_0 \, \delta_{ij}\delta_{kl}$, where K_0 is the bulk modulus. This form of the elastic moduli corresponds to a rank one 6×6 matrix $[C_{IJ}]$. The single nonzero eigenvalue is $3K_0$ since $CI = 3K_0 I$. Pentamodes obviously include isotropic acoustic fluids, for which the only stress–strain eigenmode is a hydrostatic stress, or pure pressure, and the five easy modes are all pure shear.

Given that a PM is an elastic solid with a single Kelvin modulus, the elastic stiffness must be of the form

$$C = K \, S \otimes S, \quad K > 0, \quad S \in \text{Sym}. \tag{8.2}$$

Some insight is gained by considering Hooke's law in the form $\hat{\sigma} = \hat{C}\hat{\epsilon}$, where the 6×6 matrix of moduli and the six-vectors of stress and strain are

$$\hat{C} = \begin{pmatrix} C_{11} & C_{12} & C_{13} & 2^{\frac{1}{2}}C_{14} & 2^{\frac{1}{2}}C_{15} & 2^{\frac{1}{2}}C_{16} \\ & C_{22} & C_{23} & 2^{\frac{1}{2}}C_{24} & 2^{\frac{1}{2}}C_{25} & 2^{\frac{1}{2}}C_{26} \\ & & C_{33} & 2^{\frac{1}{2}}C_{34} & 2^{\frac{1}{2}}C_{35} & 2^{\frac{1}{2}}C_{36} \\ & & & 2C_{44} & 2C_{45} & 2C_{46} \\ S & Y & M & & 2C_{55} & 2C_{56} \\ & & & & & 2C_{66} \end{pmatrix}, \quad \hat{\sigma} = \begin{pmatrix} \sigma_{11} \\ \sigma_{22} \\ \sigma_{33} \\ \sqrt{2}\sigma_{23} \\ \sqrt{2}\sigma_{31} \\ \sqrt{2}\sigma_{12} \end{pmatrix}, \quad \hat{\epsilon} = \begin{pmatrix} \epsilon_{11} \\ \epsilon_{22} \\ \epsilon_{33} \\ \sqrt{2}\epsilon_{23} \\ \sqrt{2}\epsilon_{31} \\ \sqrt{2}\epsilon_{12} \end{pmatrix}$$

Since a PM is rank one [1] the single nonzero positive eigenvalue of $\hat{C}$ is:

$$3\widetilde{K} \equiv \text{tr } \hat{C} = C_{ijij} = C_{11} + C_{22} + C_{33} + 2(C_{44} + C_{55} + C_{66}) \tag{8.3}$$

Accordingly, PM moduli have various representations

$$\mathbf{C} = K\,\mathbf{S} \otimes \mathbf{S} \Leftrightarrow \widehat{C} = K\,\mathbf{ss}^t \Leftrightarrow C_{ijkl} = K\,S_{ij}S_{kl}, \quad \mathbf{S} = \begin{pmatrix} s_1 & \frac{s_6}{\sqrt{2}} & \frac{s_5}{\sqrt{2}} \\ \frac{s_6}{\sqrt{2}} & s_2 & \frac{s_4}{\sqrt{2}} \\ \frac{s_5}{\sqrt{2}} & \frac{s_4}{\sqrt{2}} & s_3 \end{pmatrix} \tag{8.4}$$

where the tensor $\mathbf{S}$ and six-vector $\mathbf{s}$ define the single nontrivial eigenvector satisfying $\mathbf{CS} = 3\widetilde{K}\mathbf{S}$ ($\widehat{C}\mathbf{s} = 3\widetilde{K}\mathbf{s}$), and K is related to $\widetilde{K}$ by $K = 3\widetilde{K}\,(\mathrm{tr}\,\mathbf{S}^2)^{-1}$. The products in Eq. (8.2) are the important physical quantities, not K and $\mathbf{S}$ individually. This provides some freedom in defining $K > 0$ and $\mathbf{S}$. However, there is a preferred form for the latter since it can be shown that one can choose a symmetric $\mathbf{S}(\mathbf{x})$ that satisfies the static equilibrium condition [3]

$$\mathrm{div}\,\mathbf{S} = 0 \tag{8.5}$$

This identity has very important implications for TA. We say that the PM is of *canonical form* when Eq. (8.5) applies. The decomposition of Eq. (8.2) is then unique up to a multiplicative constant.

The single type of stress occurring in the PM (Eq. (8.2)) is of the form

$$\sigma = -p\mathbf{S} \quad \text{with} \quad p = -K\,\mathbf{S} : \epsilon \tag{8.6}$$

where the scalar "pseudo-pressure" p is introduced to make the stress–strain relation similar to that of an acoustic fluid ($\mathbf{S} = \mathbf{I}$). For instance, the elastic strain energy of a PM is $W = \frac{1}{2}\sigma : \epsilon = \frac{1}{2}K^{-1}p^2$, similar in form to the energy for acoustics of fluids. Note that there is no relation analogous to Eq. (8.6)$_1$ for strain in terms of stress because $\mathbf{C}$ is singular, or equivalently, only the single "component" $S : \epsilon$ is relevant; in other words, energetic. All other strains are energy-free.

Under static load and in the absence of body-force, the p in Eq. (8.6) should be chosen to be constant. The relevant surface tractions supporting the body in equilibrium are $\mathbf{t} = \sigma\mathbf{n} = -p\mathbf{S}\mathbf{n}$. Figure 8.4 illustrates the tractions required to maintain a block of PM in static equilibrium. Note that the traction vectors act obliquely to the surface, implying that shear forces are necessary. Furthermore, the tractions are not of uniform magnitude. These properties are to be compared with a normal acoustic fluid, which can be maintained in static equilibrium by constant hydrostatic pressure. This has

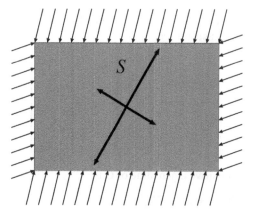

Figure 8.4 A rectangular block of PM is in static equilibrium under the action of surface tractions. The two orthogonal arrows inside the rectangle indicate the principal directions of $\mathbf{S}$ (30° from horizontal and vertical) and the relative magnitude of its eigenvalues (2:1). The equispaced arrows represent the surface loads to scale.

interesting implications for a beaker containing a PM with very low shear rigidity under gravity: the top free surface must be nonhorizontal, in contrast to the natural level surface of an inviscid fluid [6].

Writing $\mathbf{S}$ in terms of its principal directions and eigenvalues yields,

$$\mathbf{S} = \lambda_1 \mathbf{q}_1 \mathbf{q}_1 + \lambda_2 \mathbf{q}_2 \mathbf{q}_2 + \lambda_3 \mathbf{q}_3 \mathbf{q}_3 \tag{8.7}$$

where $\{\mathbf{q}_1, \mathbf{q}_2, \mathbf{q}_3\}$ is an orthonormal triad, it follows that the elastic moduli are

$$C_{IJ} = K \lambda_I \lambda_J \text{ if } I, J \in \{1, 2, 3\}; \ 0 \text{ otherwise} \tag{8.8}$$

The material symmetry displayed by PMs is therefore isotropic, transversely isotropic or orthotropic—the lowest symmetry—depending whether the triplet of eigenvalues $\{\lambda_1, \lambda_2, \lambda_3\}$ has one, two or three distinct members. The five "easy" pentamode strains correspond to the 5D space $\mathbf{S} : \epsilon = 0$ (equivalently $\mathbf{s}^t \hat{\sigma} = 0$). Three of the easy strains are pure shear: $\mathbf{q}_i \mathbf{q}_j + \mathbf{q}_j \mathbf{q}_i, i \neq j$ and the other two are $\lambda_1 \mathbf{q}_2 \mathbf{q}_2 - \lambda_2 \mathbf{q}_1 \mathbf{q}_1$ and $\lambda_2 \mathbf{q}_3 \mathbf{q}_3 - \lambda_3 \mathbf{q}_2 \mathbf{q}_2$. Any other zero-energy strain is a linear combination of these.

8.2.2 Small Rigidity and Poisson's Ratio of a PM

The stress in the PM is always proportional to the tensor $\mathbf{S}$ and only one strain element is significant: $\mathbf{S} : \epsilon$. The rank deficiency of the moduli means that there is no inverse strain–stress relation for the elements of ϵ in terms of the elements of σ. In practice, there must be some small but finite rigidity that makes $\mathbf{C}$ full rank. We can add rigidity to a PM in the following manner:

$$\mathbf{C} = K\mathbf{S} \otimes \mathbf{S} \rightarrow \mathbf{C}(K, \mu) \equiv K\mathbf{S} \otimes \mathbf{S} + 2\mu(\mathbf{I}^{(4)} - \mathbf{S} \otimes \mathbf{S}/\text{tr}\,(\mathbf{S}^2)) \tag{8.9}$$

where $\mathbf{I}^{(4)}$ is the fourth-order identity tensor, which has the property $\mathbf{I}^{(4)}\mathbf{A} = \mathbf{A}$ for all symmetric $\mathbf{A}$; that is, $I^{(4)}_{ijkl} = \frac{1}{2}(\delta_{ik}\delta_{jl} + \delta_{il}\delta_{jk})$. The form of Eq. (8.9) is chosen so that $\mathbf{CS}$ is unchanged by the addition of the extra elasticity tensor proportional to the generalized shear modulus $\mu \neq 0$. Tensor $\mathbf{C}(K, \mu)$ is positive definite in the sense of fourth-order elasticity tensors if $\mathbf{A} : \mathbf{DA} > 0$ for all nonzero symmetric $\mathbf{A}$; in other words, it is positive definite iff $K > 0$ and $\mu > 0$. In practice, a PM satisfies $\mu \ll K$. The inverse of the stiffness $\mathbf{C}$ is the compliance $\mathbf{M}$ satisfying $\mathbf{CM} = \mathbf{MC} = \mathbf{I}^{(4)}$, where, from Eq. (8.9),

$$\mathbf{M}(K, \mu) \equiv \frac{1}{K\tau^2}\mathbf{S} \otimes \mathbf{S} + \frac{1}{2\mu}\left(\mathbf{I}^{(4)} - \frac{1}{\tau}\mathbf{S} \otimes \mathbf{S}\right) \text{ with } \tau = \text{tr}\,(\mathbf{S}^2) \tag{8.10}$$

This becomes singular as the shear modulus tends to zero, indicating that PM are degenerate elastic materials with positive semi-definite strain energy.

In practice, the five soft modes of the PM are represented by $0 < \{\mu_i, i = 1, \dots, 5\} \ll K$, where the set of generalized shear moduli must be determined as part of the full elasticity tensor. An easily measured quantity that depends upon the five soft moduli is the Poisson's ratio: for a given pair of directions defined by the orthonormal vectors $\mathbf{n}$ and $\mathbf{m}$, the Poisson's ratio v_{nm} is the ratio of the contraction in the $\mathbf{m}$-direction to the extension in the $\mathbf{n}$-direction for a uniaxial applied stress along $\mathbf{n}$:

$$v_{nm} = -(\mathbf{mm} : \mathbf{Mnn})/(\mathbf{nn} : \mathbf{Mnn}) \tag{8.11}$$

As an example, consider the diamond-like structure of Figure 8.1 with Kelvin moduli given by Eq. (8.1). In the pentamode limit $K \gg \mu_1 = 3\mu_2$, with n_i, m_i as the components along the principal axes, we get (see, for example, [34])

$$v_{nm} = \frac{\frac{1}{2} - n_1^2 m_1^2 - n_2^2 m_2^2 - n_3^2 m_3^2}{n_1^4 + n_2^4 + n_3^4} \in \left[0, \frac{1}{2}\right]$$

(8.12)

The actual values of the soft moduli $\{\mu_i, i = 1, \ldots, 5\}$ are sensitive to features such as junction strength, and might not be easily calculated in comparison with the pentamode stiffness (see Section 8.4). An estimate of the Poisson effect can be obtained by assuming the five soft moduli to be equal, in which case Eq. (8.10) with $\mu/K \to 0$ gives

$$v_{nm} = \frac{(\mathbf{n} \cdot \mathbf{Sn})(\mathbf{m} \cdot \mathbf{Sm})}{\mathbf{S} : \mathbf{S} - (\mathbf{n} \cdot \mathbf{Sn})^2}$$

(8.13)

For the example of Figure 8.1, $\mathbf{S} = \mathbf{I}$ and Eq. (8.13) gives $v_{nm} = 1/2$. Generally, the values of v_{nm} from Eq. (8.13) associated with the principal axes of $\mathbf{S}$ (Eq. (8.7)) are $v_{ij} = \lambda_i \lambda_j / (\lambda_j^2 + \lambda_k^2)$, $i \neq j \neq k \neq i$. If $\lambda_1 > \lambda_2 > \lambda_3 > 0$ then the largest and smallest values are $v_{12} > \frac{1}{2}$ and $v_{32} < \frac{1}{2}$, respectively. Negative values of Poisson's ratio occur if the principal values of $\mathbf{S}$ are simultaneously positive and negative, as one might expect for re-entrant microstructures. A specific example is given in Section 8.3.2.

8.2.3 Wave Motion in a PM

The equation of motion for small-amplitude disturbances in particle velocity $\mathbf{w}$ in an elastic solid of density ρ is

$$\rho \dot{\mathbf{w}} = \mathrm{div}\sigma$$

(8.14)

Using the linear constitutive relation (Eq. $(8.6)_1$) and the fact that the PM is in canonical form, so that $\mathbf{S}$ satisfies the equilibrium condition (Eq. (8.5)), it follows that the pseudo-pressure satisfies the generalized acoustic wave equation [3]:

$$K\mathbf{S} : \nabla(\rho^{-1}\mathbf{S}\nabla p) - \ddot{p} = 0$$

(8.15)

This reduces to the acoustic equation when $\mathbf{S} = \mathbf{I}$. If the PM originates from TA, then the PM wave equation (Eq. (8.15)) is the transformed version of an acoustic equation (see the paper by the author [6], which also discusses the more general version of Eq. (8.15) that includes anisotropic inertia).

Consider plane-wave solutions for particle velocity of the form $\mathbf{w}(\mathbf{x}, t) = \mathbf{q}e^{i\kappa(\mathbf{n}\cdot\mathbf{x}-\upsilon t)}$, for $|\mathbf{n}| = 1$ and constant $\mathbf{q}$, κ and υ, and uniform PM properties. The wave equation (Eq. (8.15)) and the momentum balance (Eq. (8.14)) then imply, respectively,

$$\upsilon^2 = \rho^{-1}K\,|\mathbf{q}_L|^2, \quad \mathbf{q} = \alpha\mathbf{q}_L, \text{ where } \mathbf{q}_L \equiv \mathbf{Sn}$$

(8.16)

for $\alpha = \mathrm{const.} \neq 0$. The first equation for the nondispersive phase speed υ indicates that the slowness surface is an ellipsoid, and the second relation implies a quasi-longitudinal solution with polarization that is never orthogonal to $\mathbf{n}$ since $\mathbf{n} \cdot \mathbf{Sn} > 0$. Standard arguments for waves in anisotropic solids [35] show that the energy flux velocity is $\mathbf{c} = (\rho\upsilon)^{-1}K\,\mathbf{Sq}_L$ and satisfies $\mathbf{c} \cdot \mathbf{n} = \upsilon$, a well-known relation for generally anisotropic solids with isotropic density.

For the orthotropic PM let $c_1^2 = C_{11}/\rho$, $c_2^2 = C_{22}/\rho$, $c_3^2 = C_{33}/\rho$, then

$$\upsilon = (c_1^2 n_1^2 + c_2^2 n_2^2 + c_3^2 n_3^2)^{1/2}$$

(8.17a)

$$\mathbf{c} = v^{-1}(c_1^2 n_1 \mathbf{e}_1 + c_2^2 n_2 \mathbf{e}_2 + c_3^2 n_3 \mathbf{e}_3) \tag{8.17b}$$

$$\mathbf{q}_L = \rho^{-1/2}(c_1 n_1 \mathbf{e}_1 + c_2 n_2 \mathbf{e}_2 + c_3 n_3 \mathbf{e}_3) \tag{8.17c}$$

When a small amount of rigidity is included, as in the model of Eq. (8.9), it can be shown that the quasi-longitudinal waves are perturbed; one of the shear waves is purely transverse with speed $v_T = \sqrt{\mu/\rho}$ and polarization $\mathbf{q} \parallel \mathbf{q}_T \equiv \mathbf{n} \wedge \mathbf{q}_L$; the other has polarization approximately in the direction $\mathbf{q}_T \wedge \mathbf{q}_L$, with wave speed $v \approx (1 + |\mathbf{q}_T|^2/|\mathbf{q}_L|^2)^{1/2} v_T$.

8.3 Lattice Models for PM

PM properties can be realized from lattice microstructures that have just enough structural stiffness, but not too much; the key feature is that the coordination number (the number of nearest neighbors) is $d + 1$ [1], where $d = 2$ or 3 is the spatial dimension. Here we develop explicit results for the quasi-static effective properties of such lattice structures.

8.3.1 Effective PM Properties of 2D and 3D Lattices

We consider a unit cell of volume V in d-dimensions ($d = 2, 3$) of $d + 1$ rods. The edges of the cell are the midpoints of the rods, located at $\mathbf{a}_i$ and the single junction in the unit cell is at $\mathbf{p}$. The members interact in the static limit via combined axial forces directed along the members, and bending moments. These are associated with axial deformation and transverse flexure, respectively. We consider the limit in which the bending is small, and the deformation may be approximated by axial forces and their interaction at the junction. Physically, this corresponds to slender members with small thickness-to-length ratio.

Define $\mathbf{r}_i = r_i \mathbf{e}_i \equiv \mathbf{a}_i - \mathbf{p}$ ($i = 1, \ldots, d + 1$), where $|\mathbf{e}_i| = 1$. The elastic response of member i is characterized by its axial compliance

$$M_i = \int_0^{r_i} \frac{dx}{E_i A_i} \tag{8.18}$$

where $E_i(x)$, $A_i(x)$ are the Young's modulus and cross-sectional area, with $x = 0$ at the junction $\mathbf{p}$. We find that the effective elastic stiffness of the lattice has PM form (Eq. (8.2)); explicitly:

$$K = \left(V \sum_{k=1}^{d+1} \gamma_k \right)^{-1}, \quad \mathbf{S} = \sum_{i=1}^{d+1} \gamma_i \, \mathbf{P}_i \quad \text{where } \gamma_i = \frac{r_i^2}{M_i} - \frac{r_i}{M_i} \cdot \left(\sum_{k=1}^{d+1} \frac{\mathbf{P}_k}{M_k} \right)^{-1} \cdot \sum_{j=1}^{d+1} \frac{\mathbf{r}_j}{M_j} \tag{8.19}$$

where $\mathbf{P}_i = \mathbf{e}_i \otimes \mathbf{e}_i$. A detailed derivation is provided in Section 8.4.

Some comments are in order for the general result of Eq. (8.19). The derivation in Section 8.4 is based on the relation between the macroscopic stress and the forces $\mathbf{f}_i$ in the members of the unit cell of volume V:

$$\sigma = V^{-1} \sum_{i=1}^{d+1} \mathbf{r}_i \otimes \mathbf{f}_i \tag{8.20}$$

The values of the forces therefore follow from Eq. (8.19) and $\sigma = C\epsilon$ as

$$\mathbf{f}_i = (\Gamma : \epsilon)\frac{\gamma_i}{r_i}\mathbf{e}_i \text{ where } \Gamma = \frac{\sum_{j=1}^{d+1}\gamma_j\,\mathbf{e}_j\otimes\mathbf{e}_j}{\sum_{k=1}^{d+1}\gamma_k} \tag{8.21}$$

It may be checked from the definition of γ_i that the forces are equilibrated since

$$\sum_{i=1}^{d+1}r_i^{-1}\gamma_i\mathbf{e}_i = 0\left(\sum_{i=1}^{d+1}\mathbf{f}_i = 0\right) \tag{8.22}$$

This identity implies that $\gamma_i = 0$ for some member i only if (but not iff) the remaining d members are linearly dependent. When this unusual circumstance occurs the member i bears no load since $\mathbf{f}_i = 0$ for any applied strain. For instance, if two members are collinear in 2D, say members 1 and 2, then the third member is nonload bearing only if $r_1^{-1}\gamma_1 = r_2^{-1}\gamma_2$. When d of the members span a $(d-1)$-plane, the remaining member is nonload bearing if it is orthogonal to the plane.

8.3.2 Transversely Isotropic PM Lattice

The unit cell comprises two types of rods:

- $i = 1$ with r_1, M_1 in direction $\mathbf{e}$
- $i = 2, \ldots, d+1$ with r_2, M_2 in directions $\mathbf{e}_i$, symmetrically situated about $-\mathbf{e}$ with $-\mathbf{e}\cdot\mathbf{e}_i = \cos\theta$; see Figure 8.5.

After some simplification, Eqs. (8.2) and (8.19) give the PM elastic stiffness as

$$\mathbf{C} = ds^4r_2^2\frac{(\mathbf{I} + (\beta - 1)\mathbf{P})\otimes(\mathbf{I} + (\beta - 1)\mathbf{P})}{V(d-1)^2(dc^2M_1 + M_2)}, \quad \mathbf{P} = \mathbf{e}\otimes\mathbf{e}, \quad \begin{aligned}c &= \cos\theta\\ s &= \sin\theta\end{aligned} \tag{8.23}$$

where the nondimensional parameter β and the unit-cell volume V are

$$\beta = \frac{(d-1)c(r_1 + cr_2)}{s^2r_2}, \quad V = (sr_2)^{d-1}(r_1 + cr_2)\begin{cases}4, & d = 2\,(2D)\\ 6\sqrt{3}, & d = 3\,(3D)\end{cases} \tag{8.24}$$

Note that the elasticity of the rods enters only through the combination $dc^2M_1 + M_2$.

The nondimensional geometrical parameter β defines the anisotropy of the PM, with isotropy iff $\beta = 1$. If $\beta > 1$, the PM is stiffer along the axial or preferred direction $\mathbf{e}$ than in the orthogonal plane. Conversely, it is stiffer in the plane if $0 < \beta < 1$. The axial stiffness vanishes if $\beta = 0$, which is possible if $\theta = \frac{\pi}{2}$. The unit cell becomes re-entrant if $\theta > \frac{\pi}{2}$ $(c < 0)$. If $\theta > \frac{\pi}{2}$ then $\beta < 0$ and the principal values of $\mathbf{S}$ are simultaneously positive and negative, with the negative value associated with the axial direction. Note that $r_1 + cr_2$ must be positive since the unit-cell volume V is positive. As $r_1 + cr_2 \to 0$ the members criss-cross and the infinite lattice becomes stacked in a slab of unit thickness, hence the volume per cell tends to zero ($V \to 0$).

Let $\mathbf{e}$ be in the 1-direction. The transversely isotropic PM is then defined by the nonzero elements C_{11}, C_{22}, C_{12} in 2D, and additionally in 3D, $C_{33} = C_{22}, C_{23} = C_{22}, C_{13} = C_{12}$,

$$\begin{pmatrix}C_{11} & C_{12} & 0\\ C_{21} & C_{22} & 0\\ 0 & 0 & C_{66}\end{pmatrix} = K\begin{pmatrix}\beta & 1 & 0\\ 1 & \beta^{-1} & 0\\ 0 & 0 & 0\end{pmatrix} \tag{8.25}$$

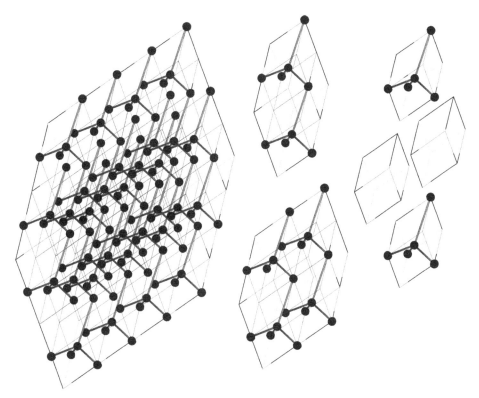

Figure 8.5 The diamond-like structure for the transversely isotropic PM lattice. The longer members are of length $2r_1$, the others of length $2r_2$.

where

$$K = \frac{d}{(d-1)} \frac{cs^2 r_2 (r_1 + cr_2)}{V(M_2 + dc^2 M_1)} \qquad (8.26)$$

The PM is isotropic if $\beta = 1$; that is, when the angle θ and r_1/r_2 are related by

$$\frac{r_1}{r_2} = \frac{1 - d\cos\theta^2}{(d-1)\cos\theta} \Leftrightarrow \text{isotropy } (\beta = 1) \qquad (8.27)$$

Hence, isotropy can be obtained if $\theta \in [\cos^{-1}\frac{1}{\sqrt{d}}, \frac{\pi}{2}]$ with the proper ratio of lengths; see Figure 8.6. At the limiting angles $r_1 \to 0$ ($r_2 \to 0$) as $\theta \to \cos^{-1}\frac{1}{\sqrt{d}}$ ($\theta \to \frac{\pi}{2}$). If the lengths are equal ($r_1 = r_2$) isotropy is obtained for $\cos\theta = \frac{1}{d}$; that is, $\theta = 60°, 70.53°$, for $d = 2, 3$, corresponding to hexagonal and tetrahedral unit cells, respectively.

The effective stiffness scales as the volume fraction of the solid material, $\mathbf{C} \propto \phi \equiv V_{\text{solid}}/V$. Consider the isotropic moduli of Eqs. (8.25) and (8.26) with uniform members of the same solid and with Young's modulus E. Then Eqs. (8.18), (8.25), and (8.26) imply

$$K = \phi E f, \quad f = s^4 \left[d - 1 + \frac{A_1}{A_2 dc}(1 - dc^2) \right]^{-1} \left[d - 1 + \frac{A_2}{A_1} dc(1 - dc^2) \right]^{-1} \qquad (8.28)$$

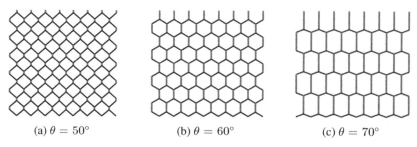

(a) $\theta = 50°$ (b) $\theta = 60°$ (c) $\theta = 70°$

Figure 8.6 Each of these 2D PM lattices has isotropic quasi-static properties. Vertical members are of length r_1; the other members are of length r_2. The ratio of r_1 to r_2 is determined by Eq. (8.27). The pure honeycomb structure is $\theta = 60°$.

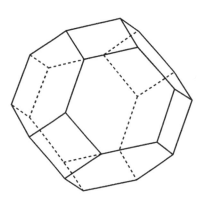

Figure 8.7 The tetrakaidecahedral unit cell [37] has low-density ($\phi \ll 1$) PM behavior similar to the diamond lattice: effective bulk modulus given by Eq. (8.29) and shear moduli of relative magnitude $\mu/K = O(\phi)$.

where A_1, A_2 are the strut thicknesses (cross-sectional areas). For a given θ and d, $f \le 1/d^2$ with equality iff $\frac{A_1}{A_2} = dc$. Hence the maximum possible isotropic effective bulk modulus for a given volume fraction ϕ is

$$K = \frac{\phi}{d^2}E \tag{8.29}$$

This agrees with the findings of Christensen [36], Equation (2.2) for $d = 2$, and with the bulk modulus for a regular lattice with tetrakaidecahedral unit cells [37, 38]; that is, an open Kelvin foam (see Figure 8.7). The latter structure, comprising joints with 4 struts and a unit cell of 14 faces (6 squares and 8 hexagons), has cubic symmetry, but the two shear moduli are almost equal so that the structure is almost isotropic. In fact, if the struts are circular and have Poisson's ratio equal to zero then the effective material is precisely isotropic with shear modulus $\mu = \frac{4\sqrt{2}}{9\pi}\phi^2 E$ [37].

Figures 8.8 and 8.9 show elastic moduli and Poisson's ratio for modified honeycomb and diamond lattices with uniform strut lengths but variable interior angles. The diamond lattice with uniform interior angle but variable strut length is considered in Figure 8.10.

Effective Moduli: 2D
While the results of Section 8.3.2 also apply to the 2D lattice, it is worth noting that in this case it is possible to include the effect of bending compliance without much

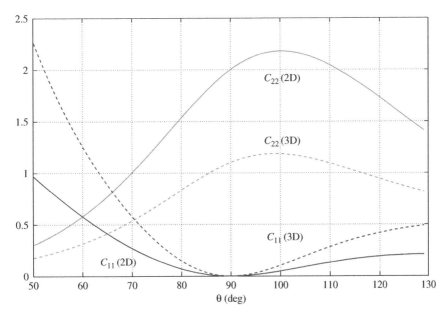

Figure 8.8 The elastic moduli for 2D and 3D PM lattices with rods of equal length ($r_1 = r_2$) and stiffness ($M_1 = M_2$) as a function of the junction angle θ. Note that the 2D (3D) moduli are identical at the isotropy angle 60° (70.53°). The axial stiffness C_{11} vanishes at $\theta = \frac{\pi}{2}$. Since $C_{12} = \sqrt{C_{11}C_{22}}$ it follows that C_{12} also vanishes at $\theta = \frac{\pi}{2}$.

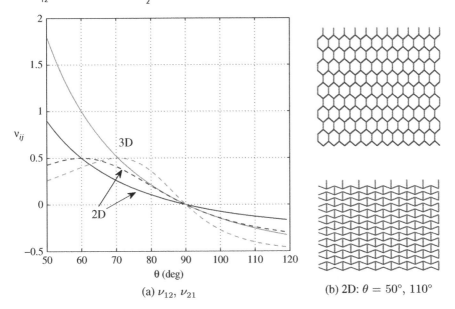

(a) ν_{12}, ν_{21}

(b) 2D: $\theta = 50°$, 110°

Figure 8.9 (a) The solid curves show ν_{12} for the type of lattice structures considered in Figure 8.6, with $r_1 = r_2$ and $M_1 = M_2$. The Poisson ratio ν_{12} describes the lateral contraction for loading along the axial e-direction. The related Poisson's ratio $\nu_{21} = \nu_{12}/(\frac{1}{2} + 2\nu_{12}^2)$ is shown by the dashed curves. (b) The 2D lattice for $\theta = 50°$ (top) and for $\theta = 110°$ (bottom).

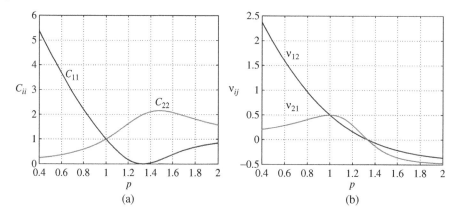

Figure 8.10 The principal stiffnesses (a) and Poisson's ratios (b) for a diamond lattice with the central "atom" shifted along the cube diagonal. The four vertices of the unit cell at (0 0 0), (0 2 2), (2 0 2), (2 2 0), and the center junction (atom) lies at $(p\ p\ p)$. Isotropy is $p = 1$.

complication. The new material parameters are the bending compliances of the individual members, N_1 and N_2 (see Eq. (8.41)), which for uniform beams are $N_j = r_j^3/(3E_jI_j)$, where I is the moment of inertia. Again taking $\mathbf{e}$ in the 1-direction, the full form of the nondegenerate 2D elastic stiffness is then (see Section 8.4.3 for details):

$$\left.\begin{matrix} C_{11} \\ C_{22} \\ C_{12} \end{matrix}\right\} = \frac{\frac{1}{2}cs}{(2c^2M_1 + M_2)N_2 + 2s^2M_1M_2} \times \begin{cases} \beta(N_2 + s^2c^{-2}M_2), \\ \frac{1}{\beta}(N_2 + s^{-2}(2M_1 + c^2M_2)) \\ (N_2 - M_2), \end{cases} \tag{8.30}$$

$$C_{66} = \frac{\frac{1}{2}sr_2(r_1 + cr_2)}{s^2(2r_2^2N_1 + r_1^2N_2) + (cr_1 + r_2)^2M_2}$$

These are in agreement with the in-plane moduli found by Zhu et al. [39]. Note that the moduli of Eq. (8.30) reduce to the PM moduli (Eq. (8.25)) as the bending compliance $N_2 \rightarrow \infty$, independent of the bending compliance N_1.

8.4   Quasi-static Pentamode Properties of a Lattice in 2D and 3D

The explicit formulas for the PM properties of 2D and 3D lattices presented in Section 8.3 are here derived in some detail using a model that includes bending effects and illustrates how the PM limit falls out as the bending compliance becomes infinite relative to the axial compliance.

8.4.1   General Formulation with Rigidity

Under the action of a static loading, the vertices of the unit cell initially located at $\mathbf{r}_i$ move to $\hat{\mathbf{r}}_i$, and the force acting at the vertex is $\mathbf{f}_i$. The forces are statically equilibrated.

The linear response produces forces assumed to be of the form

$$\mathbf{f}_i = M_i^{-1}(\hat{r}_i - r_i)\mathbf{e}_i + \sum_{j \neq i} N_{ij}^{-1} r_j (\psi_{ij} - \Psi_{ij})\hat{\mathbf{p}}_{ij} \tag{8.31}$$

where M_i is the axial compliance defined in Eq. (8.18) and $\mathbf{e}_i$ is the unit vector along $\hat{\mathbf{r}}_i$, $\mathbf{e}_i = \hat{\mathbf{r}}_i / \hat{r}_i$. The nodal compliances N_{ij} are associated with bending, and ψ_{ij}, $\Psi_{ij} \in (0, \pi)$ are the deformed and undeformed angles between members i and j. The unit vector $\hat{\mathbf{p}}_{ij}$ is perpendicular to $\mathbf{e}_i$ and lies in the plane spanned by $\mathbf{e}_i$ and $\mathbf{e}_j$ with $\hat{\mathbf{p}}_{ij} \cdot \mathbf{e}_j < 0$; that is, $\hat{\mathbf{p}}_{ij} = \csc \psi_{ij}(\cos \psi_{ij}\mathbf{e}_i - \mathbf{e}_j)$. The symmetry $N_{ij} = N_{ji}$ ensures that the moments of the shear forces are zero.

The *ansatz* Eq. (8.31) may be compared with the alternative approach for networks of beams [40] in which the total strain energy is expressed as a sum of three distinct contributions, representing:

(i) stretching of members
(ii) bending of members
(iii) the change of geometry of the nodes.

The present model ignores bending of members for several reasons. First, inclusion of flexural bending complicates the derivation considerably [24]. In particular it turns out that one needs to include extra degrees of freedom that are not required in the present model, which is more direct and leads to a semi-explicit form for the effective moduli. Also, since we are interested primarily in the PM limit it is only necessary to construct a model that yields a nondegenerate effective stiffness that has the correct leading-order properties in the PM limit. Furthermore, it turns out that Eq. (8.31) actually includes member bending in 2D if one interprets the compliances N_{ij} in terms of contributions (ii) and (iii) of the energy. This aspect is discussed further in Section 8.4.3.

Strain is introduced through the so-called *affine* kinematic assumption that the effect of deformation is to cause the cell edges to displace in a linear manner proportional to the (local) deformation gradient $\mathbf{F}$. In the absence of deformation, $\mathbf{F} = \mathbf{I}$, the edges are at $\mathbf{r}_i$ and the undeformed vertex is at the origin. In the presence of deformation the edge points are translated to $\mathbf{Fr}_i$, and the junction moves from the origin to χ, so that the vector defining the edge relative to the vertex is [41]:

$$\hat{\mathbf{r}}_i = \mathbf{Fr}_i - \chi \tag{8.32}$$

In the linear approximation $\hat{\mathbf{r}}_i$ can equally well be taken along $\mathbf{r}_i$, the approximation for the deformation is $\mathbf{F} = \mathbf{I} + \epsilon + \omega$ with $\epsilon = \epsilon^T$ and $\omega = -\omega^T$, so that $\hat{r}_i - r_i = r_i \mathbf{e}_i \cdot \epsilon \mathbf{e}_j - \mathbf{e}_i \cdot \chi$. The angle between i and j is given by $\cos \psi_{ij} = \hat{\mathbf{r}}_i \cdot \hat{\mathbf{r}}_j / (\hat{r}_i \hat{r}_j)$. The change in angle $\psi_{ij} - \Psi_{ij}$ follows by expanding about the original state, which, combined with the expression for $\hat{r}_i - r_i$, gives in the linear approximation

$$\mathbf{f}_i = \frac{r_i}{M_i} \varepsilon_{ii}\mathbf{e}_i + \sum_{j \neq i} \frac{r_j}{N_{ij}}(\mathbf{e}_i \cdot \varepsilon\hat{\mathbf{p}}_{ij} + \mathbf{e}_j \cdot \varepsilon\hat{\mathbf{p}}_{ji})\hat{\mathbf{p}}_{ij}$$

$$- \left(\frac{1}{M_i}\mathbf{e}_i \otimes \mathbf{e}_i + \sum_{j \neq i} \frac{1}{N_{ij}}\hat{\mathbf{p}}_{ij} \otimes \left(\frac{r_j}{r_i}\hat{\mathbf{p}}_{ij} + \hat{\mathbf{p}}_{ji} \right) \right) \chi \quad \text{with} \quad \varepsilon_{ij} \equiv \mathbf{e}_i \cdot \epsilon\mathbf{e}_j \tag{8.33}$$

The equilibrium condition $\sum_{i=1}^{d+1} \mathbf{f}_i = 0$ then becomes an equation for the nodal displacement χ, which can be solved as

$$\chi = \mathbf{A}^{-1} \sum_{i=1}^{d+1} \left(\frac{r_i}{M_i} \varepsilon_{ii} \mathbf{e}_i + \sum_{j \neq i} \frac{r_j}{N_{ij}} (\mathbf{e}_i \cdot \epsilon \hat{\mathbf{p}}_{ij} + \mathbf{e}_j \cdot \epsilon \hat{\mathbf{p}}_{ji}) \hat{\mathbf{p}}_{ij} \right)$$

$$\text{where} \quad \mathbf{A} = \sum_{i=1}^{d+1} \left(\frac{1}{M_i} \mathbf{e}_i \otimes \mathbf{e}_i + \sum_{j \neq i} \frac{1}{N_{ij}} \hat{\mathbf{p}}_{ij} \otimes \left(\frac{r_j}{r_i} \hat{\mathbf{p}}_{ij} + \hat{\mathbf{p}}_{ji} \right) \right)$$

(8.34)

Treating the volume of the cell as a continuum with equilibrated stress σ, integrating $\mathrm{div}\,\sigma = 0$ over V and identifying the tractions as the point forces $\mathbf{f}_i$ acting on the cell boundary, implies the well-known connection Eq. (8.20). The latter identity, combined with Eqs. (8.33) and (8.34), provides the linear relation between the strain and the stress, from which one can read off the effective elastic moduli.

8.4.2 Pentamode Limit

The pentamode limit corresponds to forces $\mathbf{f}_i$ with no transverse (bending) components. Physically, this corresponds to infinite compliance, $1/N_{ij} = 0$, and may be achieved approximately by long slender members. By ignoring bending, the stress defined by Eqs. (8.20) and (8.33) reduces to

$$\sigma = \frac{1}{V} \sum_{i=1}^{d+1} \mathbf{e}_i \otimes \mathbf{e}_i \frac{r_i}{M_i} \left[r_i \varepsilon_{ii} - \mathbf{e}_i \cdot \left(\sum_{j=1}^{d+1} \frac{1}{M_j} \mathbf{e}_j \otimes \mathbf{e}_j \right)^{-1} \sum_{k=1}^{d+1} \frac{r_k}{M_k} \varepsilon_{kk} \mathbf{e}_k \right]$$

(8.35)

Hence, $\sigma = \mathbf{C}\epsilon$ with effective elastic stiffness tensor

$$\mathbf{C} = \frac{1}{V} \sum_{i=1}^{d+1} \left[r_i^2 M_i \mathbf{B}_i \otimes \mathbf{B}_i - \sum_{k=1}^{d+1} \mathbf{r}_i \cdot \left(\sum_{j=1}^{d+1} \mathbf{B}_j \right)^{-1} \cdot \mathbf{r}_k \, \mathbf{B}_i \otimes \mathbf{B}_k \right]$$

(8.36)

where $\mathbf{B}_i = M_i^{-1} \mathbf{e}_i \otimes \mathbf{e}_i$. Although this expression for $\mathbf{C}$ possesses the symmetries expected for an elasticity tensor it is not apparent that it is actually of PM form (Eq. (8.2)). In order to see this we rewrite Eq. (8.36) as

$$\mathbf{C} = \frac{1}{V} \sum_{i,j=1}^{d+1} r_i r_j \sqrt{M_i M_j} \, P_{ij} (\mathbf{v}_i \otimes \mathbf{v}_i) \otimes (\mathbf{v}_j \otimes \mathbf{v}_j) \quad \text{where}$$

$$P_{ij} = \delta_{ij} - \mathbf{v}_i \cdot \left(\sum_{k=1}^{d+1} \mathbf{v}_k \otimes \mathbf{v}_k \right)^{-1} \cdot \mathbf{v}_j, \qquad \mathbf{v}_i \equiv M_i^{-1/2} \mathbf{e}_i$$

(8.37)

The $(d+1) \times (d+1)$ symmetric matrix $\mathbf{P}$ with elements P_{ij} has the crucial properties

$$\mathbf{P}^2 = \mathbf{P}, \quad \mathrm{tr}\,\mathbf{P} = 1$$

(8.38)

In other words, $\mathbf{P}$ is a projector, and the dimension of its projection space is one. It follows that rank $\mathbf{P} = 1$ and hence the single nonzero eigenvalue of $\mathbf{P}$ is unity:

$$\mathbf{P} = \mathbf{b}\mathbf{b}^T \quad \text{where} \quad \mathbf{b}^T \mathbf{b} = 1$$

(8.39)

Replacing P_{ij} with $b_i b_j$ in Eq. (8.37)$_1$ gives

$$\mathbf{C} = \mathbf{S} \otimes \mathbf{S} \quad \text{where} \quad \mathbf{S} = V^{-1/2} \sum_{i=1}^{d+1} r_i M_i^{1/2} \, b_i \mathbf{B}_i \tag{8.40}$$

The eigenvalue property $\mathbf{Pb} = \mathbf{b}$ implies that $\mathbf{b}$ satisfies $\sum_{i=1}^{d+1} b_i \mathbf{v}_i = 0$; that is, it is closely related with the fact that the $d+1$ vectors $\mathbf{v}_i$ are necessarily linearly dependent. Alternatively, $\mathbf{b}$ follows by setting $\mathbf{C}$ of Eq. (8.36) into PM form, $\mathbf{C} = \mathbf{S} \otimes \mathbf{S}$, then using $\mathbf{CI} = \text{Str } \mathbf{S}$ and $\mathbf{I:CI} = (\text{tr } \mathbf{S})^2$, from which we deduce that $\mathbf{C}$ of Eq. (8.2) is given explicitly by Eq. (8.19).

8.4.3 Two-dimensional Results for Finite Rigidity

Although the lattice model of Section 8.4.1 does not include beam bending explicitly, it turns out that this deformation mechanism may be mapped onto the nodal bending defined by the second term in the force expression (Eq. (8.31)). This equivalence is restricted to two dimensions; there is no analogy available for 3D. Here we just summarize the mapping, but full details may be found in the literature [24].

Consider the three ($= d + 1$) members emanating from the central node of a unit cell. Each beam bends under a shear force acting on the member end equal to $V_j = Q_j/r_j$, such that all of the moments Q_j are equilibrated. The deflection at the beam end is $w_j = V_j N_j$ where the bending compliance is

$$N_j = \int_0^{r_j} \frac{x^2 dx}{E_j I_j}, \quad j = 1, 2, 3 \tag{8.41}$$

The connection with the node-bending term in Eq. (8.31) is made by noting that the deviations in angles between neighboring members, $\theta_{ij} = \psi_{ij} - \Psi_{ij}$, may be related to the end deflections w_i, w_j and the associated angular deviations w_i/r_i, w_j/r_j. Based on this observation it can be shown that N_{ij} may be expressed in terms of the individual member bending compliances N_i as

$$N_{ij} = \left(\frac{r_1^2}{N_1} + \frac{r_2^2}{N_2} + \frac{r_3^2}{N_3} \right) \frac{N_i N_j}{r_i r_j}, \quad i \neq j \in 1, 2, 3 \text{ (no sum)} \tag{8.42}$$

Using this relation the general form for the effective moduli, implicit in Eqs. (8.20), (8.33), and (8.34), may be made explicit in the 2D case.

8.5 Conclusion

The elastic stiffness of a PM must be of the form $\mathbf{C} = K \, \mathbf{S} \otimes \mathbf{S}$, which ensures that there are five "easy" modes of deformation in 3D elasticity. The stress in the PM is then $\sigma = -p\mathbf{S}$ where the "pseudo-pressure" p is given by Eq. (8.6). Lattice frames comprised of periodic repeating unit cells provide the most natural microstructure for realizing quasi-static PM behavior. Crucially, the coordination number must be the minimal possible: $d+1$ in $d = 2$ or 3 dimensions. Our main result, Eq. (8.19), gives explicit expressions for K and $\mathbf{S}$ for lattices of this type under general assumptions on the unit-cell geometry and material properties. The key quantities are cell volume and the length, orientation and axial stiffness of each of the $d + 1$ rods in the unit cell. It is possible to predict the small but nonzero stiffnesses for the five easy modes by including the bending stiffness of the rods, see Section 8.4.1. In fact, the formulation is applicable to the entire range of stiffness possible in similarly situated lattice frameworks, from PMs corresponding to coordination number 4 in 3D with **C** of rank one, to the "stiffest" structure proposed by Gurtner and Durand [40], with coordination number 14 and full rank **C** [24].

The model and analysis presented here provide a complete description of the quasi-static PM effective behavior. A more challenging task is to understand how the unit-cell properties influence the finite-frequency phononic properties such as passband and stopbands. In order to make progress one needs to include the wave mechanics of the rods in the unit cell and to impose Bloch–Floquet boundary conditions. It is possible to adapt the model described here to include these conditions, resulting in a dispersion equation that displays PM behavior in the long-wavelength limit. This work will be described in a separate publication.

Acknowledgements

Thanks to A.J. Nagy, J. Cipolla, N. Gokhale, and X. Su for discussions and graphical assistance. This work was supported under ONR MURI grant No. N000141310631.

References

1 Milton GW, Cherkaev AV. Which elasticity tensors are realizable? *Journal of Engineering Materials and Technology* 1995;**117**(4):483–493.

2 Milton GW. *The Theory of Composites*, 1st edn. Cambridge University Press; 2001.

3 Norris AN. Acoustic cloaking theory. *Proceedings of the Royal Society A.* 2008;**464**:2411–2434.

4 Chen H, Chan CT. Acoustic cloaking in three dimensions using acoustic metamaterials. *Applied Physics Letters.* 2007;**91**(18):183518.

5 Cummer SA, Schurig D. One path to acoustic cloaking. *New Journal of Physics.* 2007;**9**(3):45.

6 Norris AN. Acoustic metafluids. *Journal of the Acoustical Society of America* 2009;**125**(2):839–849.

7 Gokhale NH, Cipolla JL, Norris AN. Special transformations for pentamode acoustic cloaking. *Journal of the Acoustical Society of America* 2012;**132**(4):2932–2941.

8 Layman CN, Naify CJ, Martin TP, Calvo DC, Orris GJ. Highly-anisotropic elements for acoustic pentamode applications. *Physical Review Letters.* 2013;**111**:024302–024306.

9 Méjica GF, Lantada AD. Comparative study of potential pentamodal metamaterials inspired by Bravais lattices. *Smart Materials and Structures.* 2013;**22**(11):115013.

10 Kadic M, Bückmann T, Schittny R, Wegener M. On anisotropic versions of three-dimensional pentamode metamaterials. *New Journal of Physics.* 2013;**15**(2):023029.

11 Layman CN, Naify CJ, Martin TP, Calvo D, Orris G. Broadband transparent peri-odic acoustic structures. *Proceedings of Meetings on Acoustics.* 2013;**19**(1):065043.

12 Kadic M, Bückmann T, Stenger N, Thiel M, Wegener M. On the feasibility of penta-mode mechanical metamaterials. *Applied Physics Letters.* 2012;**100**:191901.

13 Kadic M, Bückmann T, Stenger N, Thiel M, Wegener M. Erratum: "On the prac-ticability of pentamode mechanical metamaterials" [Appl. Phys. Lett. 100, 191901 (2012)]. *Applied Physics Letters.* 2012;**101**:049902.

14 Schittny R, Bückmann T, Kadic M, Wegener M. Elastic measurements on macro-scopic three-dimensional pentamode metamaterials. *Applied Physics Letters.* 2013;**103**(23):231905.

15 Kadic M, Bückmann T, Schittny R, Gumbsch P, Wegener M. Pentamode metama-terials with independently tailored bulk modulus and mass density. *Physical Review Applied* 2014;**2**(5):054007.

16 Martin A, Kadic M, Schittny R, Bückmann T, Wegener M. Phonon band structures of three-dimensional pentamode metamaterials. *Phyical Review B.* 2012;**86**:155116.

17 Krödel S, Delpero T, Bergamini A, Ermanni P, Kochmann DM. 3D auxetic micro-lattices with independently controllable acoustic band gaps and quasi-static elastic moduli. *Advanced Engineering Materials.* 2014;**16**(4):357–363.

18 Bückmann T, Thiel M, Kadic M, Schittny R, Wegener M. An elasto-mechanical unfeelability cloak made of pentamode metamaterials. *Nature Communications* 2014;**5**:4130.

19 Bückmann T, Schittny R, Thiel M, Kadic M, Milton GW, Wegener M. On three-dimensional dilational elastic metamaterials. *New Journal of Physics.* 2014;**16**:033032.

20 Kadic M, Bückmann T, Schittny R, Wegener M. Metamaterials beyond electromag-netism. *Reports on Progress in Physics.* 2013;**76**(12):126501.

21 Hladky-Hennion AC, Vasseur JO, Haw G, Croënne C, Haumesser L, Norris AN. Negative refraction of acoustic waves using a foam-like metallic structure. *Applied Physics Letters.* 2013;**102**(14):144103.

22 Norris AN, Nagy AJ. Metal Water: a metamaterial for acoustic cloaking. Phononics–2011–0037 in: *Proceedings of Phononics 2011*, Santa Fe, NM, USA, May 29–June 2; 2011, pp. 112–113.

23 Norris AN, Nagy AJ, Cipolla JL, Gokhale NH, Hladky-Hennion AC, Croënne C, et al. Metallic structures for transformation acoustics and negative index phononic crystals. In: *Proc. 7th Int. Congress on Advanced Electromagnetic Materials in Microwaves and Optics Metamaterials 2013*; 2013.

24 Norris AN. Mechanics of elastic networks. *Proceedings of the Royal Society A.* 2014;**470**:20140522.

25 Christensen RM. Mechanics of cellular and other low-density materials. *International Journal of Solids and Structures* 2000;**37**(1–2):93–104.

26 Fleck NA, Deshpande VS, Ashby MF. Micro-architectured materials: past, present and future. *Proceedings of the Royal Society A.* 2010;**466**(2121):2495–2516.

27 Maxwell JC. On the calculation of the equilibrium and stiffness of frames. *Philo-sophical Magazine* 1864;**27**(182):294–299.

28 Sun K, Souslov A, Mao X, Lubensky TC. Surface phonons, elastic response, and conformal invariance in twisted kagome lattices. *Proceedings of the National Academy of Sciences.* 2012;**109**(31):12369–12374.

29 Deshpande VS, Fleck NA, Ashby MF. Effective properties of the octet-truss lattice material. *Journal of the Mechanics and Physics of Solids.* 2001;**49**(8):1747–1769.

30 Deshpande VS, Ashby MF, Fleck NA. Foam topology: bending versus stretching dominated architectures. *Acta Materialia.* 2001;**49**(6):1035–1040.

31 Hutchinson RG, Fleck NA. The structural performance of the periodic truss. *Journal of the Mechanics and Physics of Solids.* 2006;**54**(4):756–782.

32 Spadoni A, Hohler R, Cohen-Addad S, Dorodnitsyn V. Closed-cell crystalline foams: self-assembling, resonant metamaterials. *Journal of the Acoustical Society of America* 2014;**135**(4):1692–1699.

33 Thomson W. Elements of a mathematical theory of elasticity. *Philosophical Transactions of the Royal Society of London* 1856;**146**:481–498.

34 Norris AN. Poisson's ratio in cubic materials. *Proceedings of the Royal Society A.* 2006;**462**:3385–3405.

35 Musgrave MJP. *Crystal Acoustics.* Acoustical Society of America; 2003.

36 Christensen RM. The hierarchy of microstructures for low density materials. In: Casey J, Crochet M, editors. *Theoretical, Experimental, and Numerical Contributions to the Mechanics of Fluids and Solids.* Birkhäuser Basel; 1995. p. 506–521.

37 Warren WE, Kraynik AM. Linear elastic behavior of a low-density Kelvin foam with open cells. *ASME Journal of Applied Mechanics.* 1997;**64**(4):787–794.

38 Zhu HX, Knott JF, Mills NJ. Analysis of the elastic properties of open-cell foams with tetrakaidecahedral cells. *Journal of the Mechanics and Physics of Solids.* 1997;**45**(3):319–343.

39 Kim HS, Al-Hassani STS. Effective elastic constants of two-dimensional cellular materials with deep and thick cell walls. *International Journal of Mechanical Sciences* 2003;**45**(12):1999–2016.

40 Gurtner G, Durand M. Stiffest elastic networks. *Proceedings of the Royal Society A..* 2014;**470**(2164):20130611.

41 Wang Y, Cuitiño A. Three-dimensional nonlinear open-cell foams with large deformations. *Journal of the Mechanics and Physics of Solids.* 2000;**48**:961–988.

9

Modal Reduction of Lattice Material Models

Dimitri Krattiger and Mahmoud I. Hussein

Department of Aerospace Engineering Sciences, University of Colorado Boulder, Boulder, Colorado, USA

9.1 Introduction

A widely used tool in structural dynamics is modal expansion, whereby a structure's vibrational behavior is represented as a linear combination of characteristic vibrations or mode shapes. The benefits of modal expansion include deeper understanding of the structure's free and forced vibration behavior, de-coupling of the system's equations of motion, and the possibility of model reduction by truncation of the mode set. Similarly, the computation of the elastic band structure of periodic materials can benefit from modal expansion. This chapter discusses how modal analysis can be used to achieve large reductions in model size and computational time for the calculation of dispersion curves of lattice materials.

Whether we are interested in a material's elastic or phonon transport properties, its electronic properties, or its electromagnetic properties, the band structure can be an extremely useful tool because it provides a fundamental understanding of what gives rise to these properties [1]. It can, however, be very time consuming to compute. For this reason, much research effort has been directed towards improving the computational efficiency of band structure calculations. Some methods take advantage of the continuously varying nature of the eigenvectors across the Brillouin zone. As one steps through the wave-vectors (κ-points), the solution from the previous step is used as a starting point for the current step in order to speed up convergence [2]. Eigenvector convergence can also be improved by multigrid methods [3], whereby a coarse mesh is solved first and then the coarse solution is used as a starting point in solving a finer mesh. Another proposed method for eigenvector solution involves minimizing the Rayleigh product using a coordinate relaxation approach [4].

Modal-expansion methods have been used to improve the computational efficiency when solving Schrödinger's equation in a crystal lattice [5]. In this approach, the eigenvectors throughout the Brillouin zone are approximated as a linear combination of the eigenvectors at just a few points in the Brillouin zone, leading to largely reduced model sizes. This concept has been extended beyond electronic structure calculations to elastic and photonic systems as well [6]. It has also been shown that the methodology of component mode synthesis from structural dynamics can be extended to dispersion computation in periodic materials [7].

Dynamics of Lattice Materials, First Edition. Edited by A. Srikantha Phani and Mahmoud I. Hussein.
© 2017 John Wiley & Sons Ltd. Published 2017 by John Wiley & Sons Ltd.

In this chapter, two distinct methods rooted in modal analysis are discussed in detail and applied to a 2D lattice material. First, a modal-expansion method [6] is presented in the context of elastic band structure calculations. Then a mode-synthesis method [7] is described and demonstrated. Both methods achieve large model reductions and large savings in computational time with relatively low error in the band structure.

9.2 Plate Model

The modal-reduction methods presented in this chapter are demonstrated on a 2D periodic lattice material that can be treated as a plate. The plate is modeled using Mindlin–Reissner theory [8, 9] and a corresponding descretized representation is formed using the finite-element method. For completeness, a brief discussion of the model's creation is included here. More details on the development of Mindlin–Reissner plate elements may be found in finite-element textbooks [10–13].

9.2.1 Mindlin–Reissner Plate Finite Elements

Consider a plate whose normal is oriented in the z-direction, as shown in Figure 9.1. The governing equation of Mindlin–Reissner plate theory is

$$\left(\nabla^2 - \frac{\rho}{\mu\alpha}\frac{\partial^2}{\partial t^2}\right)\left(D\nabla^2 - \frac{\rho h^3}{12}\frac{\partial^2}{\partial t^2}\right)w + \rho h\frac{\partial^2 w}{\partial t^2} = 0 \tag{9.1}$$

where h is the plate thickness, μ is the shear modulus, α is the shear correction factor, $D = (Eh^3)/(12 - 12v^2)$ is the plate stiffness, E is the Young's modulus, v is the Poisson ratio, h is the plate thickness, and w is the transverse displacement of the plate. We will assume the following displacement field:

$$u = z\theta_x(x, y), \quad v = z\theta_y(x, y), \quad w = w(x, y) \tag{9.2}$$

where θ_x and θ_y represent rotations of the normal to the plate. We may then use a set of polynomial shape functions, N_i, to describe the full displacement field in terms of the nodal rotations and displacements θ_{xi}, θ_{yi}, and w_i:

$$\theta_x = \sum_{i=1}^{k} N_i(x, y)\theta_{xi}, \quad \theta_y = \sum_{i=1}^{k} N_i(x, y)\theta_{yi}, \quad w = \sum_{i=1}^{k} N_i(x, y)w_i \tag{9.3}$$

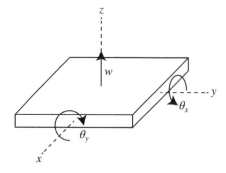

Figure 9.1 Plate schematic.

The element stiffness matrix is derived from Eq. (9.1) and is evaluated by integrating over the element domain, A^e [12]:

$$K^e = \underbrace{\frac{h^3}{12} \int_{A^e} B_B^{\mathrm{T}} D_B B_B dA^e}_{K_B^e} + \underbrace{\alpha h \int_{A^e} B_S^{\mathrm{T}} D_S B_S dA^e}_{K_S^e} \tag{9.4}$$

where we choose $\alpha = 5/6$. This shear correction factor is intended for use with simply supported boundary conditions, but it is also adequate for the present purpose. The first term in Eq. (9.4) is the bending stiffness, and the second term is the shear stiffness. The bending strain-displacement matrix is given by

$$B_B = \begin{bmatrix} \begin{bmatrix} 0 & N_{1,x} & 0 \\ 0 & 0 & N_{1,y} \\ 0 & N_{1,y} & N_{1,x} \end{bmatrix} \cdots \begin{bmatrix} 0 & N_{k,x} & 0 \\ 0 & 0 & N_{k,y} \\ 0 & N_{k,y} & N_{k,x} \end{bmatrix} \end{bmatrix} \tag{9.5}$$

and the shear strain-displacement matrix is given by

$$B_S = \begin{bmatrix} \begin{bmatrix} N_{1,x} & N_1 & 0 \\ N_{1,y} & 0 & N_1 \end{bmatrix} \cdots \begin{bmatrix} N_{k,x} & N_k & 0 \\ N_{k,y} & 0 & N_k \end{bmatrix} \end{bmatrix} \tag{9.6}$$

The constitutive matrices for bending and shear are, respectively,

$$D_B = \frac{E}{1-v^2} \begin{bmatrix} 1 & v & 0 \\ v & 1 & 0 \\ 0 & 0 & \frac{1-v}{2} \end{bmatrix}, \quad D_S = \frac{E}{2(1+v)} \begin{bmatrix} 1 & 0 \\ 0 & 1 \end{bmatrix} \tag{9.7}$$

The element mass matrix may also be found by integrating the shape functions over the element domain:

$$M^e = \int_{A^e} N^{\mathrm{T}} GN dA^e, \tag{9.8}$$

$$N = \begin{bmatrix} \begin{bmatrix} N_1 & 0 & 0 \\ 0 & N_1 & 0 \\ 0 & 0 & N_1 \end{bmatrix} \cdots \begin{bmatrix} N_k & 0 & 0 \\ 0 & N_k & 0 \\ 0 & 0 & N_k \end{bmatrix} \end{bmatrix} \tag{9.9}$$

$$G = \rho \begin{bmatrix} h & 0 & 0 \\ 0 & \frac{h^3}{12} & 0 \\ 0 & 0 & \frac{h^3}{12} \end{bmatrix} \tag{9.10}$$

where ρ is the material density. The element degree-of-freedom vector contains the nodal displacements:

$$\{q^e\} = \{w_1 \ \theta_{x1} \ \theta_{y1} \ \cdots \ w_k \ \theta_{xk} \ \theta_{yk}\}^{\mathrm{T}} \tag{9.11}$$

A well documented problem with Mindlin–Reissner plate elements is shear locking [11]. Shear locking occurs when low-order elements encounter bending. In pure bending along one axis, a plate will form a circular arc, as shown in Figure 9.2a. A linear element, however, will not be able to take on a circular displacement. Thus it will enter a state of pure shear, as shown in Figure 9.2b, in order to accommodate the nodal rotations. This is termed *shear locking* and results in much higher strain energy for the same rotational displacements. This leads to overprediction of vibration frequencies. The problem

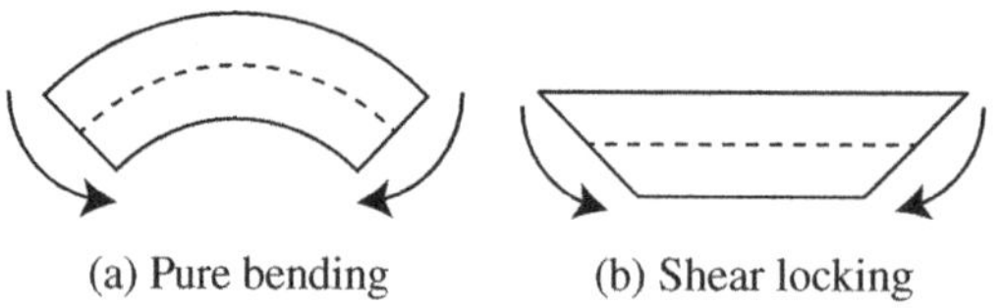

(a) Pure bending   (b) Shear locking

Figure 9.2 Demonstration of (a) a pure bending state and (b) a linear element under bending load.

occurs in higher-order elements as well, but to a much smaller degree. The most common treatment of this issue is underintegration of the shear stiffness matrix [14]. For example, in a four-node quadrilateral element, a single-point Gauss quadrature is performed, rather than the typical 2×2 Gauss quadrature.

While this underintegration allows for reasonable solutions to be obtained using Mindlin–Reissner elements, it allows for some zero-energy vibration modes. These zero-energy modes have a highly oscillatory displacement profile, similar to the folds in an accordian, and may be prevented by adding a stabilization matrix to the shear stiffness matrix [15]. The stabilization matrix is simply a fully integrated shear stiffness matrix. For the 2×2 case, the shear stiffness matrix may be found as follows:

$$K_S^e = (1 - \varepsilon)K_S^{e[1\times1]} + \varepsilon K_S^{e[2\times2]} \tag{9.12}$$

where ε is a perturbation parameter that is normalized by the thickness squared divided by the area; in other words:

$$\varepsilon = rh^2/A \tag{9.13}$$

The bracketed superscripts in Eq. (9.12) represent the Gauss quadrature rule being employed. The parameter r is taken to be 0.1 in the current model. This is large enough to suppress the zero-energy modes, but small enough to avoid significant shear-locking effects. At this point, a model may be generated by meshing the domain and applying the direct stiffness method [10]. The resulting global mass and stiffness matrices, M and K, are referred to as the "free" mass and stiffness matrices because no boundary conditions have been applied. Neglecting damping and assuming zero forcing, the free equations of motion are:

$$M\ddot{q} + Kq = 0 \tag{9.14}$$

where q is the free global degree-of-freedom vector.

9.2.2 Bloch Boundary Conditions

Bloch's theorem states that a field in a periodic material can be represented as a periodic function multiplied by a plane wave [16]:

$$u(x, \kappa, t) = \underbrace{\tilde{u}(x, \kappa)}_{\substack{\text{periodic} \\ \text{function}}} \underbrace{e^{i\kappa^T x}}_{\substack{\text{plane-} \\ \text{wave} \\ \text{term}}} \underbrace{e^{-i\omega t}}_{\substack{\text{harmonic} \\ \text{term}}}, \quad \tilde{u}(x, \kappa) = \tilde{u}(x + g, \kappa) \tag{9.15}$$

where x is the position vector, κ is the wave number, ω is the temporal frequency, and g is a lattice vector. From Eq. (9.15) we see that the degrees of freedom on the boundary are related to each other through a plane-wave term. The degrees of freedom on the left boundary are related to those on the right boundary, and the degrees of freedom on the bottom boundary are related to those on the top boundary:

$$q_R = q_L \lambda_x, \quad q_T = q_B \lambda_y \tag{9.16}$$

where $\lambda_x = e^{i\kappa_x L_x}$ and $\lambda_y = e^{i\kappa_y L_y}$. Similarly, the corner degrees of freedom are related to each other:

$$\mathbf{q}_{BR} = \mathbf{q}_{BL}\lambda_x, \quad \mathbf{q}_{TR} = \mathbf{q}_{BL}\lambda_x\lambda_y, \quad \mathbf{q}_{TL} = \mathbf{q}_{BL}\lambda_y \tag{9.17}$$

Equations (9.16) and (9.17) may be collected into a transformation matrix that relates the periodic degree-of-freedom vector to the free degree-of-freedom vector:

$$
\overbrace{\begin{bmatrix} \mathbf{q}_I \\ \mathbf{q}_L \\ \mathbf{q}_R \\ \mathbf{q}_B \\ \mathbf{q}_T \\ \mathbf{q}_{BL} \\ \mathbf{q}_{BR} \\ \mathbf{q}_{TR} \\ \mathbf{q}_{TL} \end{bmatrix}}^{q} = \overbrace{\begin{bmatrix} I & 0 & 0 & 0 \\ 0 & I & 0 & 0 \\ 0 & \lambda_x I & 0 & 0 \\ 0 & 0 & I & 0 \\ 0 & 0 & \lambda_y I & 0 \\ 0 & 0 & 0 & I \\ 0 & 0 & 0 & \lambda_x I \\ 0 & 0 & 0 & \lambda_x\lambda_y I \\ 0 & 0 & 0 & \lambda_y I \end{bmatrix}}^{P} \overbrace{\begin{bmatrix} \mathbf{q}_I \\ \mathbf{q}_L \\ \mathbf{q}_B \\ \mathbf{q}_{BL} \end{bmatrix}}^{\overline{q}} \tag{9.18}
$$

Note that Eqs. (9.16)–(9.18) consider 2D wave propagation, but the ideas are easily extended to three dimensions. We may now substitute Eq. (9.18) into Eq. (9.14), and pre-multiply by $P^\dagger$ to arrive at the periodic equations of motion

$$\underbrace{P^\dagger M P}_{\overline{M}(\kappa)}\ddot{\overline{q}} + \underbrace{P^\dagger K P}_{\overline{K}(\kappa)}\overline{q} = 0 \tag{9.19}$$

where $\overline{M}(\kappa)$ and $\overline{K}(\kappa)$ are the wave-vector-dependent periodic mass and stiffness matrices, and $(\cdot)^\dagger$ indicates the complex conjugate transpose. A common implementation mistake made with this method is incorrect sorting of the degrees of freedom. Special attention must be paid in Eq. (9.19) to ensure that the rows of P are sorted to match the degree-of-freedom order of M and K.

There are other methods of forming models for band structure calculations. Here we have enforced the Bloch plane-wave solution using boundary conditions. The result is that we compute the entire displacement field shown in Eq. (9.15). Another approach is to consider the plane-wave solution implicitly [6]. The resulting eigenvector solution provides just the periodic function in Eq. (9.15). It is very easy to go between the two solutions by multiplying each element in the displacement vector by the plane-wave component, $e^{i\kappa^T x}$, corresponding to its physical location in the unit cell, x. An important characteristic of the Bloch boundary-condition formulation considered here is that the interior partitions of the resulting mass and stiffness matrices are independent of the wave vector. We take advantage of this feature in the Bloch mode synthesis (BMS) model-reduction method.

9.2.3 Example Model

The modal-reduction methods of this chapter will be demonstrated using the plate-based periodic lattice material shown in Figure 9.3a. This model contains 1408 elements and has 4698 degrees of freedom. The material properties and model

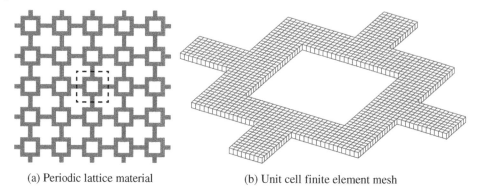

(a) Periodic lattice material   (b) Unit cell finite element mesh

Figure 9.3 (a) Portion of a periodic lattice material with a single unit cell outlined; (b) finite-element mesh of the unit cell for this material. This model is featured on the cover of the book.

dimensions are summarized as follows:

$$E = 2.4 \text{ GPa}, \quad \rho = 1040 \text{ kg}/m^3, \quad v = 0.33, \quad L_x = L_y = 10 \text{ cm}, \quad h = 1 \text{ mm} \quad (9.20)$$

9.3 Reduced Bloch Mode Expansion

The goal of modal reduction is to use a set of mode shapes to approximate the system's vibrational behavior within the frequency range of interest. For a structure this usually involves keeping mode shapes with natural frequencies up to 1.5–2 times the highest frequency of interest. For a periodic material, however, the wave behavior depends not only on the frequency, but also on the wave-vector. Thus, the set of modes used to reduce a periodic model must capture the behavior, not only in the frequency range of interest, but also in the wave-vector range of interest. Hussein presented a method capable of doing exactly that, called reduced Bloch mode expansion (RBME) [6]. By combining a set of Bloch mode shapes from wave-vector κ_a with a set of mode shapes from wave-vector κ_b, we can approximate the wave behavior of the system for any wave-vector *between* κ_a and κ_b.

9.3.1 RBME Formulation

Suppose we have a model of a lattice material. In other words, we have the mass and stiffness matrices, $\overline{M}(\kappa)$ and $\overline{K}(\kappa)$, as functions of the wave-vector. The band structure is typically computed by discretizing the wave-vector path and solving the following eigenvalue problem for every κ-point:

$$\overline{K}(\kappa_j)\Phi_j = \overline{M}(\kappa_j)\Phi_j\Lambda_j \quad (9.21)$$

where the subscript j denotes the jth κ-point, Λ is a matrix containing the eigenvalues on its diagonal, and Φ is a matrix whose columns are the eigenvectors. As mentioned earlier, we may use the endpoints of the wave-vector segment as expansion points. By collecting the important modes from each expansion point, we can form a transformation matrix that reduces the model size but still retains the important dynamic information:

$$\underbrace{\Phi_a}_{n \times n} \Rightarrow \underbrace{\hat{\Phi}_a}_{n \times m} , \quad \underbrace{\Phi_b}_{n \times n} \Rightarrow \underbrace{\hat{\Phi}_b}_{n \times m} \tag{9.22}$$

$$R = [\hat{\Phi}_a, \hat{\Phi}_b] \tag{9.23}$$

This transformation matrix is then used to reduce the mass and stiffness matrices of the periodic model as follows:

$$\overline{M}^R(\kappa_j) = R^\dagger \overline{M}(\kappa_j)R, \quad \overline{K}^R(\kappa_j) = R^\dagger \overline{K}(\kappa_j)R \tag{9.24}$$

The mode truncation in Eq. (9.22) keeps the m lowest frequency mode shapes at each expansion point. The parameter m is chosen such that all modes up to 1.5–2 times the highest frequency of interest are included in the transformation. As long as the highest frequency of interest is low enough, m will be significantly smaller than n.

An eigenvalue solution of the reduced system at all of the intermediate κ-points yields the approximate band structure:

$$\overline{K}^R(\kappa_j)\Phi_j^R = \overline{M}^R(\kappa_j)\Phi_j^R \Lambda_j^R \tag{9.25}$$

The approximate system mode shapes can be recovered from the reduced mode shapes using the RBME transformation matrix

$$\Phi_j \approx R\Phi_j^R \tag{9.26}$$

As with modal reduction in structures, the accuracy of the reduced band structure can be improved by including more mode shapes. With the RBME method, this is achieved either by increasing the number of modes, m, kept at each expansion point, or by increasing the number of expansion points. The expansion points thus far have simply been the end-points of the wave-vector segment, but we may also add extra expansion points between κ_a and κ_b, as shown in Figure 9.4. As additional expansion points are added, the mode shapes computed at each additional point may simply be concatenated with the existing mode shapes in the transformation matrix.

9.3.2 RBME Example

Figure 9.5 shows a comparison of the full-system dispersion diagram with the 2- and 3-point RBME dispersion diagrams for the lattice-material model described in the previous section. Note that for this example we are interested in three different wave-vector

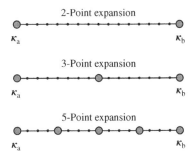

Figure 9.4 Expansion point locations for 2-point, 3-point, and 5-point RBME schemes.

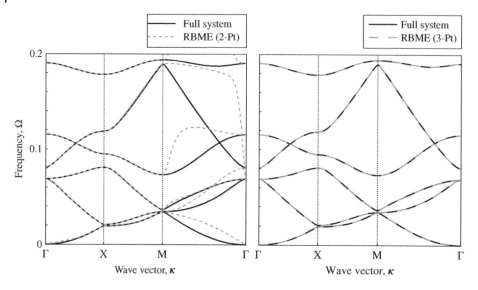

Figure 9.5 Comparison of full system dispersion with 2-point RBME dispersion (left) and 3-point RBME dispersion (right). Frequencies are normalized as follows: $\Omega = \omega L_x \sqrt{\rho/E}$.

segments: Γ–X, X–M, and M–Γ. Thus we compute a new RBME transformation matrix for each segment.

As mentioned earlier, when more Bloch modes are kept in the reduction, the error in the dispersion frequencies decreases. From Eq. (9.26), we can see that the mode shapes predicted by the RBME model are a linear combination of the expansion modes contained in the RBME transformation matrix. The accuracy of the frequencies predicted by the RBME model depend on how well the approximate mode shapes match the full-model mode shapes. Thus, given that the RBME transformation matrix contains mode shapes from the expansion points, the reduced model is able to exactly recover the mode shapes and frequencies at the expansion points. As we move away from the expansion points, the error in mode shapes and frequencies increases. This is illustrated in Figure 9.6, which shows the error in the 6th dispersion branch over the Brillouin zone. We can see that incorporating additional expansion points can significantly reduce the error.

From Figures 9.5 and 9.6, we can see that the RBME reduction is least accurate in the M–Γ segment. This indicates that the mode shapes in this region are more difficult to approximate with the modes in the RBME transformation. There is also unexpectedly high error near the M-point when $m = 6$. This occurs because the 6th branch is degenerate at the M-point, so the 6th mode used in the RBME transformation is actually a linear combination of the 6th and 7th system modes. This problem is resolved once additional modes are employed in the RBME transformation.

In practice, it is often most useful to know the maximum error in a dispersion branch as the number of expansion modes is increased. This is shown for the 6th dispersion branch in Figure 9.7, using 2-point, 3-point, and 5-point expansions. All three expansion schemes show a linear trend in the logarithmic error plot as the number of expansion points is increased.

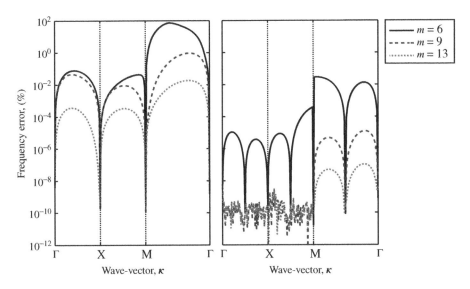

Figure 9.6 Error in 6th dispersion branch computed with 2-point (left) and 3-point (right) RBME models as the number of modes, *m*, per expansion point is increased.

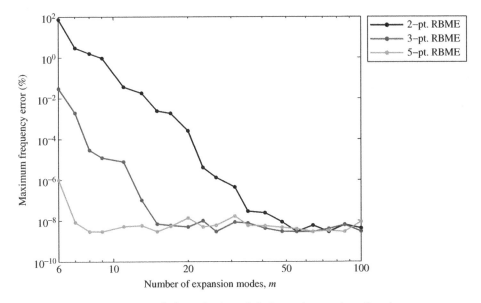

Figure 9.7 Maximum error in 6th dispersion branch for increasing number of modes, *m*, per expansion point.

9.3.3 RBME Additional Considerations

We have seen that increasing the number of modes in the RBME transformation can quickly reduce the error. When we do so, however, we risk creating a set of reduction modes that is linearly dependent or nearly linearly dependent in a numerical sense. This will create numerical ill-conditioning in the resulting reduced equations of motion, causing large errors and introducing spurious modes. In order to avoid this, a Gram-Schmidt orthogonalization should be applied to R in order to improve the numerical conditioning of the reduced mass and stiffness matrices.

9.4   Bloch Mode Synthesis

In structural dynamics analysis it is very common to split a structure into several smaller structures, called substructures. Each substructure model connects to the other substructures via some interface degrees of freedom. The full model may be recovered from the substructure models by reconnecting the interface degrees of freedom from each substructure. Methods have been developed that allow for modal analysis and subsequent reduction of the substructure models while retaining the ability to connect the interface degrees of freedom. These are commonly referred to as component-mode-synthesis methods because the system modes are synthesized by connecting the modes of each component [17].

Component mode synthesis has several advantages over considering an entire structure at once. Structures are often designed as a system of substructures, so it is natural to separately analyze individual substructures throughout the design process. It also allows for a divide-and-conquer approach to the full-level modal solution. Eigenvalue analysis of several small systems is usually much less costly than analysis of a single large system formed by combining all of these small systems. Thus the computational effort associated with modal analysis of the full system is greatly reduced by performing modal reduction on the individual substructures before connecting them.

The same idea may be applied to the analysis of a unit cell. Although we do not typically have multiple substructures to connect when modeling a lattice material, we do have many different sets of boundary conditions. In fact, we have slightly different boundary conditions for every κ-point we consider. We saw earlier that these boundary conditions are enforced by connecting the interface to itself through some phase term. The Bloch modal solution is thus obtained by synthesizing the modes of the free unit cell *to themselves* through the Bloch boundary conditions. This method is termed Bloch mode synthesis (BMS) [7].

As mentioned earlier, there are substructure representations that allow for modal reduction while retaining the ability to connect to other substructures. The most common is the Craig–Bampton substructure representation [18]. This representation uses a set of fixed-interface mode shapes to describe the vibrational motion of the substructure's interior, and a complementary set of constraint modes to describe the static motion in the structure due to deflection of the boundary degrees of freedom.

9.4.1   BMS Formulation

A constraint mode may be created by enforcing a static unit deflection of a single boundary degree of freedom while holding the rest of the boundary fixed. If we consider just the static portion of Eq. (9.19) and partition it into interior and boundary partitions, we can write the following expression containing the constraint modes:

$$\begin{bmatrix} K_{II} & K_{IA} \\ K_{AI} & K_{AA} \end{bmatrix} \begin{bmatrix} \Psi \\ I \end{bmatrix} = \begin{bmatrix} 0 \\ f_A \end{bmatrix} \tag{9.27}$$

where the subscript I denotes the interior degrees of freedom, and the subscript A denotes the boundary degrees of freedom. Note that the boundary degrees of freedom are replaced with the identity matrix. This is equivalent to perturbing a single boundary degree of freedom at a time while holding the rest fixed. We can solve for the constraint modes, Ψ, by considering just the upper half of Eq. (9.27):

$$\Psi = -K_{II}^{-1} K_{IA} \tag{9.28}$$

A few constraint modes are shown on the left-hand side of Figure 9.8.

The fixed-interface modes are found by setting the boundary motion in Eq. (9.14) to zero, and solving the following eigenvalue problem for the dynamic modes of the interior:

$$K_{II} \Phi_I = M_{II} \Phi_I \Lambda_I \tag{9.29}$$

A few fixed-interface modes are shown on the right-hand side of Figure 9.8. The set of fixed-interface modes may be truncated so that only those n_{FI} modes with frequencies

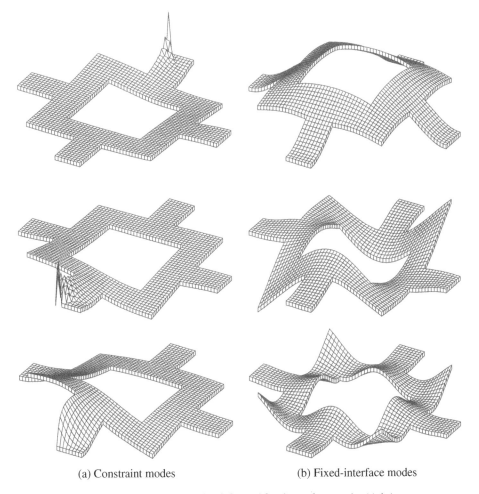

(a) Constraint modes   (b) Fixed-interface modes

Figure 9.8 A selection of constraint modes (left), and fixed-interface modes (right).

less than 1.5-2 times the highest frequency of interest are kept; in other words,

$$\underbrace{\Phi_I}_{n_I \times n_I} \Rightarrow \underbrace{\hat{\Phi}_I}_{n_I \times n_{FI}} \tag{9.30}$$

The Craig–Bampton transformation matrix may be formed by collecting the fixed-interface modes and the constraint modes together as follows:

$$T = \begin{bmatrix} \Phi_I & \Psi \\ 0 & I \end{bmatrix} \tag{9.31}$$

This is used to create the BMS reduced system representation:

$$\mathcal{M}^B = T^\dagger M T, \quad \mathcal{K}^B = T^\dagger K T \tag{9.32}$$

Now we apply the periodicity transformation to the BMS reduced mass and stiffness matrices:

$$\overline{\mathcal{M}}^B(\kappa_j) = P^\dagger \mathcal{M}^B P, \quad \overline{\mathcal{K}}^B(\kappa_j) = P^\dagger \mathcal{K}^B P \tag{9.33}$$

Note that the periodicity transformation must be modified slightly because we have replaced the interior degrees of freedom with some interior mode shapes. As in the previous section, an eigenvalue solution of the reduced system is used to approximate the band structure frequencies,

$$\overline{\mathcal{K}}^B(\kappa_j)\Phi_j^B = \overline{\mathcal{M}}^B(\kappa_j)\Phi_j^B \Lambda_j^B \tag{9.34}$$

and the approximate unit-cell mode shapes are recovered from the reduced mode shapes using the Craig–Bampton transformation matrix,

$$\Phi_j \approx T\Phi_j^B \tag{9.35}$$

9.4.2 BMS Example

For the Mindlin plate model described earlier, Figure 9.9 shows a comparison of the full-system dispersion diagram with the dispersion diagrams computed with BMS reduced models containing 6 fixed-interface modes and 24 fixed-interface modes, respectively.

We can see from Figure 9.9 that the error in frequency computed with the BMS model seems to grow as we increase in frequency. Increasing the number of fixed-interface modes kept in the reduction will improve the accuracy. Figure 9.10 shows the error in the 6th dispersion branch over the Brillouin zone for BMS models with increasing number of fixed-interface modes. Unlike the RBME method, we can see that the error is relatively consistent over the entire Brillouin zone.

Figure 9.11 shows the maximum error in the 6th dispersion branch as the number of fixed-interface modes is increased. We can observe an approximately linear trend in the logarithm of the error versus the logarithm of the number of fixed-interface modes.

9.4.3 BMS Additional Considerations

In Section 9.2.2, we noted that as a result of the Bloch boundary-condition formulation, the interior partition of the mass and stiffness matrices was κ-independent over the unit cell. Because of this, the fixed-interface mode computation in Eq. (9.29) must be

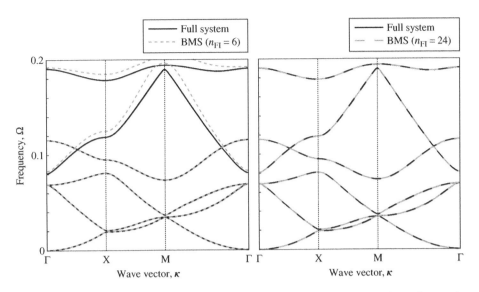

Figure 9.9 Comparison of full-system dispersion with BMS dispersion using 6 fixed-interface modes (left), and 24 fixed-interface modes (right). Frequencies are normalized as follows: $\Omega = \omega L_x \sqrt{\rho/E}$.

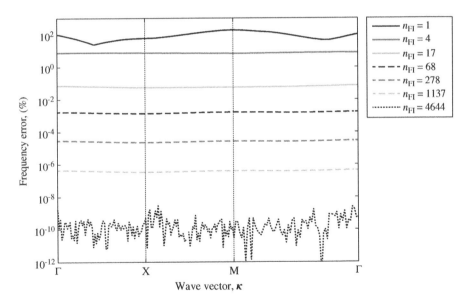

Figure 9.10 Error in 6th dispersion branch computed with BMS models for increasing numbers of fixed-interface modes, n_{FI}.

performed only once. For very large models, the fixed-interface mode computation often represents the majority of the BMS band structure computation time, so it is worth using the Bloch boundary-condition formulation.

The size of the reduced BMS model includes the reduced set of fixed-interface modes as well as all of the constraint modes. There is a constraint mode for every boundary

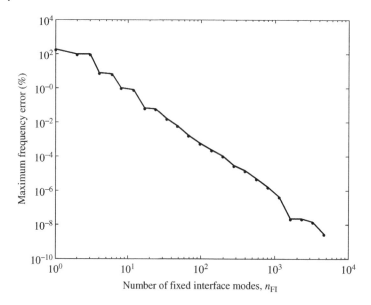

Figure 9.11 Maximum error in 6th dispersion branch computed with BMS models as the number of fixed-interface modes is increased.

degree of freedom, so this may lead to a reduced model that is still quite large. In these situations, a boundary reduction may be possible in order to further reduce the model size [19].

9.5 Comparison of RBME and BMS

Both of the methods discussed in this chapter have been shown to be capable of producing accurate elastic dispersion diagrams with greatly reduced model sizes. In this section we examine the benefits and drawbacks of each method.

9.5.1 Model Size

The first point of comparison is model size. For comparable error levels, RBME models will be considerably smaller than BMS models. A significant number of fixed-interface modes must be retained in a BMS model to achieve low error. This is because individually the fixed-interface modes do not necessarily resemble the Bloch eigenvectors, so they do not correspond directly to a dispersion branch. Rather, the fixed-interface modes span the cumulative vibrational behavior of the system within some frequency range. This is analogous to the way a set of trigonometric functions can be used to approximate a square wave by Fourier series. No individual cosine function resembles a square wave, but the summation of many cosine functions can lead to a close approximation of a square wave. In contrast, the modes used in the RBME model correspond directly to the dispersion branches, albeit at different κ-points, so very few are needed in order to provide a good approximation of the dispersion.

Another factor that increases BMS model size is the constraint mode set. This set contains one constraint mode for every boundary degree of freedom. This means that the BMS model will have at least as many degrees of freedom as there are boundary degrees of freedom in the full model.

9.5.2 Computational Efficiency

Computational efficiency measures how much faster the dispersion curves are computed using a reduced model compared to the full model. This can be described using the computational time percent, R, which gives the reduced computation time, t_r, as a percentage of the full-model computation time, t_f:

$$R = 100 * \frac{t_r}{t_f} \tag{9.36}$$

Figure 9.12 presents flowcharts for the dispersion computation using the full model, the RBME reduction, and the BMS reduction. The most expensive computations in each section are highlighted. This figure shows that the computation time associated with each model reduction has two components: an up-front computational cost, t_u, and a per-κ-point computational cost, t_κ,

$$t_r = t_u + n_\kappa t_\kappa, \tag{9.37}$$

where n_κ is the number of κ-points. We can see that as n_κ becomes very large, t_κ becomes more significant. The main factor that influences t_κ is the reduced model size. As discussed earlier, RBME has an advantage when it comes to model size, so it also has a smaller per-κ-point computational cost.

For both RBME and BMS, t_u rises quickly as the full-model size is increased. Thus for very large full-model sizes, the up-front computational cost may be so large that

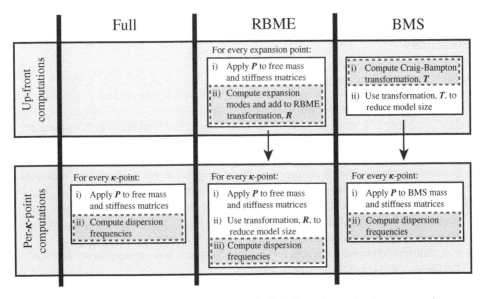

Figure 9.12 Flowcharts showing computation steps for full-dispersion evaluation compared to reduced methods. The computationally limiting step in each section is highlighted.

the per-κ-point computational cost becomes negligible. The up-front computational cost associated with RBME is the time necessary to compute the expansion modes. This involves solution of the full complex eigenvalue problem for a few low-frequency eigenvectors at multiple expansion points. The up-front computational cost of the BMS method is the time necessary to compute the fixed-interface mode set and the constraint mode set. These involve a single real eigenvalue solution and a system solution respectively. In terms of up-front computational cost, BMS has an advantage over RBME because it avoids ever solving the full complex eigenvalue problem.

9.5.3 Ease of Implementation

Here we make a few comments on the ease of implementation and flexibility of both RBME and BMS. In this category, RBME has the advantage. For any model that can be used to compute band structure, applying RBME simply involves computing expansion modes and then using these to reduce the model at all of the other κ-points. It does not matter how the model is created or what type of basis functions are used. For example, RBME is equally effective with models created using finite elements where the degrees of freedom represent real space displacements, or plane-wave expansions where the degrees of freedom represent contributions of basis functions over the entire domain.

The implementation of BMS is slightly more involved. It requires that the model be created in such a way that the degrees of freedom represent physical values of the displacement field at the nodes. This is the case with finite-element and finite-difference models, but not plane-wave expansion models. Nevertheless, the Bloch boundary-condition formulation allows for models created with conventional finite-element approaches to be converted into Bloch periodic models. This makes available the vast finite-element libraries from commercial software packages.

References

1 M. I. Hussein, M. J. Leamy, and M. Ruzzene, "Dynamics of phononic materials and structures: Historical origins, recent progress, and future outlook," *Applied Mechanics Reviews*, vol. 66, no. 4, p. 040802, 2014.

2 D. C. Dobson, "An efficient method for band structure calculations in 2D photonic crystals," *Journal of Computational Physics*, vol. 149, no. 2, pp. 363–376, 1999.

3 R. L. Chern, C. C. Chang, C. C. Chang, and R. R. Hwang, "Large full band gaps for photonic crystals in two dimensions computed by an inverse method with multigrid acceleration," *Physical Review E*, vol. 68, no. 2, p. 026704, 2003.

4 W. Axmann and P. Kuchment, "An efficient finite element method for computing spectra of photonic and acoustic band-gap materials: I. Scalar case," *Journal of Computational Physics*, vol. 150, no. 2, pp. 468–481, 1999.

5 E. L. Shirley, "Optimal basis sets for detailed Brillouin-zone integrations," *Physical Review B*, vol. 54, no. 23, p. 16464, 1996.

6 M. I. Hussein, "Reduced Bloch mode expansion for periodic media band structure calculations," *Proceedings of the Royal Society A: Mathematical, Physical and Engineering Science*, vol. 465, no. 2109, pp. 2825–2848, 2009.

7 D. Krattiger and M. I. Hussein, "Bloch mode synthesis: Ultrafast method for elastic band-structure calculations," *Physical Review E*, vol. 29, no. 3, pp. 313–327, 2014.

8 R. Mindlin, "Influence of rotatory inertia and shear on flexural motions of isotropic, elastic plates," *Journal of Applied Mechanics*, vol. 18, p. 3138, 1951.

9 E. Reissner, "The effect of transverse shear deformation on the bending of elastic plates," *Journal of Applied Mechanics*, vol. 12, p. 6877, 1945.

10 R. D. Cook, D. S. Malkus, M. E. Plesha, and R. J. Witt, *Concepts and Applications of Finite Element Analysis*. Wiley, 2001.

11 K. J. Bathe, *Finite Element Procedures*. Klaus-Jurgen Bathe, 2006.

12 A. J. M. Ferreira, *MATLAB Codes for Finite Element Analysis: Solids and Structures*. Solid mechanics and its applications, Springer, 2008.

13 T. J. R. Hughes, *The Finite Element Method: Linear Static and Dynamic Finite Element Analysis*. Courier Corporation, 2012.

14 O. C. Zienkiewicz, R. L. Taylor, and J. M. Too, "Reduced integration technique in general analysis of plates and shells," *International Journal for Numerical Methods in Engineering*, vol. 3, no. 2, pp. 275–290, 1971.

15 T. Belytschko, C. S. Tsay, and W. K. Liu, "A stabilization matrix for the bilinear mindlin plate element," *Computer Methods in Applied Mechanics and Engineering*, vol. 29, no. 3, pp. 313–327, 1981.

16 F. Bloch, "Über die quantenmechanik der elektronen in kristallgittern," *Zeitschrift für Physik*, vol. 52, no. 7–8, pp. 555–600, 1929.

17 R. R. Craig and A. J. Kurdila, *Fundamentals of Structural Dynamics*. Wiley, 2011.

18 M. C. C. Bampton and R. R. Craig, "Coupling of substructures for dynamic analyses," *AIAA Journal*, vol. 6, no. 7, pp. 1313–1319, 1968.

19 M. P. Castanier, Y.-C. Tan, and C. Pierre, "Characteristic constraint modes for component mode synthesis," *AIAA Journal*, vol. 39, no. 6, pp. 1182–1187, 2001.

10

Topology Optimization of Lattice Materials

Osama R. Bilal and Mahmoud I. Hussein

Department of Aerospace Engineering Sciences, University of Colorado Boulder, Colorado, USA

10.1 Introduction

Lattice materials are periodic material systems composed of a basic building block (the unit cell) that repeats in space. A lattice material features intriguing properties both statically [1] and dynamically [2]. The dynamical characteristics of periodic materials in general describe the nature of wave propagation through the internal structure of the material, whether it is electromagnetic waves [3, 4] or mechanical (elastodynamic) waves [5, 6]. Unique frequency characteristics are observed in this broad class of materials, among which is the presence of band gaps: frequency ranges where waves are prohibited from propagation.

Lattice materials can be classified into two subcategories based on the band gap formation mechanism: phononic crystals (PnCs) and acoustic/elastic metamaterials (MMs) [7]. Bragg scattering (scattering due to periodicity) results in wave interferences in PnCs. For MMs, band gaps exist as a result of a hybridization mechanism of the underlying dispersion curves due to coupling with local resonances. Owing to the fundamental difference in their mechanisms of band gap creation, the constraints on unit-cell size and other features are also very distinct. MMs with resonators can influence waves that are orders of magnitudes larger in wavelength than the unit-cell size; while a PnC unit cell must roughly be the same size as, or larger than, the targeted wavelength [8]. Lattice materials are modeled and analyzed in either continuous (for example, rod, beam, plate, shell, or bulk material) or discrete forms (such as lumped-mass systems), as seen in Figure 10.1. However, it is sufficient to model a simple 1D spring–mass model to capture the important dynamical characteristics of a lattice material (see Section 10.5).

Applications of lattice materials based on their interactions with mechanical waves include elastic or acoustic waveguiding [9], focusing [10], vibration minimization [11], sound collimation [12], frequency sensing [13, 14], acoustic cloaking [15, 16], acoustic rectification [17], opto-mechanical wave coupling in photonic devices [18], thermal conductivity lowering in semiconductors [19–23] and, most recently, flow control [24]. The reader is referred to the literature for others [7, 25–27].

Dynamics of Lattice Materials, First Edition. Edited by A. Srikantha Phani and Mahmoud I. Hussein.
© 2017 John Wiley & Sons Ltd. Published 2017 by John Wiley & Sons Ltd.

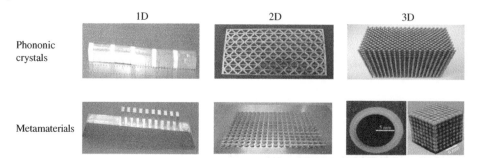

Figure 10.1 Realization of PnCs and MMs in one-, two-, and three-dimensions.

All of these applications require well-defined characteristics of the material band structure; hence unit-cell design and optimization is of profound importance in improving performance. Since many of these applications utilize band gaps, a common design objective is to widen the size of band gaps, as discussed further in the next section.

10.2 Unit-cell Optimization

Designing a lattice material at the unit-cell level is advantageous in many aspects. A unit-cell analysis provides a complete picture of the intrinsic local dynamics, which is obscured when analyzing the structure as a whole. Moreover, unit-cell design is computationally less expensive than designing an entire structure. An optimized unit cell is scalable and can be used to attenuate waves at a broad spectrum of frequencies. In order to keep the unit-cell's dimensions small, the band gap central frequency should be lowered as much as possible.

10.2.1 Parametric, Shape, and Topology Optimization

Unit-cell design optimization for a desired band structure objective can be carried out using three approaches, outlined in Figure 10.2. One approach is a parametric optimization, where the unit-cell attributes are varied systematically while preserving the unit-cell shape. Another approach is to start with a certain topological distribution of holes (fixed number of holes), and then to vary the shape of the holes to improve the design without altering the general topological layout. The third approach involves no assumptions about the number of holes *a priori*, and uses the optimization algorithm to fully define the topology. The third method is the most general and is the focus of this chapter. Since both shape and topology approaches involve varying design parameters throughout the optimization process, modern classifications consider parametric sweeping a subclass of either one.

10.2.2 Selection of Studies from the Literature

Phononic crystal unit-cell optimization has been investigated in the literature for both one dimension [28–30] and two dimensions [31–38], using gradient-based [31–34] as well as non-gradient-based [29, 30, 35–38] techniques. However, previous studies

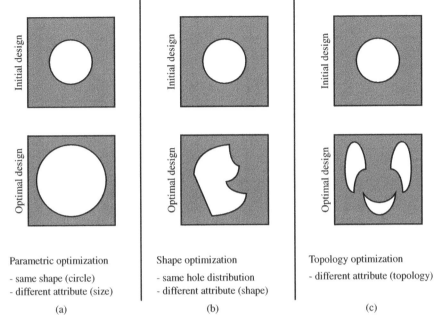

Parametric optimization
- same shape (circle)
- different attribute (size)

(a)

Shape optimization
- same hole distribution
- different attribute (shape)

(b)

Topology optimization
- different attribute (topology)

(c)

Figure 10.2 Unit-cell optimization approaches.

[29, 39, 40] and preliminary investigations [30, 38, 41] have observed that the general band gap optimization problem is multi-modal in nature and manifests a complex design landscape. While there are many studies for band gap maximization for PnCs, less work has been done for locally resonant acoustic/elastic MMs [42, 43].

10.2.3 Design Search Space

The design space for the unit-cell topology optimization problem is intractable, so analysis of all the alternatives is unfeasible. For example, when considering a simple 2D unit cell with 64 × 64 elements, the number of possible designs equals $2^{528} \approx 8.7 \times 10^{158}$, where the base "2" is the number of possible states (that is, material A or B) and the power "528" is the number of unique elements ($u_{ele} = ((n + 1)^2 - 1)/8, n = 64$) in the unit cell, assuming a high degree of symmetry within it. If a computer is capable of evaluating a single design in a second, it would take the current age of the universe to compute only 1% of a search of this size; Figure 10.3 illustrates this remarkable scalability relationship.

In addition to the exponential growth of the search space with respect to the number of lattice unit-cell design variables, and the non-convex nature of the problem, the search space is discontinuous. For solid-void unit cells, having a solid region that is not connected to the lattice is inadmissible. Furthermore, having a fully connected unit cell without proper connectivity at the boundaries to form a lattice is also inadmissible. Examples of disconnected or inadmissible lattice-material unit cells are shown in Figure 10.4.

Exploring the search space of the problem before carrying out an optimization study is most beneficial. Our initial study begins with a unit cell that is discretized with a coarse

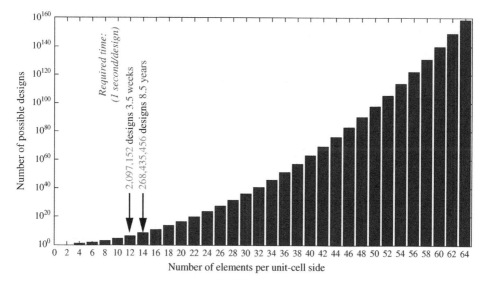

Figure 10.3 Search-space size for a simple $n \times n$ 2D unit cell.

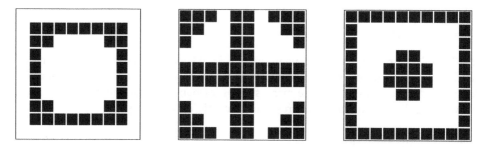

Figure 10.4 Example of disconnected unit cells.

mesh of 10×10 elements. All the possible designs (32,768 in number) are generated and evaluated.

Due to the presence of eigenvalue multiplicities and the non-convex nature of the objective function considered, gradient-based methods usually suffer from low regularity and non-differentiability. Therefore, gradient-free algorithms are a natural choice for the problem.

10.3   Plate-based Lattice Material Unit Cell

We consider a lattice material with a square unit cell extended repeatedly in the x–y plane and with small thickness across the z-direction. This configuration creates a square lattice, as shown in Figure 10.5a. The material distribution does not vary in the z-direction, so the design problem is two-dimensional. A hexagonal lattice is constructed in a similar manner, as presented in Figure 10.5b.

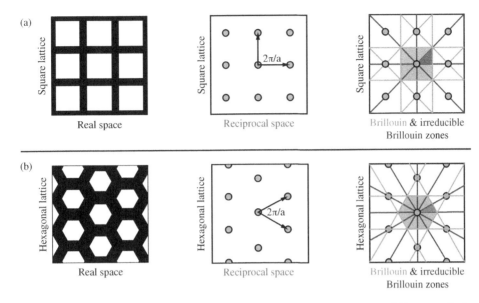

Figure 10.5 Schematic representation of the most common lattices—(a) square; (b) hexagonal—in real and reciprocal space. The first Brillouin zone of each is shown; the irreducible Brilliouin zone is also highlighted.

10.3.1 Equation of Motion and FE Model

The governing equation of motion for thin plates, based on Mindlin theory [44, 45], is

$$\left(\nabla^2 - \frac{\rho}{\mu\alpha}\frac{\partial^2}{\partial t^2}\right)\left(D\nabla^2 - \frac{\rho h^3}{12}\frac{\partial^2}{\partial t^2}\right)w + \rho h\frac{\partial^2 w}{\partial t^2} = 0 \tag{10.1}$$

where h is the plate thickness, μ is the shear modulus, α is the shear correction factor, $D = (Eh^3)/[12(1 - v^2)]$ is the plate stiffness, E is the Young's modulus, v is the Poisson's ratio, and $w = w(x, y, t)$ is the plate's transverse displacement.

Due to lattice symmetry, the analysis is restricted to the first Brillouin zone (BZ). We consider square lattices and impose C_{4v} symmetry, where only one-eighth of the unit cell (a triangular portion) is unique at the unit-cell level [46]. Subsequently, design representation is only needed in a portion of the unit cell and the band structure calculation is limited to the corresponding irreducible Brillouin zone (IBZ) (Figure 10.5). Furthermore, we model only the solid portion of the unit cell since we permit only contiguous distribution of solid material. In this manner, the lattice material considered exhibits geometric periodicity (with free internal in-plane surfaces) rather than material periodicity. In practice, the voids will be either in a vacuum or filled with air. This model presents an adequate representation of both cases because elastic waves propagating in the solid will be dominant despite having holes filled with air [47]. In fact, this suggests that the results shown are practically independent of the choice of the solid material.

In topology optimization, the most common representation of solid-void models is to assume a configuration consisting of a regular solid and a very soft solid that may be approximated as a void. This "almost void" phase is introduced to avoid re-meshing or re-numbering of the finite-element nodes for different designs [48]. However, here we utilize a different approach: the void is modeled by setting $\rho = 0, E = 0$ and the same

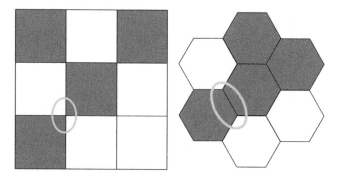

Figure 10.6 Single-node connectivity when using four-node quadratic elements versus hexagonal elements. The gray pixels represent "material" and the white pixels represent "void".

mesh is used for all designs. Once the mass and stiffness matrices are assembled, the zero rows and columns are eliminated to remove singularities. Owing to the nature of the resulting topologies, limiting the meshing to the solid part only (that is, no air or highly compliant material) reduces the computational time by a significant factor with no additional cost of re-meshing. Next, we utilize four-node bilinear quadrilateral elements to numerically solve the emerging eigenvalue problem. When using square elements in a solid-void design problem, single-node connectivity between elements may occur (Figure 10.6). This can produce erroneous results, so it is crucial to penalize two elements connected by only one node. One way to avoid this problem is to substitute hexagonal elements for square elements [49]. However, since quadrilateral elements are more common in meshing algorithms, they are used throughout this chapter.

10.3.2 Mathematical Formulation

A square unit cell Y, of side length a, is represented by $n \times n$ pixels, forming a binary matrix $\mathbf{g}$ for a more compact representation. In this chapter, the matrix and vector representations are used interchangeably. Following the underlying unit-cell symmetry, this matrix may be converted to a vector, $\mathbf{G}$. Each of the pixels can be assigned to a value corresponding to either no-material (void) or material (silicon is chosen as an example); that is, $g_s \in \{0, 1\}$. The formulation for designing a 2D lattice in a square unit cell with a C_{4v} symmetry condition can be represented mathematically as follows [50]:

Decision variables:

$\mathbf{G} = \{G_{ij}\}$, $\mathbf{G}$ is an $n \times n$ matrix, and G_{ij} denotes a pixel in row i and column j; $i, j = 1, 2, \ldots n$

Objective function:

$$\text{Maximize } f(\mathbf{G}) \tag{10.2}$$

Subject to:

$$G_{(i+\frac{n}{2})(j+\frac{n}{2})} = G_{(j+\frac{n}{2})(i+\frac{n}{2})} \quad \forall i, j = 1, 2, \ldots, \frac{n}{2} \tag{10.3}$$

$$G_{(i+\frac{n}{2})(j)} = G_{(i+\frac{n}{2})(n-j+1)} \quad \forall i, j = 1, 2, \ldots, \frac{n}{2} \tag{10.4}$$

$$G_{(i)(j)} = G_{(n-i+1)(j)}; i = 1, 2, \ldots, \frac{n}{2} \quad \forall j = 1, 2, \ldots, n \tag{10.5}$$

$$n = 2k + 1^1 \text{ where } k \in Z \tag{10.6}$$

$$G_{ij} \in \{0, 1\} \tag{10.7}$$

Equation (10.3) sets the rule for the top right-quarter square to consist of two mirror-reflected triangles of the IBZ, while Eq. (10.4) defines the mirror reflection for the upper-left quarter square as being equivalent to its counterpart on the right. The last symmetry condition, in Eq. (10.5), reflects the upper-half rectangle of the unit cell to the lower half. Figure 10.7 shows an example vector **g** and its usage in constructing the corresponding unit cell.

10.4 Genetic Algorithm

A genetic algorithm (GA) is a search heuristic that emulates the process of natural evolution. A GA starts with a population of different design topologies (solutions), which are evaluated according to their performance based on a given objective function (fitness value). A solution selection is then carried out to reproduce other topologies or new designs. To further mimic the natural evolution process, a mutation is implemented on these topologies according to a certain probability. Afterwards, the new "offspring" creates a new population. This process is repeated until a certain number of generations is reached or convergence (no further improvement of the objective value) is reached [52].

10.4.1 Objective Function

The objective function is formulated in terms of the size of a particular band gap width, normalized with respect to its midpoint frequency. In general, it is most advantageous to have the frequency range of a band gap maximized while positioning its midpoint

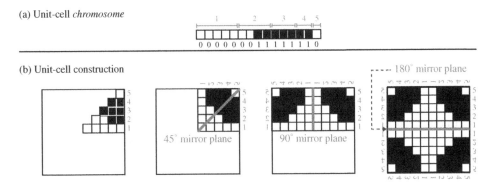

(a) Unit-cell *chromosome*

(b) Unit-cell construction

Figure 10.7 Unit-cell construction.

---

1 Since Bloch boundary conditions are used, n must be an odd number in order to avoid the shear-locking problem [51].

as low as possible in order to minimize unit-cell size. The function takes the following form:

$$f(\mathbf{G}) = \frac{\max(\min_{j=1}^{n_\kappa}(\omega_{i+1}^2(\kappa_j, \mathbf{G})) - \max_{j=1}^{n_\kappa}(\omega_i^2(\kappa_j, \mathbf{G})), 0)}{(\min_{j=1}^{n_\kappa}(\omega_{i+1}^2(\kappa_j, \mathbf{G})) + \max_{j=1}^{n_\kappa}(\omega_i^2(\kappa_j, \mathbf{G})))/2} \tag{10.8}$$

where $\min_{j=1}^{n_\kappa}(\omega_i^2(\kappa_j, \mathbf{G}))$ and $\max_{j=1}^{n_\kappa}(\omega_i^2(\kappa_j, \mathbf{G}))$ denote the minimum and maximum, respectively, of the ith frequency ω_i over the entire discrete wave-vector set, $\kappa_j, j = 1, \ldots, n_\kappa$, which traces the border of the IBZ. Selecting the topology distribution of the material phases inside the unit cell provides a powerful means of reaching this target; this has been the focus of numerous research studies on PnCs and photonic crystals. The band gap exists only when the minimum of the $(i + 1)$th branch is greater than the maximum of the ith branch.

10.4.2 Fitness Function

For the initialization step, most randomly generated designs at high unit-cell resolution exhibit no band gap. To address this problem, the area between the two dispersion branches of interest is used as an indicator for the fitness of the unit-cell design:

$$\text{Fitness} = \phi f(\mathbf{G}) + F \tag{10.9}$$

where ϕ is set to a large number, say $\phi = 10^4$, in order to guarantee that any design that has a band gap is selected over a design that has no band gap, and F is a step function defined as

$$F = \begin{cases} 0 & \text{if } f(\mathbf{G}) > 0 \quad \text{(Band gap exists)} \\ A & \text{if } f(\mathbf{G}) = 0 \quad \text{(No band gap)} \end{cases} \tag{10.10}$$

where A is defined as the distance between each of the two consecutive points in the two dispersion branches; that is,

$$A = \sum_{j=1}^{n_k} [(\omega_{i+1}^2(\kappa_j, \mathbf{G})) - (\omega_i^2(\kappa_j, \mathbf{G}))] \tag{10.11}$$

10.4.3 Selection

Based on the fitness function mentioned above, a tournament selection operator is used to choose the two parents; this is done by randomly selecting N designs from the current population and utilizing the best two. The advantage of tournament selection is that when a design has a low fitness value it can still contribute to the reproduction operation. In other words, the best design of the population does not dominate the selection process. In addition, because of the tournament between the N selected designs, the worst candidate in the population cannot be involved in reproduction.

10.4.4 Reproduction

Since binary chromosomes are used, the crossover operator implemented is a simple single-point crossover between the two selected designs. The offspring mutates according to a specific probability using the rule given by Bilal et al. [50]: select a random pixel x if $\sum_{r=-1}^{1} g_{x+r} > 1$, set the three pixels to ones, otherwise to zeros (see Table 10.1).

Table 10.1 Possible design segments before and after mutation.

Design segment	Mutated segment
0 0 0	0 0 0
0 0 1	0 0 0
0 1 0	0 0 0
0 1 1	1 1 1
1 0 0	0 0 0
1 0 1	1 1 1
1 1 0	1 1 1
1 1 1	1 1 1

10.4.5 Initialization and Termination

The initial population of the GA is set up randomly, but, without any loss of generality, each pair of adjacent pixels in a row has the same material type. The GA terminates when the generation counter reaches the maximum number of generations or after convergence. At the end of the search, the final design topology passes through a simple one-point flip local search, where changing one pixel to its opposite material state results in fine tuning the solution.

10.4.6 Implementation

In applying the specialized GA described above, we consider the following properties for isotropic silicon (λ and μ denote Lame's coefficients): $\rho_s = 2330\,\text{kg/m}^3$, $\lambda_s = 85.502\,\text{GPa}$, $\mu_s = 72.835\,\text{GPa}$, and a mesh resolution of $n = 32$. With this resolution, the total number of possible material distributions within the unit-cell domain is 8.7×10^{40}. This highlights the tremendously large search space that the GA must navigate. At the end of each complete GA iteration, we double the resolution of the emerged topology to 64×64 pixels, and then smooth the topology (while keeping it pixelated) by following a few simple rules.

Figure 10.8 presents the unit-cell topology and band structure of the optimized unit cell for thin-plate flexural waves, featuring a first band gap with a relative size of nearly 58%. The optimized design features a frame-like topology that is simple and manufacturable. It is observed that the design exhibits very thin connecting beams. One might expect that such a configuration would yield a locally resonant band gap. However, this is not the case. Local resonance band gaps are difficult to obtain and generally require relatively heavy resonators [8, 42].

To compare the performance of this design with a corresponding optimized, manufacturable, thin-plate configuration reported by Halkjær et al. [34], we recomputed the frequency band structure using the same material ($\rho_p = 1200\,\text{kg/m}^3$, $E_p = 2.3\,\text{GPa}$, $v_p = 0.35$) and geometric properties ($h = 0.0909a$) given in that study. We found the normalized band gap size of the new design to be 0.56, which is more than three times higher in value than the design reported by Halkjær et al. This is despite the fact that

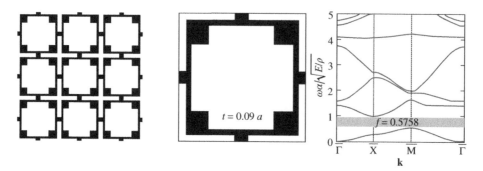

Figure 10.8 Optimized unit-cell design.

the design reported here is based on a square lattice and the design provided by Halkjær et al. is based on a hexagonal lattice, which is generally known to allow larger band gaps.

10.5 Appendix

A 1D model of a lattice material is represented by two masses, m_1 and m_2, connected in series by springs k_1 and k_2, as shown in Figure 10.9a. The equations of motion for the system are presented in Eqs. (10.12) and (10.13) for m_1 and m_2, respectively.

$$m_1 \ddot{u}_1^j + k_1(u_1^j - u_2^{j-1}) + k_2(u_1^j - u_2^j) = 0 \tag{10.12}$$

$$m_2 \ddot{u}_2^j + k_1(u_2^j - u_1^{j+1}) + k_2(u_2^j - u_1^j) = 0 \tag{10.13}$$

where u is the displacement, j is the unit-cell index number, and the double dots denote second derivative with respect to time.

For an MM, the same number of masses and springs are used, albeit connected differently (Figure 10.9b). The equations of motion for the system are presented in Eqs. (10.14) and (10.15) for m_1 and m_2, respectively.

$$m_1 \ddot{u}_1^j + k_1(2u_1^j - u_1^{j-1} - u_1^{j+1}) + k_2(u_1^j - u_2^j) = 0 \tag{10.14}$$

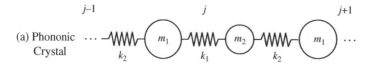

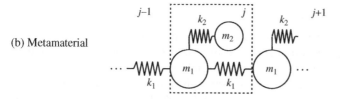

Figure 10.9 Schematic models for PnCs and MMs.

$$m_2 \ddot{u}_2^j + k_2(u_2^j - u_1^j) = 0 \tag{10.15}$$

We assume a generalized Bloch solution of the form $\mathbf{u}^{j+n} = U e^{i(\omega t + n\kappa a)}$, where ω is the frequency, κ is the wavenumber, and a is the unit-cell size. Combining the equations of motion for both systems yields a complex, generalized eigenvalue problem in the form

$$(\omega^2 \mathbf{M} + \mathbf{K}(\kappa))\mathbf{u} = \mathbf{0} \tag{10.16}$$

where the mass and stiffness matrices are written as

$$\mathbf{M}_{PnC} = \begin{bmatrix} m_1 & 0 \\ 0 & m_2 \end{bmatrix}, \mathbf{K}_{PnC}(\kappa) = \begin{bmatrix} k_1 + k_2 & -(k_1 e^{-i\kappa a} + k_2) \\ -(k_1 e^{i\kappa a} + k_2) & k_1 + k_2 \end{bmatrix} \tag{10.17}$$

$$\mathbf{M}_{MM} = \begin{bmatrix} m_1 & 0 \\ 0 & m_2 \end{bmatrix}, \mathbf{K}_{MM}(\kappa) = \begin{bmatrix} 2k_1(1 - \cos(\kappa a)) & -k_2 \\ -k_2 & k_2 \end{bmatrix} \tag{10.18}$$

In order to obtain statically equivalent models for PnCs and MMs, the longitudinal speed of sound, c, should match at the long-wavelength limit [53]:

$$c = \lim_{\kappa \to 0} \frac{d\omega}{d\kappa} \tag{10.19}$$

By using the values in Table 10.2 for both systems, we solve the eigenvalue problem in Eq. (10.16), and obtain the dispersion curves (Figure 10.10). The two dispersion curves have the same slope (group velocity) in the first branch for small κ values. However, the group velocity decays quicker in the MM as κ increases, due to the coupling between the first branch (also known as the acoustic branch) and the local resonance mode. The

Table 10.2 Summary of parameters for PnC and MM unit cells.

Cell	$m_1[kg]$	$m_2[kg]$	$k_1[N/m]$	$k_2[N/m]$	$c(\kappa \to 0)[m/s]$
MM	1	2	10 000	10 000	57.73
PnC	1	2	20 000	20 000	57.73

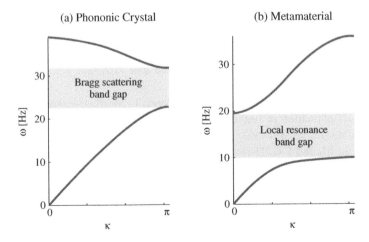

Figure 10.10 Dispersion curves for PnC and MM systems.

band gaps are highlighted in gray. The PnC band gap ranges from 22–31 Hz, while for the statically equivalent MM system, the band gap occupies a lower frequency range of 10–20 Hz.

References

1 X. Zheng, H. Lee, T. H. Weisgraber, M. Shusteff, J. DeOtte, E. B. Duoss, J. D. Kuntz, M. M. Biener, Q. Ge, J. A. Jackson, *et al.*, "Ultralight, ultrastiff mechanical metamaterials," *Science*, vol. 344, no. 6190, pp. 1373–1377, 2014.

2 A. S. Phani, J. Woodhouse, and N. Fleck, "Wave propagation in two-dimensional periodic lattices," *The Journal of the Acoustical Society of America*, vol. 119, no. 4, pp. 1995–2005, 2006.

3 S. John, "Strong localization of photons in certain disordered dielectric super lattices," *Physical Review Letters*, vol. 58, no. 23, pp. 2486–2489, 1987.

4 E. Yablonovitch, "Inhibited spontaneous emission in solid-state physics and electronics," *Physical Review Letters*, vol. 58, no. 20, pp. 2059–2062, 1987.

5 M. Sigalas and E. Economou, "Elastic and acoustic wave band structure," *Journal of Sound and Vibration*, vol. 158, no. 2, pp. 377–382, 1992.

6 M. S. Kushwaha, P. Halevi, L. Dobrzynski, and B. Djafari-Rouhani, "Acoustic band structure of periodic elastic composites," *Physical Review Letters*, vol. 71, no. 13, pp. 2022–2025, 1993.

7 M. I. Hussein, M. J. Leamy, and M. Ruzzene, "Dynamics of phononic materials and structures: Historical origins, recent progress, and future outlook," *Applied Mechanics Reviews*, vol. 66, no. 4, p. 040802, 2014.

8 Z. Liu, X. Zhang, Y. Mao, Y. Zhu, Z. Yang, C. Chan, and P. Sheng, "Locally resonant sonic materials," *Science*, vol. 289, no. 5485, pp. 1734–1736, 2000.

9 A. Khelif, A. Choujaa, S. Benchabane, B. Djafari-Rouhani, and V. Laude, "Guiding and bending of acoustic waves in highly confined phononic crystal waveguides," *Applied Physics Letters*, vol. 84, pp. 4400–4402, 2004.

10 S. Yang, J. H. Page, Z. Liu, M. L. Cowan, C. T. Chan, and P. Sheng, "Focusing of sound in a 3D phononic crystal," *Physical Review Letters*, vol. 93, no. 2, p. 024301, 2004.

11 M. I. Hussein, G. M. Hulbert, and R. A. Scott, "Dispersive elastodynamics of 1D banded materials and structures: Design," *Journal of Sound and Vibration*, vol. 307, pp. 865–893, 2007.

12 J. Christensen, A. I. Fernandez-Dominguez, F. de Leon-Perez, L. Martin-Moreno, and F. J. Garcia-Vidal, "Collimation of sound assisted by acoustic surface waves," *Nature Physics*, vol. 3, no. 12, pp. 851–852, 2007.

13 I. El-Kady, R. H. Olsson III, and J. G. Fleming, "Phononic band-gap crystals for radio frequency communications," *Applied Physics Letters*, vol. 92, p. 233504, 2008.

14 S. Mohammadi, A. A. Eftekhar, W. D. Hunt, and A. Adibi, "High-q micromechanical resonators in a two-dimensional phononic crystal slab," *Applied Physics Letters*, vol. 94, p. 051906, 2009.

15 S. A. Cummer and D. Schurig, "One path to acoustic cloaking," *New Journal of Physics*, vol. 9, no. 3, pp. 45–53, 2007.

16 D. Torrent and J. Sánchez-Dehesa, "Acoustic cloaking in two dimensions: a feasible approach," *New Journal of Physics*, vol. 10, p. 063015, 2008.

17 X. F. Li, X. Ni, L. A. Feng, M. H. Lu, C. He, and Y. F. Chen, "Tunable unidirectional sound propagation through a sonic-crystal-based acoustic diode," *Physical Review Letters*, vol. 106, p. 084301, 2011.

18 M. Eichenfield, J. Chan, R. M. Camacho, K. J. Vahala, and O. Painter, "Optomechanical crystals," *Nature*, vol. 462, no. 7269, pp. 78–82, 2009.

19 N. Cleland, D. R. Schmidt, and C. S. Yung, "Thermal conductance of nanostructured phononic crystals," *Physical Review B*, vol. 64, p. 172301, 2001.

20 E. S. Landry, M. I. Hussein, and A. J. H. McGaughey, "Complex superlattice unit cell designs for reduced thermal conductivity," *Physical Review B*, vol. 77, p. 184302, 2008.

21 J. K. Yu, S. Mitrovic, D. Tham, J. Varghese, and J. R. Heath, "Reduction of thermal conductivity in phononic nanomesh structures," *Nature Nanotechnology*, vol. 5, no. 10, pp. 718–721, 2010.

22 P. E. Hopkins, C. M. Reinke, M. F. Su, R. H. Olsson III, E. A. Shaner, Z. C. Leseman, J. R. Serrano, L. M. Phinney, and I. El-Kady, "Reduction in the thermal conductivity of single crystalline silicon by phononic crystal patterning," *Nano Letters*, vol. 11, no. 1, pp. 107–112, 2010.

23 B. L. Davis and M. I. Hussein, "Nanophononic metamaterial: Thermal conductivity reduction by local resonance," *Physical Review Letters*, vol. 112, no. 5, p. 055505, 2014.

24 M. Hussein, S. Biringen, O. Bilal, and A. Kucala, "Flow stabilization by subsurface phonons," in *Proceedings of the Royal Society of London A: Mathematical, Physical and Engineering Sciences*, vol. 471, p. 20140928, 2015.

25 R. H. Olsson III and I. El-Kady, "Microfabricated phononic crystal devices and applications," *Measures in Science and Technology*, vol. 20, p. 012002, 2009.

26 P. A. Deymier, *Acoustic Metamaterials and Phononic Crystals*, vol. 173. Springer Science & Business Media, 2013.

27 M. Maldovan, "Sound and heat revolutions in phononics," *Nature*, vol. 503, no. 7475, pp. 209–217, 2013.

28 M. I. Hussein, G. M. Hulbert, and R. A. Scott, "Tailoring of wave propagation characteristics in periodic structures with multilayer unit cells," in *Proceedings of 17th American Society for Composites Technical Conference, West Lafayette, Indiana*, 2002.

29 M. I. Hussein, K. Hamza, G. M. Hulbert, R. A. Scott, and K. Saitou, "Multiobjective evolutionary optimization of periodic layered materials for desired wave dispersion characteristics," *Structural and Multidisciplinary Optimization*, vol. 31, no. 1, pp. 60–75, 2006.

30 O. R. Bilal, M. A. El-Beltagy, and M. I. Hussein, "Optimal design of periodic Timoshenko beams using genetic algorithms," in *52nd AIAA/ASME/ASCE/AHS/ASC Structures, Structural Dynamics and Materials Conference*, 2011.

31 O. Sigmund, "Microstructural design of elastic band gap structures," in *Proceedings of the 2nd World Congress on Structural Multidisciplinary Optimization, Dalian, China*, 2001.

32 O. Sigmund and J. Jensen, "Systematic design of phononic band-gap materials and structures by topology optimization," *Philosophical Transactions: Mathematical, Physical and Engineering Sciences*, pp. 1001–1019, 2003.

33 A. R. Diaz, A. G. Haddow, and L. Ma, "Design of band-gap grid structures," *Structural and Multidisciplinary Optimization*, vol. 29, no. 1, pp. 418–431, 2005.

34 S. Halkjær, O. Sigmund, and J. S. Jensen, "Maximizing band gaps in plate structures," *Structural and Multidisciplinary Optimization*, vol. 32, pp. 263–275, 2006.

35 G. A. Gazonas, D. S. Weile, R. Wildman, and A. Mohan, "Genetic algorithm optimization of phononic bandgap structures," *International Journal of Solids and Structures*, vol. 43, no. 18–19, pp. 5851–5866, 2006.

36 M. I. Hussein, K. Hamza, G. Hulbert, and K. Saitou, "Optimal synthesis of 2D phononic crystals for broadband frequency isolation," *Waves in Random and Complex Media*, vol. 17, no. 4, pp. 491–510, 2007.

37 O. R. Bilal and M. I. Hussein, "Ultrawide phononic band gap for combined in-plane and out-of-plane waves," *Physical Review E*, vol. 84, no. 6, p. 065701, 2011.

38 O. R. Bilal and M. I. Hussein, "Topologically evolved phononic material: breaking the world record in band gap size," in *Society of Photo-Optical Instrumentation Engineers (SPIE) Conference Series*, vol. 8269, p. 12, 2012.

39 L. Shen, Z. Ye, and S. He, "Design of two-dimensional photonic crystals with large absolute band gaps using a genetic algorithm," *Physical Review B*, vol. 68, p. 035109, 2003.

40 S. Preble, M. Lipson, and H. Lipson, "Two-dimensional photonic crystals designed by evolutionary algorithms," *Applied Physics Letters*, vol. 86, pp. 061111–061111, 2005.

41 M. El-Beltagy and M. Hussein, "Design space exploration of multiphase layered phononic materials via natural evolution," in *Proceedings ASME International Mechanical Engineering Congress and Exposition, Chicago, Illinois*, 2006.

42 O. R. Bilal and M. I. Hussein, "Trampoline metamaterial: Local resonance enhancement by springboards," *Applied Physics Letters*, vol. 103, no. 11, pp. 111901–4, 2013.

43 L. Lu, T. Yamamoto, M. Otomori, T. Yamada, K. Izui, and S. Nishiwaki, "Topology optimization of an acoustic metamaterial with negative bulk modulus using local resonance," *Finite Elements in Analysis and Design*, vol. 72, pp. 1–12, 2013.

44 R. D. Mindlin, "Influence of rotatory inertia and shear on flexural motions of isotropic, elastic plates," *Journal of Applied Mechanics*, vol. 18, pp. 31–38, 1951.

45 E. Reissner, "The effect of transverse shear deformation on the bending of elastic plates," *Journal of Applied Mechanics*, vol. 12, pp. 68–77, 1945.

46 O. R. Bilal, M. A. El-Beltagy, and M. I. Hussein, "Topologically evolved photonic crystals: breaking the world record in band gap size," in *SPIE Smart Structures and Materials & Nondestructive Evaluation and Health Monitoring*, p. 834609, 2012.

47 C. M. Reinke, M. F. Su, R. H. Olsson III, and I. El-Kady, "Realization of optimal bandgaps in solid-solid, solid-air, and hybrid solid-air-solid phononic crystal slabs," *Applied Physics Letters*, vol. 98, pp. 061912–3, 2011.

48 O. Sigmund and K. Maute, "Topology optimization approaches," *Structural and Multidisciplinary Optimization*, vol. 48, no. 6, pp. 1031–1055, 2013.

49 C. Talischi, G. H. Paulino, A. Pereira, and I. F. Menezes, "Polytop: a Matlab implementation of a general topology optimization framework using unstructured

polygonal finite element meshes," *Structural and Multidisciplinary Optimization*, vol. 45, no. 3, pp. 329–357, 2012.

50 O. R. Bilal, M. A. El-Beltagy, M. H. Rasmy, and M. I. Hussein, "The effect of symmetry on the optimal design of two-dimensional periodic materials," in *Informatics and Systems (INFOS), 2010 The 7th International Conference on*, pp. 1–7, IEEE, 2010.

51 T. Belytschko, C. Tsay, and W. Liu, "A stabilization matrix for the bilinear Mindlin plate element," *Computer Methods in Applied Mechanics and Engineering*, vol. 29, no. 3, pp. 313–327, 1981.

52 D. Goldberg, *Genetic Algorithms in Search, Optimization and Machine Learning*. Addison-Wesley, 1989.

53 M. I. Hussein and M. J. Frazier, "Metadamping: An emergent phenomenon in dissipative metamaterials," *Journal of Sound and Vibration*, vol. 332, no. 20, pp. 4767–4774, 2013.

11

Dynamics of Locally Resonant and Inertially Amplified Lattice Materials

Cetin Yilmaz[1] and Gregory M. Hulbert[2]

[1] *Bogazici University, Department of Mechanical Engineering, Bebek, Istanbul, Turkey*
[2] *University of Michigan, Department of Mechanical Engineering, Ann Arbor, Michigan, USA*

11.1 Introduction

Lattice materials (periodic materials and structures) have received considerable attention in the last two decades because they can be used to prevent propagation of acoustic or elastic waves in certain frequency ranges [1–3]. These frequency ranges appear as gaps in the phononic band structure of an infinite periodic lattice, and hence they are known as *phononic band gaps.*

Phononic band gaps in periodic lattices are commonly generated by two means: Bragg scattering and local resonances [4–6]. In Bragg scattering, a band gap appears as a result of destructive interference of the wave reflections from the periodic inclusions within the lattice. Band gaps can also be formed via local resonators that hinder wave propagation around their resonance frequencies. Moreover, a band gap can be generated by the combination of these two effects [4–6].

The lowest-frequency band gap formed by Bragg scattering is of the order of the wave speed (longitudinal or transverse) of the lattice divided by the lattice constant [6, 7]. However, with the local resonance method, gaps can be obtained at much lower frequencies [7–10]. Bragg gaps and local resonance gaps appear qualitatively different in complex band structure plots [11, 12]. For each gap in the real part of the band structure plot of a lattice, the attenuation characteristics within the gap can be seen in the imaginary part of the plot. The attenuation properties of a Bragg gap vary smoothly within the gap and the maximum attenuation generally occurs close to the mid-gap frequency. On the other hand, local resonance gaps display high attenuation near the resonance frequency of the local resonator and the attenuation profile over the gap range is generally asymmetric [12]. Differences in frequency response function (FRF) plots for finite periodic lattices are also observed between Bragg gaps and local resonance gaps. Similar to the infinite periodic case, attenuation properties of a Bragg gap vary smoothly within the gap and the maximum attenuation generally occurs close to the mid-gap frequency, whereas local resonance gaps display sharp attenuation near the resonance frequency of the local resonator and the attenuation profile over the gap range is generally asymmetric [9, 13, 14].

In finite periodic structures, the frequencies where sharp attenuation occurs are known as *antiresonance frequencies.* Generally, antiresonance frequencies in a structure

Dynamics of Lattice Materials, First Edition. Edited by A. Srikantha Phani and Mahmoud I. Hussein.
© 2017 John Wiley & Sons Ltd. Published 2017 by John Wiley & Sons Ltd.

can be engendered by two different methods [15]. The first method involves resonating substructures: the well-known local resonance effect. The second method involves inertial coupling and amplification within the structure [15–17]. Both methods are capable of producing low-frequency band gaps in finite and infinite periodic lattices. However, these two methods produce qualitatively different results in terms of wave-energy localization characteristics. In local resonance induced gaps, a considerable fraction of the wave energy propagates through the lattice because local resonances are active throughout the lattice. On the other hand, the energy is localized near the excitation source for inertial amplification induced gaps [17]. In other words, since the antiresonance generation phenomena are different in locally resonant and inertially amplified lattices, their wave-propagation characteristics are qualitatively different.

In this chapter, lattices that owe their band gaps to antiresonance effects will be investigated. The differences and similarities of locally resonant and inertially amplified lattices will be shown through 1D, 2D, and 3D lattice examples from the literature.

11.2 Locally Resonant Lattice Materials

There are many studies in the literature regarding 1D, 2D, and 3D locally resonant lattices [6–14, 18–36]. However, only one study per lattice type will be examined, for brevity and clarity, but without loss of generality.

11.2.1 1D Locally Resonant Lattices

In this section, 1D locally resonant lattices, such as rods, shafts, strings and beams, will be investigated. The governing equation for the longitudinal excitations in rods (bars), torsional excitations in shafts, and transverse excitations in strings is the 1D wave equation:

$$\frac{d^2u}{dt^2} = c^2 \frac{d^2u}{dx^2} \tag{11.1}$$

where u represents axial displacement in rods, angular displacement in shafts, or transverse displacement in strings and c represents the wave speed in the medium. These three structures are therefore dynamically equivalent. The governing equation for flexural waves in slender (Euler–Bernoulli) beams is

$$\frac{d^2w}{dt^2} + c^2 \frac{d^2w}{dx^4} = 0 \tag{11.2}$$

where w represents transverse displacement of the beam and c again represents the wave speed.

First, we consider longitudinal (axial) vibrations in rods. Figure 11.1 shows an elastic rod with periodically attached resonators [13]. The elastic modulus, density, and cross-sectional area of the rod are E, ρ, and A, respectively. The stiffness and mass of each resonator are k and m, respectively. The lattice constant is L_x. Using the transfer matrix method and the Bloch theorem, the dispersion equation of the lattice can be obtained [13]:

$$\cos(\pm\varepsilon) = \cosh(\pm\mu) = \cos(\beta L_x) - \frac{km\omega^2}{2\beta EA(k - m\omega^2)} \sin(\beta L_x) \tag{11.3}$$

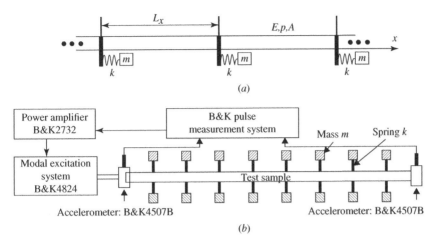

Figure 11.1 Longitudinal (axial) vibrations in rods: (a) infinite periodic elastic rod with resonators, with longitudinal waves propagating along the x-direction considered here; (b) finite periodic rod with eight resonators used in the experiment. Source: Wang et al. [13], with permission of ASME.

where ω is the angular frequency, $\beta = \omega/c$ is the longitudinal wave number of the uninterrupted uniform rod at frequency ω, and $c = \sqrt{E/\rho}$ is the longitudinal elastic wave speed in the rod, and μ and ε are the attenuation and phase constants, respectively.

Figure 11.2a shows the calculated attenuation and phase constants (μ and ε) of the 1D locally resonant lattice made up of an organic glass rod with periodic resonators. Here, $L_x = 0.05$ m, $A = 5 \times 10^{-6}$ m^2, $E = 1.5 \times 10^{10}$ Pa, and $\rho = 1200$ kg/m^3. Each resonator is composed of a pair of steel beams (acting as springs) and masses. They are placed symmetrically on the rod to counteract moments exerted on the rod. The stiffness and mass of the resonators are $k = 5.12 \times 10^6$ N/m and $m = 47.6$ g, respectively [13]. In Figure 11.2a, a band gap is formed in the range 1584–3047 Hz. One can obtain the normalized band gap frequency range by multiplying these frequencies by L_x/c. For the parameter values above, $c = 3536$ m/s and the normalized band gap is obtained in the range 0.0224–0.0431. This range is quite low compared to that of a Bragg gap, which is generally centred around 0.5 in terms of normalized frequency. Moreover, unlike a Bragg gap, the attenuation is highly asymmetric within this locally resonant band gap, in the sense that attenuation is very high close to the lower limit of the band gap, whereas it is quite low close to the upper limit. Figure 11.2b depicts the FRF of 1D locally resonant lattices comprising six and eight unit cells. Calculations and measurements match quite well for the finite lattices and the band gap is formed in the same frequency range as in Figure 11.2a [13]. In Figure 11.2a, the maximum attenuation occurs at $\sqrt{k/m}/2\pi = 1651$ Hz, which corresponds to the natural frequency of the resonators. The local resonances of the resonators cause an antiresonance notch at the same frequency in the FRF plot shown in Figure 11.2b.

Analogous to these locally resonant rods that are excited longitudinally, there are pipe systems that have periodic Helmholtz resonators [18]. Therefore, 1D locally resonant lattices exist for acoustic waves, as well.

Secondly, we consider torsional vibrations in shafts. Figure 11.3 shows a shaft with periodically attached torsional resonators [14]. The torsional resonators consist of a soft

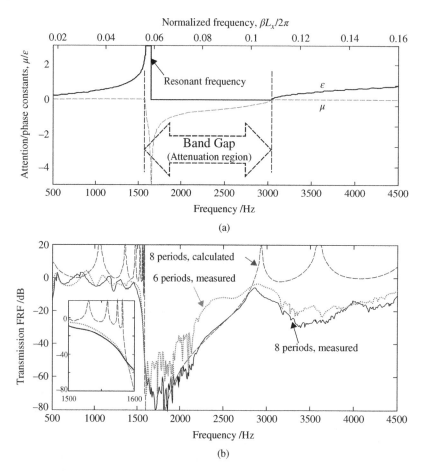

Figure 11.2 Infinite 1D locally resonant lattice made up of organic glass rod: (a) calculated attenuation (μ) and phase (ε) constants; (b) calculated and measured FRF for six and eight unit cells. Source: Wang et al. [13], with permission of ASME.

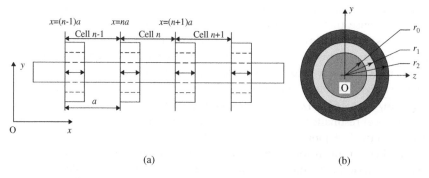

Figure 11.3 Infinite periodic elastic shaft with torsional resonators: (a) schematic of the shaft, with torsional waves propagating along the x-direction considered here; (b) cross-section. Source: Yu et al. [14], with permission of Elsevier.

rubber ring enclosed with an outer metal ring. The outer radius of the rubber ring is r_1 and the outer radius of metal ring is r_2. The common thickness of the two rings is l. The lattice constant is a. G_0, ρ_0, G_1, ρ_1, and G_2, ρ_2, are the shear modulus and density of the shaft, rubber ring, and metal ring, respectively. Using the transfer matrix method and the Bloch theorem, the dispersion equation of the lattice can be obtained [14]:

$$cos(ka) = cos(\beta a) - \frac{KI\omega^2}{2\beta G_0 J_t(K - I\omega^2)} sin(\beta a) \tag{11.4}$$

where ω is the angular frequency, $\beta = \omega/c$ is the wave number of the uninterrupted uniform shaft at frequency ω, $c = \sqrt{G_0/\rho_0}$ is the torsional wave speed in the circular shaft, k is the wave vector, K is the torsional stiffness of the rubber rings in the resonator, I is the mass moment of inertia of the metal rings in the resonator, and J_t is the polar moment of area of the shaft.

Figure 11.4 shows the complex band structure of the lattice. The shaft is made up of epoxy ($G_0 = 1.59 \times 10^9$ Pa, $\rho_0 = 1180$ kg/m³) and the resonators are made up of rubber and lead ($G_1 = 3.4 \times 10^5$ Pa, $\rho_1 = 1300$ kg/m³ and $G_2 = 1.49 \times 10^{10}$ Pa, $\rho_2 = 11600$ kg/m³). Moreover, the dimensions of the lattice are as follows: $r_0 = 5$ mm, $r_1 = 8$ mm, $r_2 = 10$ mm, $l = 25$ mm and $a = 75$ mm. As seen in Figure 11.4a a band gap is formed in the range 198–1138 Hz. One can obtain the normalized band gap frequency range by multiplying these frequencies by a/c. For the parameter values above, $c = 1161$ m/s and the normalized band gap is obtained in the range 0.0128–0.0735. This range is quite low compared to a Bragg gap, which is generally centred around 0.5 in terms of the normalized frequency. Again, the asymmetric attenuation profile is obtained within the band gap (as seen in Figure 11.4b), which is typical for a locally resonant gap.

Figure 11.5 depicts the FRF of the locally resonant shaft consisting of eight unit cells, obtained using the finite-element method. The band gap frequency range agrees quite well with that of the infinite periodic case shown in Figure 11.4 [14]. In Figure 11.4b, the maximum attenuation occurs at $\sqrt{K/I}/2\pi = 203$ Hz, which corresponds to the torsional natural frequency of the resonators. The local resonances of the resonators

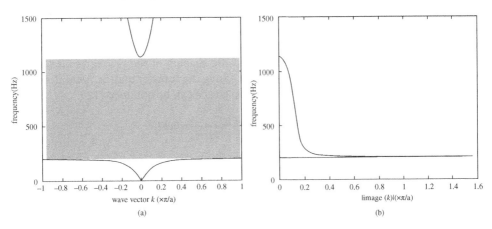

Figure 11.4 Complex band structure of the elastic shaft with torsional resonators. (a) Real wave vector. (b) Absolute value of the imaginary part of the complex wave vector. Source: Yu et al. [14], with permission of Elsevier.

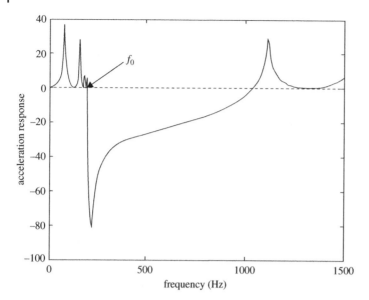

Figure 11.5 Frequency response function of the locally resonant shaft with eight unit cells. Source: Yu et al.[14], with permission of Elsevier.

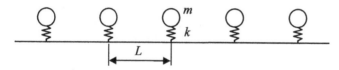

Figure 11.6 Taut string with periodically attached resonators. Source: Xiao et al. [12], with permission of Elsevier.

cause an antiresonance notch at the same frequency in the FRF plot, shown in Figure 11.5. Other examples of 1D locally resonant shafts with torsional resonators can be found in the paper by Song et al. [19] and the references therein.

As a third case, we consider transverse vibrations in strings. Figure 11.6 shows a taut string under tension, T, with periodically attached identical resonators [12]. The mass density per unit length of the string is ρ and the spacing of the resonators is L. Each resonator consists of a spring k and a mass m. Using the transfer matrix method and the Bloch theorem, the dispersion equation of the lattice can be obtained [12]:

$$cos(qL) = cosh(iqL) = cos(\beta L) - \frac{km\omega^2}{2\beta T(k - m\omega^2)} sin(\beta L) \tag{11.5}$$

where ω is the angular frequency, $\beta = \omega/c$ is the wave number of the uninterrupted uniform string at frequency ω, $c = \sqrt{T/\rho}$ is the wave speed in the string, and q is the complex wave vector.

To describe cogently the band structure of the lattice, we define:

- the nondimensional mass of the resonators, $\gamma = m/(\rho L)$
- nondimensional frequency, $\Omega = \beta L/\pi$
- nondimensional resonance frequency of the resonator, $\Omega_0 = (\Omega/\omega)\sqrt{k/m}$

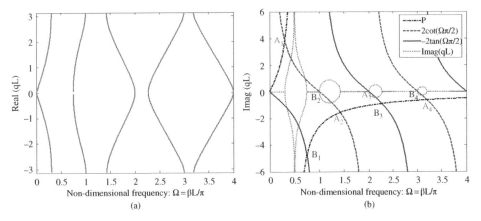

Figure 11.7 Complex band structure of the string with local resonators for $\Omega_0 = 0.5$ and $\gamma = 2.5$: (a) real wave vector; (b) absolute value of the imaginary part of the complex wave vector. Source: Xiao et al. [12], with permission of Elsevier.

- negative of the nondimensional dynamic stiffness of the resonator, $P = (\gamma \pi \Omega_0^2 \Omega)/(\Omega_0^2 - \Omega^2)$.

Figure 11.7 shows the complex band structure of the lattice for $\Omega_0 = 0.5$ and $\gamma = 2.5$. In Figure 11.7b, imag(qL) is the imaginary part of the wave vector. Intersection of the curves P, $2 \cot(\Omega \pi /2)$, and $-2 \tan(\Omega \pi /2)$ define the band edge frequencies (labelled A_i and B_i in Figure 11.7b) [12]. The band gap centred at $\Omega = 0.5$ is due to the local resonance effect as it shows sharp attenuation characteristics at the natural frequency of the resonators. Moreover, the other band gaps have integer band edge frequencies and the attenuation characteristics within these band gaps vary smoothly. Thus, they are Bragg type gaps (Here, note that the normalized frequency $\Omega = 2L/c$, which is twice the normalized frequency of the earlier cases). As discussed in Ref. [12], if the resonance frequency of the resonator is placed at 0.5 or some integer plus 0.5, then the attenuation profile within the gap is symmetrical. However, in general the natural frequency of the resonators can be at any frequency, which results in an asymmetrical attenuation profile.

To summarize, since the governing equation for the longitudinal excitations in rods, torsional excitations in shafts and transverse excitations in strings is the 1D wave equation, their dispersion equations have the same form; compare Eqs. (11.3), (11.4), and (11.5).

As a fourth case, we consider flexural vibrations in slender (Euler–Bernoulli) beams. Figure 11.8 shows a beam with periodically attached resonators [20]. The governing equation for this case is different to the earlier cases; compare Eqs. (11.1) and (11.2). The dispersion relation of Wang et al. [20] is not repeated here. However, the band structure and the FRF of the lattice are shown in Figure 11.9 for the following dimensions and material parameter values: $a = 0.05$ m, $b = h = 0.01$ m, $d = 0.04$ m, $h_{rub} = 0.002$ m; $\rho_{Al} = 2799$ kg/m^3, $\lambda_{Al} = 589.55 \times 10^8$ Pa, $\mu_{Al} = 268.12 \times 10^8$ Pa for duralumin; $\rho_{rub} = 1300$ kg/m^3, $\lambda_{rub} = 15.32 \times 10^6$ Pa, $\mu_{rub} = 1.02 \times 10^6$ Pa for silicone rubber; $\rho_{Cu} = 8356$ kg/m^3, $\lambda_{Cu} = 1.726 \times 10^{10}$ Pa, $\mu_{Cu} = 7.527 \times 10^{10}$ Pa for copper. Here, λ and μ are the Lamé constants.

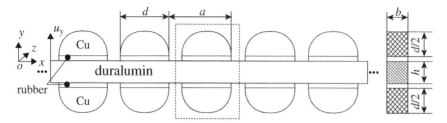

Figure 11.8 Beam with periodically attached resonators. Source: Wang et al. [20], with permission of APS.

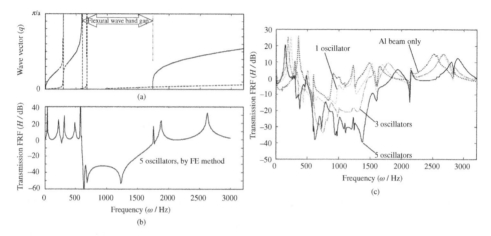

Figure 11.9 (a) Band structure of the locally resonant beam; (b) calculated frequency response function of the locally resonant beam with five unit cells; (c) measured frequency response function of the finite periodic locally resonant beams. The solid, dash-dot, dashed and dotted lines represent the measured results corresponding to samples of five, three, one, and no oscillators, respectively. Source: Wang et al. [20], with permission of APS.

Figure 11.9a shows the band structure of the locally resonant beam. Unlike the earlier cases, there are many branches. Consequently, the band structure is qualitatively different. In Figure 11.9a a band gap is formed in the range 596–1734 Hz, which matches well with those of the calculated and measured finite periodic cases shown in Figure 11.9b and c, respectively. One can obtain the normalized band gap frequency range by multiplying the band edge frequencies by a/c. For the parameter values above, the normalized band gap is obtained in the range 0.1272–0.2170 [20]. This range is quite low compared to a Bragg gap, which is generally centred around 0.5 in terms of normalized frequency.

Finally, unlike the earlier cases, there are multiple antiresonance frequencies within the band gap of the FRF plot shown in Figure 11.9b. These antiresonance frequencies correspond to the different vibration modes of the resonators. Multiple antiresonance frequencies within the band gap of locally resonant beams can also be seen in other studies [21–23], in particular, in thick beams (Timoshenko beams), where shear deformations and rotary inertia are taken into consideration [22, 23]. Finally, Liu and Hussein [24] have compared the band structures of Euler–Bernoulli and Timoshenko beams.

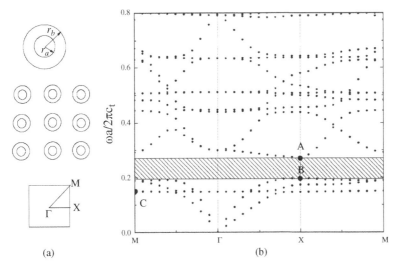

Figure 11.10 Square array of rubber coated lead cylinders in glass matrix: (a) cross-section and the Brillouin zone of the square lattice; (b) band structure of the lattice for a filling fraction of 0.4 and r_a/r_b = 0.722. Source: Zhang et al. [25], with permission of Elsevier.

11.2.2 2D Locally Resonant Lattices

In this section, 2D locally resonant lattices will be investigated. These lattices can be designed to achieve band gaps for in-plane or out-of-plane excitations.

First, we consider in-plane response of 2D locally resonant lattices. Figure 11.10a shows a three-component (ternary) lattice with square unit cell and the Brillouin zone of the lattice, in which the irreducible zone is shown as a triangle. In this lattice, there are infinitely long rubber coated lead cylinders in a glass matrix and the in-plane response is calculated on the edges of the irreducible Brillouin zone using the plane-wave expansion method [25]. The material parameters of the lattice are as follows: $\rho = 11.4$ g/cm^3, $c_l = 2.16$ km/s, $c_t = 0.86$ km/s for lead; $\rho = 1.0$ g/cm^3, $c_l = 1.83$ km/s, $c_t = 0.5$ km/s for rubber; $\rho = 2.6$ g/cm^3, $c_l = 5.84$ km/s, $c_t = 3.37$ km/s for glass; where ρ, c_l and c_t are the density, and the longitudinal and transverse wave speeds, respectively. Moreover, r_a/r_b is the radius ratio of the core and the coating layer. Figure 11.10b shows the band structure of the locally resonant square lattice for a filling fraction of 0.4 and $r_a/r_b = 0.722$. One can see that a band gap is present below the Bragg limit, which is attained when $\omega a/2\pi c_t = 0.5$. Here, a is the lattice constant and c_t is the transverse wave speed in the glass matrix. This band gap is formed because of the local resonances of the lead cylinders within the glass matrix [25].

In Figure 11.11, the normalized gap width ($\Delta\omega/\omega_g$) versus radius ratio of the core and the coating layer (r_a/r_b) is shown. Here, ω_g represents the mid-gap frequency. Note that for every filling fraction, the gap opens for a certain range of radius ratios. Moreover, for every filling ratio, the largest gap is attained at a different radius ratio. Furthermore, wider band gaps are obtained by using larger filling fractions [25].

The dependence of the normalized gap width on the radius ratio of the core and the coating layer or the filling fraction can be found in other studies, as well [6, 7]. Still other

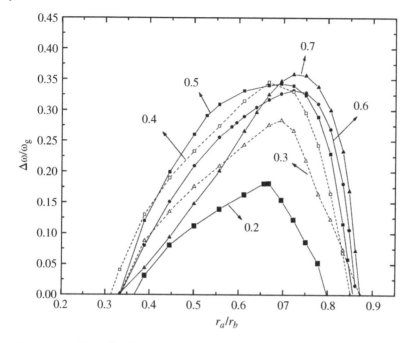

Figure 11.11 Normalized gap width ($\Delta\omega/\omega_g$) versus radius ratio of the core and the coating layer (r_a/r_b). Here, the numbers on each curve represent the filling fraction for that curve. Source: Zhang et al. [25], with permission of Elsevier.

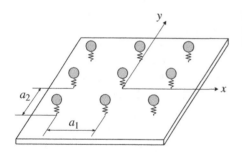

Figure 11.12 Thin plate with 2D periodic array of attached mass-spring resonators. Source: Xiao et al. [11], with permission of IOP.

papers investigate the in-plane response of 2D locally resonant mass-spring lattices [9, 26].

Secondly, we consider the out-of-plane response of 2D locally resonant lattices. Figure 11.12 shows a thin aluminium plate with a 2D periodic array of attached mass-spring resonators. The material parameters of the plate are: $E = 70$ GPa, $\rho = 2700$ kg/m^3, and $\nu = 0.3$. The plate thickness, $h = 0.002$ m, and the resonators are equally spaced along the in-plane x- and y-directions ($a_1 = a_2 = a = 0.1$ m). Finally, the resonator stiffness and mass are $k_R = 9.593 \times 10^4$ N/m and $m_R = 0.027$ kg, respectively [11]. The resonance frequency of each resonator is therefore $f_R = \sqrt{k_R/m_R}/2\pi = 300$ Hz.

Figure 11.13a shows the real part of the band structure of the locally resonant plate and the bare plate without resonators. Note that two band gaps (g_1 and g_2) are formed along the ΓX direction and a complete band gap (G_1) is formed for the locally

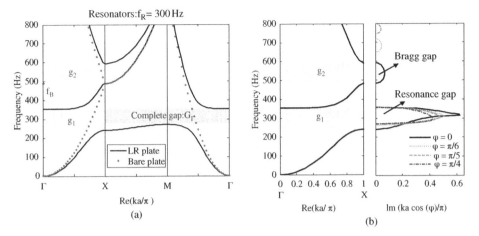

Figure 11.13 (a) Real band structures of the locally resonant plate (solid lines) and the bare plate without resonators (dotted lines). (b) Real and imaginary parts of the complex band structure of the locally resonant plate along the ΓX direction ($\varphi = 0$). The imaginary part of the plot also contains solutions along other directions. Here, $\varphi = \pi/4$ corresponds to the ΓM direction. Source: Xiao et al. [11], with permission of IOP.

resonant plate. Figure 11.13b shows the real and imaginary parts of the complex band structure of the locally resonant plate along the ΓX direction. Note that the resonance frequency of the resonators ($f_R = 300$ Hz) lies within the first band gap, g_1. Moreover, for this lattice, the lowest-frequency Bragg gap along the ΓX direction is obtained at $f_B = (\pi/a)^2 \sqrt{D/\rho h}/2\pi$, where $D = Eh^3/12(1-v^2)$ is the flexural rigidity of the base plate. For the given parameters, the Bragg limit is $f_B = 484$ Hz [11]. This frequency marks the lowest limit of the second gap (g_2) in the band structure plots shown in Figures 11.13a and b. Moreover, the imaginary part of the complex band structure of the lattice in Figure 11.13b also shows that the attenuation profile within the second gap (g_2) is smooth, which is typical for a Bragg gap. On the other hand, attenuation is very sharp within the first gap (g_1) when the excitation frequency matches the resonance frequency of the resonators ($f_R = 300$ Hz).

There are also experimental studies on the out-of-plane response of locally resonant plates, in which the local resonators are realized by depositing silicone rubber on an aluminium base plate [27]. By using lead caps on the silicone rubber resonators, wider band gaps at lower frequencies are obtained [28]. Alternatively, wide band gaps can be obtained by cutting out regions from the base plate [29].

Locally resonant band gaps can also be generated for surface acoustic waves. In order to generate this type of band gap, 2D arrays of resonators are placed on 3D solid structures [30, 31]. Finally, local resonance induced band gaps can also be seen in 2D acoustic systems [32]. Here, local resonances are generated by Helmholtz resonators.

11.2.3 3D Locally Resonant Lattices

In this section, 3D locally resonant lattices will be investigated. Typical 3D configurations achieve local resonances using three-component (ternary) lattices in which heavy masses are coated with soft layers and embedded in stiff matrices [6, 8, 33–35].

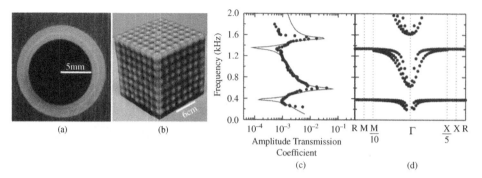

Figure 11.14 Silicone rubber coated lead sphere in an epoxy matrix: (a) cross section; (b) 8×8×8 ternary locally resonant lattice; (c) calculated (solid line) and measured (circles) amplitude transmission coefficient along the [100] direction; (d) band structure of the 3D lattice. Source Liu et al. [8], with permission of AAAS.

Conversely, local resonances can also be generated by forming holes in a 3D solid structure [36].

First, we consider three-component (ternary) locally resonant lattices. Figure 11.14a shows the cross section of a silicone rubber coated lead sphere in an epoxy matrix, which is the unit cell for the $8 \times 8 \times 8$ SC crystal with a lattice constant of 1.55 cm (see Figure 11.14b). Here, the lead balls have 1 cm diameter and the silicone coatings are 2.5 mm thick. Figure 11.14c shows the measured sound transmission from the center of the specimen to one of the faces; effectively, four layers of the lattice are utilized. The ratio of the amplitude measured at the center to the incident wave shows two dips (antiresonances) at 380 Hz and 1350 Hz followed by two peaks (resonances). Figure 11.14d shows the band structure for an infinite periodic structure with a SC arrangement of coated spheres, which was obtained by using the multiple-scattering theory [8]. Note that the antiresonances in Figure 11.14c define the lower limits of the locally resonant band gaps in Figure 11.14d, while the resonances define the upper limits. The mid-gap frequency of the first gap is around 500 Hz and the lattice constant of this lattice is 300 times smaller than that of longitudinal wavelength in epoxy. Therefore, this local resonance induced gap is formed at a frequency that is two orders of magnitude lower than the Bragg limit [8].

Figure 11.15 shows the calculated displacements of the unit cell at the first and second antiresonance frequencies [8]. At the first antiresonance frequency (Figure 11.15a), the lead balls displace within the epoxy matrix by straining the silicone rubber. Here, the lead ball acts like a heavy mass and the silicone rubber acts like a soft spring. At the second antiresonance frequency, the maximum displacement occurs inside the silicone rubber (Figure 11.15b). The displacement of the lead ball is small but nonzero. This is analogous to the "optical mode" in molecular crystals with two atoms per unit cell, where one of the atoms is much heavier than the other [8].

Secondly, we consider local resonance induced gaps in 3D solid structures with holes. Unlike three-component (ternary) lattices, there is only one material in this type of lattice. Figure 11.16a shows the unit cell of the 3D structure with a SC arrangement of holes. In this structure, the resonators are periodically arranged cubic lumps in the cubic holes, connected to the matrix by narrow connectors. These narrow connectors

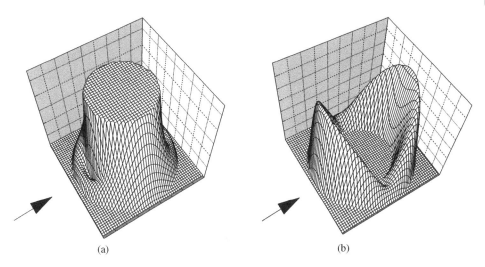

(a) (b)

Figure 11.15 Calculated displacements of the unit cell at the first (a) and second (b) antiresonance frequencies. The displacement shown is for a cross-section through the center of one coated sphere, located at the front surface. The arrows indicate the direction of the incident wave. Source Liu et al. [8], with permission of AAAS.

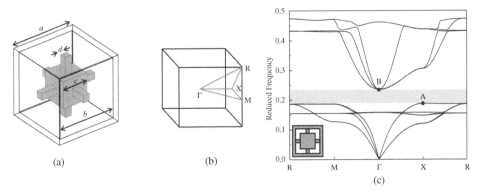

(a) (b) (c)

Figure 11.16 3D structure with SC arrangement of holes: (a) unit cell; (b) corresponding first Brillouin zone of the periodic lattice; (c) band structure of the 3D lattice. Source: Wang and Wang [36], with permission of ASME.

act as soft springs (analogous to the soft coatings in ternary 3D lattices). The band structure of the lattice is shown in Figure 11.16c for the following dimensions: $b/a = 0.9$, $c/a = 0.8$ and $d/a = 0.1$. As reduced frequencies ($\omega a/2\pi c_t$) are used in this plot, only Poisson's ratio ($\nu = 0.33$) is given as a material parameter. The band gap frequency range is 0.19–0.23 in terms of the reduced frequency, which is below the Bragg limit [36].

Figure 11.17 shows the vibration modes of the unit cell at the band edges in Figure 11.16c. For the lower-edge mode (point A in Figure 11.16c), the lump oscillates like a rigid body, with the connectors acting as springs; the external frame is almost still (see Figures 11.17a–c). For the upper-edge mode (point B in Figure 11.16c), the lump and the external frame vibrate in opposite phases along the body diagonal of the unit cell (see Figures 11.17d–f). So the emergence of the complete band gap is due to

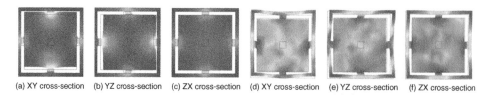

(a) XY cross-section   (b) YZ cross-section   (c) ZX cross-section   (d) XY cross-section   (e) YZ cross-section   (f) ZX cross-section

Figure 11.17 Vibration modes of the unit cell at the band edges in Figure 11.16(c). Here, (a)–(c) show the modes corresponding to the lower-edge of the band gap (point A in Figure 11.16c), whereas (d)–(f) show the modes corresponding to the upper-edge of the band gap (point B in Figure 11.16c). Source: Wang and Wang [36], with permission of ASME.

the local resonances of the resonators, which is similar to the case of ternary locally resonant lattices [36].

Finally, local resonance induced band gaps can be seen in 3D acoustic systems as well [37]. Here, local resonances are generated by Helmholtz resonators embedded in a fluid matrix.

11.3 Inertially Amplified Lattice Materials

Inertial amplification is a new method of generating band gaps in lattice materials. Unlike locally resonant lattices, there are relatively few studies in the literature regarding 1D, 2D, and 3D inertially amplified lattices. The differences and similarities between locally resonant and inertially amplified lattices will be highlighted below.

11.3.1 1D Inertially Amplified Lattices

The idea of generating band gaps in lattice structures through inertial coupling and amplification was first proposed by Yilmaz et al. [16], who studied a 2D lattice. Figure 11.18a shows a unit cell that can be used in a 1D mass-spring lattice with inertial amplification. Here, the thick links that connect the masses m to the mass m_a are rigid. Hence, the motion of the mass m_a is coupled to the two ends of the unit cell and there is no additional degree of freedom associated with this mass. As the two ends of the unit cell displace, amplified inertial forces are generated by the mass m_a, provided that the angle θ is small.

To highlight the effect of amplified inertial forces, the equation of motion of the system in Figure 11.18a is given by

$$[m + m_a(\cot^2\theta + 1)/4]\ddot{x} + kx = [m_a(\cot^2\theta - 1)/4]\ddot{y} + ky \tag{11.6}$$

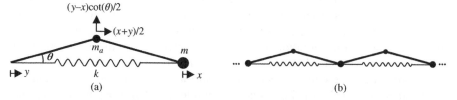

Figure 11.18 1D mass-spring lattice with inertial amplification: (a) the unit cell; (b) unit cells connected in series to form the lattice.

In Eq. (11.6), y is considered as the input displacement and x is considered as the output displacement, the resonance and the antiresonance frequencies of this single degree-of-freedom system are obtained as, respectively,

$$\omega_p = \sqrt{\frac{k}{m + m_a(\cot^2\theta + 1)/4}} \tag{11.7}$$

$$\omega_z = \sqrt{\frac{k}{m_a(\cot^2\theta - 1)/4}} \tag{11.8}$$

In the denominator of the resonance and the antiresonance frequencies of the system, the mass m_a is multiplied by $\cot^2\theta$. Hence, the effective inertia of the system is amplified.

The unit cell shown in Figure 11.18a is connected in series to form a 1D mass-spring lattice with inertial amplification, depicted in Figure 11.18b. The band structure of this lattice will be compared to locally resonant mass-spring lattice shown in Figure 11.19. Unit cells of both lattices will have the same masses (m, m_a) and same static stiffness (k) between x and y. These lattices will also be compared with a mass-spring lattice composed of alternating masses m and m_a connected in series by springs with stiffness $2k$ as shown in Figure 11.20. Hence, the static stiffness of the unit cell of this third lattice between x and y is also k.

It is assumed that the lattice constant is a for all the three lattices. Using the transfer matrix method and the Bloch theorem, the dispersion equations of the inertially amplified, locally resonant, and alternating mass-spring lattices are given respectively, by

$$\cos(\gamma a) = \frac{2k - \omega^2[m + m_a(\cot^2\theta + 1)/2]}{2k - \omega^2 m_a(\cot^2\theta - 1)/2} \tag{11.9}$$

$$\cos(\gamma a) = 1 - \frac{k_a^2}{2k(k_a - \omega^2 m_a)} + \frac{k_a - \omega^2 m}{2k} \tag{11.10}$$

$$\cos(\gamma a) = 1 + \frac{\omega^4 m m_a - 4k(m + m_a)\omega^2}{8k^2} \tag{11.11}$$

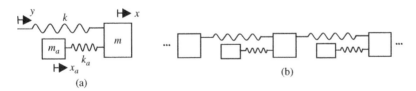

(a)　　　(b)

Figure 11.19　1D locally resonant mass-spring lattice: (a) the unit cell; (b) unit cells connected in series to form the lattice.

(a)　　　(b)

Figure 11.20　1D alternating mass-spring lattice: (a) the unit cell; (b) unit cells connected in series to form the lattice.

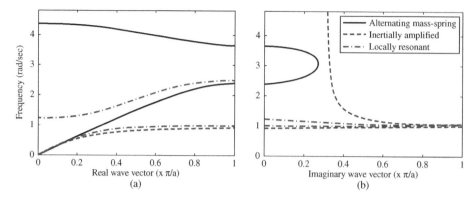

Figure 11.21 Complex band structures of the alternating mass-spring, inertially amplified, and locally resonant lattices: (a) real wave vector; (b) absolute value of the imaginary part.

where ω is the angular frequency and γ is the complex wave vector. Figure 11.21 shows the complex band structures of these three lattices for $m = 0.7$, $m_a = 0.3$, and $k = 1$. For the inertially amplified lattice, $\theta = \pi/12$. For the locally resonant lattice, $k_a = 4k/(\cot^2\theta - 1)$ so that the antiresonance frequency of the locally resonant lattice $(\sqrt{k_a/m_a})$ is equal to that of the inertially amplified lattice (see Eq. (11.8)). As seen in Figure 11.21a, there is only one branch for the inertially amplified lattice, because its unit cell has one degree of freedom while the other two lattices have two branches. Consequently, the inertially amplified lattice has a semi-infinite band gap. Figure 11.21b shows the imaginary parts of the wave vectors for the three lattices, focusing on the lowest-frequency band gaps. Note that the attenuation profiles of these three lattices are qualitatively different. The alternating mass-spring lattice shows a Bragg gap. The inertially amplified and locally resonant lattices show sharp attenuation near their antiresonance frequencies which have values around one-third of the mid-gap frequency of the Bragg gap. However, inertially amplified lattice offers quite high values of attenuation above the antiresonance frequency whereas the locally resonant lattice offers high attenuation only near the antiresonance frequency.

One-dimensional inertially amplified lattices have also been studied in the case of distributed parameter systems composed of beams [38,41]. Taniker and Yilmaz [41] shows that ultra wide gaps can be realized by the inertial amplification method, which surpass the ones that can be obtained by Bragg scattering or local resonances.

11.3.2 2D Inertially Amplified Lattices

In this section, 2D inertially amplified lattices will be described. These are designed to achieve band gaps for in-plane excitations.

First, we consider a 2D inertially amplified mass-spring lattice [17]. Figure 11.22 shows the 2D mass-spring lattice with inertial amplification and its square unit cell. Here, the horizontal and vertical thin lines have stiffness k and the large dots have mass m. Each central mass m_c is attached to the neighboring masses m via lines with stiffness k_c. Consequently, k, k_c, m, and m_c form the structural backbone of the lattice. Thus the nodes with masses m and m_c are denoted as structural nodes. The thick lines with stiffness k_a and the small dots with mass m_a form the amplification mechanisms. The angle θ determines the amplification within the mechanisms. If the lattice is excited at a frequency

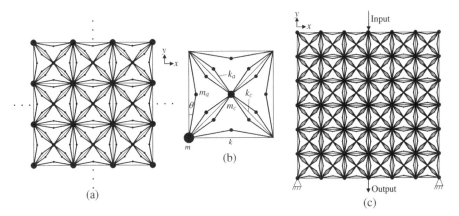

Figure 11.22 2D mass-spring lattice with inertial amplification: (a) the lattice; (b) the unit cell; (c) 6×6 finite periodic lattice. Source: Yilmaz and Hulbert [17], with permission of Elsevier.

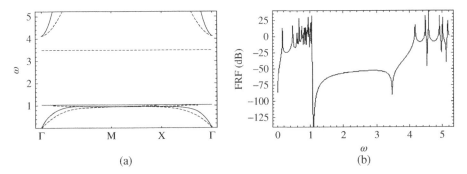

Figure 11.23 2D mass-spring lattice with inertial amplification: (a) band structure; (b) FRF plot of the 6×6 lattice. Source: Yilmaz and Hulbert [17], with permission of Elsevier.

less than the resonance frequencies of the amplification mechanisms and θ is small, the relative motion of the structural nodes will cause amplified motion for the masses m_a. Consequently, amplified inertial forces will be generated by the masses m_a [17].

Figure 11.23a shows the band structure of the 2D inertially amplified lattice for $k = 1$, $m = 0.2$, $k_c = 1$, $m_c = 0.2$, $k_a = 10$, $m_a = 0.05$ and $\theta = \pi/18$. Here, longitudinal and transverse branches are represented with continuous and dashed lines, respectively. Note that there are two band gaps separated by a flat branch corresponding to the transverse resonance modes of the amplification mechanisms. The flat branch is at $\omega = \sin(\theta)\sqrt{2k_a/m_a} = 3.47$. The first band gap between 1.04 and 3.47 is due to the inertial amplification effect, whereas the second band gap between 3.47 and 4.14 is due to local resonances. Figure 11.23b shows the FRF plot of the 6×6 lattice depicted in Figure 11.22c. Notice that the antiresonance notch generated by the inertial amplification effect at $\omega = 1.09$ is quite deep, and there is a relatively flat and deep interval until the second antiresonance frequency at $\omega = 3.47$, which is due to the local resonance effect. Since the FRF values (output over input accelerations) are computed at a structural node (see Figure 12.22c), instead of a resonance peak, there is an antiresonance notch in the FRF plot at $\omega = 3.47$. As a result, in Figure 11.23b, rather than two gaps, there is a combined gap [17].

The FRF plot only gives information about a specific (in this case, the output) node and it does not provide insight regarding the two different antiresonance generation methods in this lattice. However, when we consider the whole lattice, the perspectives are quite different. Figure 11.24a shows the contour plot of acceleration at the inertial amplification induced antiresonance frequency, $\omega = 1.09$. One can see that the energy is localized near the excitation source. Figure 11.24b shows the contour plot of acceleration at the local resonance induced antiresonance frequency, $\omega = 3.47$. At this frequency, the structural nodes have very small acceleration and appear as black dots. However, due to local resonances, energy propagates throughout the lattice, exhibited by light gray areas throughout the whole lattice. Therefore, the two types of antiresonance-frequency-generating methods produce qualitatively different results in terms of energy localization [17].

Secondly, we consider inertial amplification induced band gaps in 2D solid structures. Figure 11.25 shows an inertial amplification mechanism that is formed by combining different size beam sections [38]. Here, l_i, t_i, and m_i are the length, thickness, and mass of the ith beam section that form the mechanism. t_2 and t_4 are much smaller than t_1 and t_3,

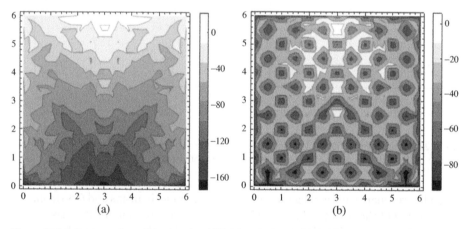

Figure 11.24 Contour plots of acceleration (dB): (a) at the inertial amplification induced antiresonance frequency, $\omega = 1.09$; (b) at the local resonance induced antiresonance frequency, $\omega = 3.47$. Source: Yilmaz and Hulbert [17], with permission of Elsevier.

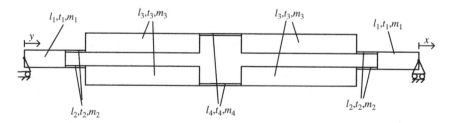

Figure 11.25 Distributed parameter model of the inertial amplification mechanism that is formed by combining different-sized beam sections. Source: Acar and Yilmaz [38], with permission of Elsevier.

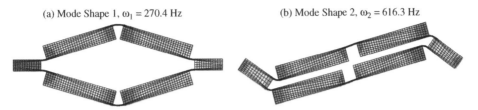

Figure 11.26 First two mode shapes of the optimized inertial amplification mechanism. Source: Acar and Yilmaz [38], with permission of Elsevier.

so beams with thicknesses t_2 and t_4 act like flexure hinges. This mechanism will be used to form a 2D inertially amplified lattice. To have a wide inertial amplification induced band gap, the ratio of the first two resonance frequencies of this mechanism (ω_{p1}/ω_{p2}) is minimized. Figure 11.26 shows the first two mode shapes of the optimized mechanism made of steel ($E = 205$ GPa, $\nu = 0.29$, $\rho = 7800$ kg/m^3). The dimensions of the optimized mechanism are $t_1 = 4.7$ mm, $t_2 = t_4 = 0.5$ mm, $t_3 = 6$ mm, $l_1 = 12$ mm, $l_2 = 2.5$ mm, $l_3 = 30$ mm, and $l_4 = 5$ mm [38].

In order to obtain a 2D square lattice, as shown in Figure 11.27, a second mechanism is needed, the length of which is $\sqrt{2}$ times the length of the optimized mechanism. In addition, this large mechanism should be optimized so that its first two resonance frequencies (ω_{p1}, ω_{p2}) and the first antiresonance frequency (ω_{z1}) are close to the

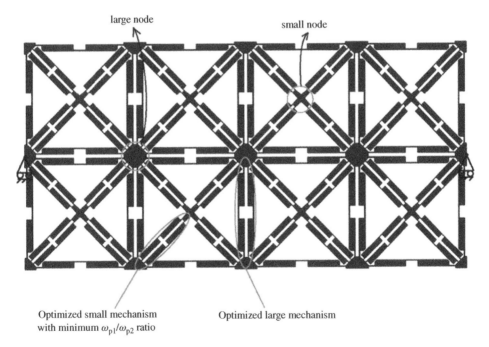

Figure 11.27 Two-dimensional distributed parameter lattice with embedded inertial amplification mechanisms. Here, four small mechanisms are connected to each small node and four small and four large mechanisms are connected to each large node. Source: Acar and Yilmaz [38], with permission of Elsevier.

corresponding ones of the first optimized mechanism, which ensures that these two mechanisms have similar stop-band widths and depths. The dimensions of the large optimized mechanism are $t_1 = 6.5$ mm, $t_2 = t_4 = 1.05$ mm, $t_3 = 7.66$ mm, $l_1 = 13.5$ mm, $l_2 = 7.35$ mm, $l_3 = 38.3$ mm, and $l_4 = 14.7$ mm [38].

Modal analysis of the finite-element model of the lattice reveals 43 resonance frequencies below 282 Hz. However, the next resonance frequency is obtained at 619 Hz. A phononic gap (stop band) is generated between these two resonance frequencies. The 43rd and 44th mode shapes are shown in Figure 11.28. In Figure 11.28a, the 1D building blocks (inertial amplification mechanisms) deform close to their first mode shapes (see Figure 11.26a), and in Figure 11.28b, they deform close to their second mode shapes (see Figure 11.26b). Thus the first two modes of the 1D building blocks define the gap limits of the 2D lattice.

In Figure 11.29a, the experimental and numerical FRF plots of the 2D lattice in the longitudinal direction are shown. One can see that these two results closely match and both show quite deep stop bands in the frequency range 300–600 Hz. Figure 11.29b shows the experimental FRF plots of the 2D lattice in the longitudinal and transverse

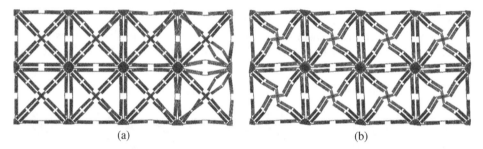

(a)   (b)

Figure 11.28 2D distributed parameter lattice with embedded inertial amplification mechanisms: (a) 43rd mode shape (282.3 Hz); (b) 44th mode shape (619.2 Hz). Source: Acar and Yilmaz [38], with permission of Elsevier.

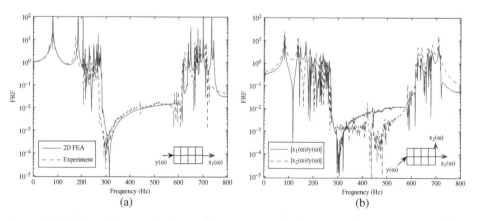

(a)   (b)

Figure 11.29 2D inertially amplified lattice: (a) experimental and numerical FRF plots for the longitudinal direction: $|x_1(\omega)/y(\omega)|$; (b) experimental FRF plots for the longitudinal (x_1) and transverse (x_2) directions. Source: Acar and Yilmaz [38], with permission of Elsevier.

directions. Again, for both cases, quite deep stop bands are formed in the frequency range 300–600 Hz.

As shown in Yuksel and Yilmaz [42], shape optimization can be used to enhance the depth and width of the inertial amplification induced stop bands. Finally, Frandsen et al. [43] shows that inertial amplification effect can be potentially realized in the form of a surface coating, to be used for sound and vibration control.

11.3.3 3D Inertially Amplified Lattices

In this section, 3D inertially amplified lattices will be considered. First, we look at inertially amplified BCC and FCC mass-spring lattices [39]. Figure 11.30 shows the unit cells of BCC and FCC lattices, as well as the inertial amplification mechanism that will be placed around each spring k and k_c in these lattices. Band structures of the BCC and FCC lattices without amplification mechanisms are shown in Figure 11.31. Here, the spring stiffnesses are inversely proportional to their lengths, hence $k/k_c = \sqrt{3}/2$ in the BCC lattice and $k/k_c = \sqrt{2}/2$ in the FCC lattice. Note that the lowest-frequency band gaps are formed in the normalized frequency ranges 0.540–0.712 and 0.635–1.191 for the BCC and FCC lattices, respectively. In Figure 11.31b, there are two flat branches (each composed of three coinciding branches) at $\omega = 0$ and $\omega l/2\pi c_t = 1.191$ corresponding to the out-of-plane and in-plane local resonances of the masses in the face centers, respectively. The in-plane resonance frequency occurs at $\omega = \sqrt{2k_c/m_c}$ [39].

Figure 11.32 shows the band structures of the BCC and FCC lattices with embedded amplification mechanisms. If inertial amplification mechanisms are placed around each

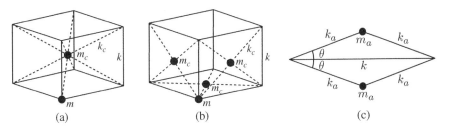

(a)   (b)   (c)

Figure 11.30 3D inertially amplified lattices: (a) BCC unit cell; (b) FCC unit cell; (c) inertial amplification mechanism. Source: Taniker and Yilmaz [39], with permission of Elsevier.

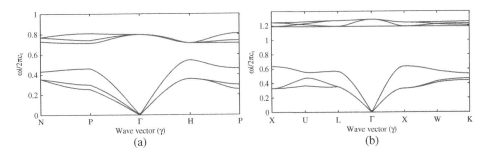

Figure 11.31 Phononic band structures without amplification mechanisms: (a) BCC lattice for $m_c/m_{total} = 1/5, k/k_c = \sqrt{3}/2$; (b) FCC lattice for $3m_c/m_{total} = 1/5, k/k_c = \sqrt{2}/2$. Source: Taniker and Yilmaz [39], with permission of Elsevier.

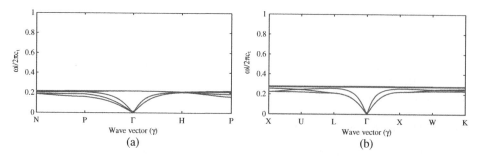

Figure 11.32 Phononic band structures with inertial amplification mechanisms: (a) BCC lattice, for $m_c/m_{total} = 1/5, m/m_{total} = 1/5, 22m_a/m_{total} = 3/5, k_a/k \to \infty, \theta = \pi/18$; (b) FCC lattice, for $3m_c/m_{total} = 1/5, m/m_{total} = 1/5, 30m_a/m_{total} = 3/5, k_a/k \to \infty, \theta = \pi/18$. Source: Taniker and Yilmaz [39], with permission of Elsevier.

spring k and k_c in the BCC unit cell, there will be 22 m'_as within the unit cell whereas there will be 30 m'_as within the inertially amplified FCC unit cell. Here, $k_a/k \to \infty$ for both lattices. Therefore, the motions of the m'_as are coupled to the masses m and m_c. Consequently, no extra branch can be seen in the band structure plots. In Figure 11.31a the highest frequency is 0.807, and the band gap is formed between the third and fourth branches. In Figure 11.32a, due to inertial amplification, all six branches are suppressed to lower frequencies so that a semi-infinite band gap starts at 0.218. In Figure 11.31b the highest frequency is 1.287, and the band gap is formed between the sixth and seventh branches. In Figure 11.32b, due to inertial amplification, all twelve branches are suppressed to lower frequencies so that a semi-infinite band gap starts at 0.281 [39].

The effects of finite stiffness amplification mechanisms on the stop-band limits are seen in Figure 11.33. Here, FRF plots are generated for $8 \times 8 \times 8$ BCC and FCC lattices, in which the k_a/k ratio is taken as 10. For the BCC and FCC lattices, the stop bands are formed in the normalized frequency ranges 0.207–0.661 and 0.263–0.759, respectively. The upper limits of these stop bands are dictated by the transverse

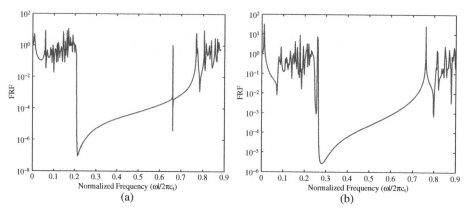

Figure 11.33 FRF plots of the $8\times8\times8$ lattices with inertial amplification: (a) BCC, for $m_c/m_{total} = 1/5, m/m_{total} = 1/5, 22m_a/m_{total} = 3/5, k_a/k \to 10, \theta = \pi/18$; (b) FCC, for $3m_c/m_{total} = 1/5, m/m_{total} = 1/5, 30m_a/m_{total} = 3/5, k_a/k \to 10, \theta = \pi/18$. Source: Taniker and Yilmaz [39], with permission of Elsevier.

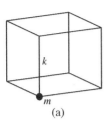
(a)

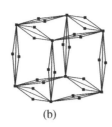
(b)

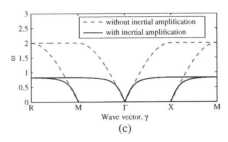
(c)

Figure 11.34 Inertial amplification in SC lattices: (a) SC unit cell; (b) SC unit cell with embedded inertial amplification mechanisms; (c) phononic band structures of the SC lattices with and without embedded inertial amplification mechanisms. For the SC lattice, $k = 1$ and $m = 1$; for the SC lattice with inertial amplification mechanisms, $k = 1$, $m = 0.5$, $m_a = 0.5/6$, and $\theta = \pi/18$. Source: Taniker and Yilmaz [40], with permission of ASME.

resonance frequencies of the inertial amplification mechanisms ($\omega = \sin(\theta)\sqrt{2k_a/m_a}$). Therefore, wide and deep stop bands can be obtained with the use of moderately stiff inertial amplification mechanisms in these lattices [39].

Secondly, we consider a similar study that compares SC and BCC lattices with and without inertial amplification mechanisms [40]. As BCC lattices were analyzed above, only the effect of inertial amplification on a SC lattice will be considered here. Figure 11.34a shows the unit cell of a SC lattice. In Figure 11.34b, inertial amplification mechanisms (see Figure 11.30c) are placed around each spring k, so there are 6 m'_as within the SC unit cell with embedded inertial amplification mechanisms. Figure 11.35c shows the band structures of the SC lattices with and without embedded inertial amplification mechanisms. Here, it is assumed that $k_a/k \to \infty$. Therefore, the motions of the m'_as are coupled to the masses m. Consequently, the band structures of the SC lattices with and without inertial amplification mechanism have the same number of branches. Moreover, the unit cells of both lattices have the same stiffness and total mass. However, the semi-infinite band gap in the SC lattice decreases from 2 to 0.826 with the addition of the inertial amplification mechanism [40].

Finally, inertial amplification induced band gaps can be seen in 3D solid structures as well [44]. Here, wide stop bands are achieved using a 3D printed structure.

11.4 Conclusions

In this chapter, it has been shown that local resonance induced band gaps can be obtained:

- in 1D lattices such as rods, shafts, strings, or beams
- in 2D lattices that are designed for in-plane, out-of-plane, or surface waves
- in ternary 3D lattices or 3D solid structures with holes.

Additionally, local resonance induced band gaps can be seen in 1D, 2D, and 3D acoustic systems. Despite the differences in these various lattice types, local resonance induced gaps always form around the resonance frequencies of the local resonators in the lattice. Unlike Bragg gaps, high attenuation is observed at the resonance frequencies of the local resonators and the attenuation profile over the gap range is generally asymmetric. Furthermore, local resonance induced gaps can be placed well below the Bragg limit, which depends on the lattice constant and the wave speed (transverse or longitudinal) within the lattice.

Inertial amplification induced band gaps can also be obtained in 1D, 2D, and 3D lattices. Similar to local resonance gaps, inertial amplification gaps can be placed well below the Bragg limit. However, "local resonances" and "inertial amplification" produce qualitatively different results in terms of energy localization characteristics. In local resonance induced gaps, a considerable fraction of energy propagates through the lattice because local resonances are generated throughout the lattice. In contrast, the energy is localized near the excitation source for inertial amplification induced gaps. In other words, as the antiresonance generation means are different in locally resonant and inertially amplified lattices, their wave-propagation characteristics are qualitatively different. It has been also shown that local resonance gaps always have some upper limit because the insertion of the local resonators causes resonances besides antiresonances within the lattice. However, when inertial amplification mechanisms with rigid links are embedded in a lattice, they only generate antiresonances. Consequently, the number of branches in the band structure changes, but the highest frequency in the lattice decreases considerably to yield a low-frequency semi-infinite gap. Finally, quite wide and deep low-frequency band gaps can be obtained with the use of moderately stiff inertial amplification mechanisms in a lattice.

References

1 Sigalas MM, Economou EN. Elastic and acoustic wave band structure. *Journal of Sound and Vibration*. 1992; **158**(2): 377–82.

2 Sigmund O, Jensen JS. Systematic design of phononic band–gap materials and structures by topology optimization. Philosophical Transactions of the Royal Society of London. *Series A: Mathematical, Physical and Engineering Sciences*. 2003; **361**(1806): 1001–19.

3 Khelif A, Aoubiza B, Mohammadi S, Adibi A, Laude V. Complete band gaps in two-dimensional phononic crystal slabs. *Physical Review E*. 2006; **74**(4): 046610.

4 Kushwaha MS. Classical band structure of periodic elastic composites. *International Journal of Modern Physics B*. 1996; **10**(9): 977–1094.

5 Kushwaha MS, Djafari-Rouhani B, Dobrzynski L, Vasseur JO. Sonic stop-bands for cubic arrays of rigid inclusions in air. *The European Physical Journal B*. 1998; **3**(2): 155–61.

6 Liu Z, Chan CT, Sheng P. Three-component elastic wave band-gap material. *Physical Review B*. 2002; **65**: 165116.

7 Goffaux C, Sanchez-Dehesa J. Two-dimensional phononic crystals studied using a variational method: Application to lattices of locally resonant materials. *Physical Review B*. 2003; **67**: 144301.

8 Liu Z, Zhang X, Mao U, Zhu YY, Yang Z, Chan CT, Sheng P. Locally resonant sonic materials. *Science*. 2000; **289**(5485):1734–36.

9 Jensen JS. Phononic band gaps and vibrations in one- and two-dimensional mass-spring structures. *Journal of Sound and Vibration*. 2003; **266**(5): 1053–78.

10 Hirsekorn M, Delsanto PP, Batra NK, Matic P. Modelling and simulation of acoustic wave propagation in locally resonant sonic materials. *Ultrasonics.* 2004; **42**(1–9): 231–5.

11 Xiao Y, Wen J, Wen X. Flexural wave band gaps in locally resonant thin plates with periodically attached spring–mass resonators. *Journal of Physics D.* 2012; **45**(19): 195401.

12 Xiao Y, Mace BR, Wen J, Wen X. Formation and coupling of band gaps in a locally resonant elastic system comprising a string with attached resonators. *Physics Letters A.* 2011; **375**(12): 1485–91.

13 Wang G, Wen X, Wen J, Liu Y. Quasi-one-dimensional periodic structure with locally resonant band gap. *Journal of Applied Mechanics.* 2006; **73**(1): 167–170.

14 Yu D, Liu Y, Wang G, Cai L, Qiu J. Low frequency torsional vibration gaps in the shaft with locally resonant structures. *Physics Letters A.* 2006; **348**(3-6): 410–5.

15 Yilmaz C, Kikuchi N. Analysis and design of passive low-pass filter-type vibration isolators considering stiffness and mass limitations. *Journal of Sound and Vibration.* 2006; **293**(1–2): 171–95.

16 Yilmaz C, Hulbert GM, Kikuchi N. Phononic band gaps induced by inertial amplification in periodic media. *Physical Review B.* 2007; **76**(5): 054309.

17 Yilmaz C, Hulbert GM. Theory of phononic gaps induced by inertial amplification in finite structures. *Physics Letters A.* 2010; **374**(34): 3576–84.

18 Cheng Y, Xu JY, Liu XJ. One-dimensional structured ultrasonic metamaterials with simultaneously negative dynamic density and modulus. *Physical Review B.* 2008; **77**(4): 045134.

19 Song Y, Wen J, Yu D, Wen X. Analysis and enhancement of torsional vibration stopbands in a periodic shaft system. *Journal of Physics D.* 2013; **46**(14): 145306.

20 Wang G, Wen J, Wen X. Quasi-one-dimensional phononic crystals studied using the improved lumped-mass method: Application to locally resonant beams with flexural wave band gap. *Physical Review B.* 2005; **71**(10): 104302.

21 Yu D, Liu Y, Zhao H, Wang G, Qui J. Flexural vibration band gaps in Euler–Bernoulli beams with locally resonant structures with two degrees of freedom. *Physical Review B.* 2006; **73**(6): 064301.

22 Yu D, Liu Y, Wang G, Zhao H, Qiu J. Flexural vibration band gaps in Timoshenko beams with locally resonant structure. *Journal of Applied Physics.* 2006; **100**(12): 124901.

23 Raghavan L, Phani AS. Local resonance bandgaps in periodic media: theory and experiment. *Journal of the Acoustical Society of America.* 2013; **134**(3): 1950–59.

24 Liu L, Hussein MI. Wave motion in periodic flexural beams and characterization of the transition between Bragg scattering and local resonance. *Journal of Applied Mechanics.* 2012; **79**(1): 011003.

25 Zhang X, Liu Y, Wu F, Liu Z. Large two-dimensional band gaps in three-component phononic crystals. *Physics Letters A.* 2003; **317**(1–2): 144–9.

26 Martinsson PG, Movchan AB. Vibrations of lattice structures and phononic band gaps. *Quarterly Journal of Mechanics and Applied Mathematics.* 2003; **56**(1): 45–64.

27 Oudich M, Senesi M, Assouar MB, Ruzenne M, Sun JH, Vincent B, Hou Z, Wu TT. Experimental evidence of locally resonant sonic band gap in two-dimensional phononic stubbed plates. *Physical Review B.* 2011; **84**(16): 165136.

28 Assouar MB, Oudich M. Enlargement of a locally resonant sonic band gap by using double-sides stubbed phononic plates. *Applied Physics Letters*. 2012; **100**(12): 123506.

29 Bilal OR, Hussein MI. Trampoline metamaterial: Local resonance enhancement by springboards. *Applied Physics Letters*. 2013; **103**(11): 111901.

30 Oudich M, Assouar MB. Surface acoustic wave band gaps in a diamond-based two-dimensional locally resonant phononic crystal for high frequency applications. *Journal of Applied Physics*. 2012; **111**(1): 014504.

31 Khelif A, Achaoui Y, Benchabane S, Laude V, Aoubiza B. Locally resonant surface acoustic wave band gaps in a two-dimensional phononic crystal of pillars on a surface. *Physical Review B*. 2010; **81**(21): 214303.

32 Hu XH, Chan CT, Zi J. Two-dimensional sonic crystals with Helmholtz resonators. *Physical Review E*. 2005; **71**(5): 055601.

33 Zhang X, Liu Z, Liu Y. The optimum elastic wave band gaps in three dimensional phononic crystals with local resonance. *The European Physical Journal B*. 2004; **42**(4): 477–82.

34 Liu Z, Chan CT, Sheng P. Analytic model of phononic crystals with local resonances. *Physical Review B*. 2005; **71**(1): 014103.

35 Wang G, Shao LH, Liu YZ. Accurate evaluation of lowest band gaps in ternary locally resonant phononic crystals. *Chinese Physics*. 2006; **15**(8): 1843–48.

36 Wang YF, Wang YS. Complete bandgap in three-dimensional holey phononic crystals with resonators. *Journal of Vibration and Acoustics*. 2013; **135**(4): 041009.

37 Li J, Wang YS, Zhang C. Tuning of acoustic bandgaps in phononic crystals with Helmholtz resonators. *Journal of Vibration and Acoustics*. 2013; **135**(3): 031015.

38 Acar G, Yilmaz C. Experimental and numerical evidence for the existence of wide and deep phononic gaps induced by inertial amplification in two-dimensional solid structures. *Journal of Sound and Vibration*. 2013; **332**(24): 6389–404.

39 Taniker S, Yilmaz C. Phononic gaps induced by inertial amplification in BCC and FCC lattices. *Physics Letters A*. 2013; **377**(31-33): 1930–36.

40 Taniker S, Yilmaz C. Inertial-amplification-induced phononic band gaps in SC and BCC lattices. IMECE2013–62674 in Proceedings of the ASME 2013 International Mechanical Engineering Congress and Exposition. 2013 Nov 15–21; San Diego, California, USA:.

41 Taniker S, Yilmaz C. Generating ultra wide vibration stop bands by a novel inertial amplification mechanism topology with flexure hinges. *International Journal of Solids and Structures*. 2017; **106–107**: 129–138.

42 Yuksel O, Yilmaz C. Shape optimization of phononic band gap structures incorporating inertial amplification mechanisms. *Journal of Sound and Vibration*. 2015; **355**: 232–245.

43 Frandsen NM, Bilal OR, Jensen JS, Hussein MI. Inertial amplification of continuous structures: Large band gaps from small masses. *Journal of Applied Physics*. 2016; **119**(12): 124902.

44 Taniker S, Yilmaz C. Design, analysis and experimental investigation of three-dimensional structures with inertial amplification induced vibration stop bands. *International Journal of Solids and Structures*. 2015; **72**: 88–97.

12

Dynamics of Nanolattices: Polymer-Nanometal Lattices

Craig A. Steeves[1], Glenn D. Hibbard[2], Manan Arya[1] and Ante T. Lausic[2]

[1]*Institute for Aerospace Studies, University of Toronto, Ontario, Canada*
[2]*Department of Materials Science and Engineering, University of Toronto, Ontario, Canada*

12.1 Introduction

Periodic lattices are generated by tessellating a reference cell consisting of a network of struts over two or three dimensions. They are ubiquitous in engineering applications, used as cores in sandwich panels, sandwich beams, and as space trusses. Plane-wave propagation through such materials has been studied extensively [1]. Recently, the propagation of waves through two-dimensional periodic lattices was characterized using a finite-element approach that models struts in the reference cell as Euler–Bernoulli beams [2] or as Timoshenko beams [3]. Three-dimensional structures, such as the octet truss lattice [4] and the pyramidal truss lattice have applications as structural materials, or as cores in sandwich panels. However, similar wave propagation analysis for 3D periodic structures is sparse [5]. In part, this is because the difficulty in fabricating such lattices on a large scale makes moot such an analysis. Herein we discuss both a practical method for manufacturing 3D lattices, as well as the analysis of wave propagation through them.

12.2 Fabrication

The challenge of fabricating optimal lattices possessing millimeter- to micron-scale features is not trivial. Each fabrication route has a particular combination of advantages and disadvantages. This section discusses those associated with fabricating metal/polymer truss structures.

Rapid prototyping offers a promising pathway to fabricate microtrusses and other complex lattice shapes [6]. While the polymeric build material lacks the mechanical performance and thermal stability required for many applications, a secondary coating step can be used to electrodeposit a thin sleeve of metal encapsulating the polymer core. Moreover, if the deposition conditions are suitably controlled, it is possible to drive the grain size of the deposited metal down to the nanometer scale, creating a nanocrystalline material having ultra-high strength, and which is optimally positioned away from the neutral bending axis of the internal struts. These metal–polymer lattices have critical

Dynamics of Lattice Materials, First Edition. Edited by A. Srikantha Phani and Mahmoud I. Hussein.
© 2017 John Wiley & Sons Ltd. Published 2017 by John Wiley & Sons Ltd.

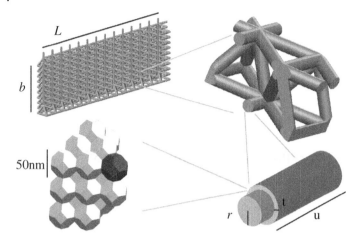

Figure 12.1 Multi-scale breakdown of metal–polymer microtruss lattices. The starting beam can be divided into unit cells which are composed of a varying number of struts based on geometry choice. Finally, the grain size of the coating allows for modifications to the mechanical properties of the hybrid structure.

internal length scales ranging from the level of the beam down to the nanometer-scale grain size of the metal coatings (Figure 12.1). In between, the unit-cell geometry can be selected from regular, space-filling polyhedra, while the struts can be tailored to control the degree of hybridization.

The starting point for creating metal–polymer lattice materials is the additive man-ufacturing of the polymer template. This subset of rapid prototyping, also colloquially known as *3D printing* is differentiated by how the build material is laid down in each successive step. Currently, the most common and low-cost of these methods is *fused deposition modeling* (FDM), where a filament is run through a heating element, is lique-fied, and then the layers are "drawn" on the base platform, each layer resting upon the previous one. The precision of the nozzle and the nozzle extrusion diameter define the resolution limits. With very little waste and a wide variety of thermoplastic polymer fila-ments available, it is popular for its ease of use. However, the drawback of FDM lies in its inability to make certain types of complex structures, since the liquefied filament needs to be deposited on a base of sufficient stability to prevent sagging [7]. Furthermore, the inherent packing of cylindrical tubes introduces porosity in the finished structure and leaves a rougher surface finish [8, 9].

Immersion printing bypasses this limitation by having the assembly and platform within a vat of photopolymerizable liquid polymer. The stage is lowered just below the level of the liquid and a UV laser draws out the first layer. Wherever the laser illuminates liquid polymer, the photoinitiator additives radicalize and set off a photopolymerization reaction, thus solidifying the area. After a slice is complete, the stage descends, allowing the now-hard area to submerge and a further layer is cured. Upon completion, the part is removed, rinsed of the resin and thermally cured to its full hardness. While this process alleviates the concern of unsupported parts of the microtruss, the movement from vat to curing is difficult because the green part is delicate and flexible, potentially deforming the desired shape [7]. A modification of this technique was developed by

Schaedler et al. [10], who did the photoinitiating and curing process simultaneously by shining the UV source through a mask of holes to create a complex 3D lattice material.

Stereolithography (SLA) combines the layer-by-layer slicing of FDM and the photopolymerization of immersion printing with the ability to print two different materials at the same time. By using a second material that can be easily removed (for example, a low-melting-point wax), SLA is able to build virtually any desired shape [6, 7]. Each layer is cured individually and, once finished, the part is placed in an oven where the wax is melted away leaving the final part with the desired internal complexity.

Several options are available for reinforcing the polymer lattice after printing. From an aqueous environment, electroless deposition can be used, either as a thin sleeve of reinforcement [10] or as a seed layer for subsequent electrodeposition [11]. In general terms, electrodeposited layers can be thicker, higher strength, and more ductile than electroless layers [12, 13]. Nanocrystalline electrodeposition is particularly attractive, since grain-size control of the deposited layer allows the mechanical properties to be tuned.

Nanocrystalline materials are physically distinct from their conventional polycrystalline counterparts because of the significant volume fraction of atoms associated with their intercrystalline interfaces [14]. The mechanical properties of nanocrystalline metals can be divided into those that are grain-size sensitive and those that are largely not [15]. Young's modulus, for example, is relatively insensitive to grain size. Nanoindentation experiments on electrodeposited nickel–phosphorus alloys having essentially the same chemistry but with grain sizes (d) ranging from 4 nm to 29 nm showed that in fully dense nanocrystalline materials, the elastic modulus is essentially independent of grain size until below approximately 15 nm [16].

Strength, however, is strongly dependent on grain size. Compared to their conventional polycrystalline counterparts, the yield strength and hardness of nanocrystalline electrodeposits can be improved by up to 5 to 10 times. This strength increase can be understood in terms of the finer spacing between grain boundaries, which acts to create a barrier to dislocation motion and is generally described through the Hall–Petch relationship, in which the yield strength increases with decreasing grain size as $d^{-1/2}$ [17, 18]. Beyond a certain point of grain-size reduction, however, the Hall–Petch model breaks down and there is a transition toward inverse Hall–Petch (softening) behavior with decreasing grain size [19]. Depending on the alloy system in question, this transition generally occurs over the range of 10 to 30 nm.

The overall effect of grain size on the mechanical properties of nanocrystalline materials can be seen in Figure 12.2 where typical tensile curves of electrodeposited nickel having grain sizes of 10 μm, 40 nm, and 20 nm are shown. Also shown in Figure 12.2 is the tensile curve for a type of photopolymer used in SLA-type rapid prototyping. The tensile ductility of the photopolymer is comparable to that of the 10-μm nickel, but with a yield strength five times lower, and 34 times lower than the strength of the 20-nm nanocrystalline nickel. Alloying nickel during the deposition stage (with, for example, iron or tungsten) allows an additional degree of control over the final properties. Much like the polymer matrix in a fibre-reinforced composite, the function of the 3D printed polymer is to enable the most advantageous positioning of the high-performance material, in this case the nanocrystalline metal. The difference in mechanical behavior between polymer core and electrodeposited sleeve is best illustrated by plotting Young's modulus and tensile strength against density for both material types (Figure 12.3). The following section

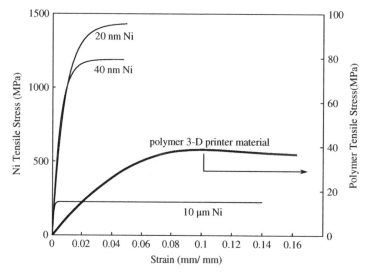

Figure 12.2 Tensile stress–strain curves for as-deposited 20 nm, 40 nm, and 10 μm electrodeposited Ni. As a comparison, the tensile stress–strain curve of the as-printed polymer material is included on the secondary ordinate axis.

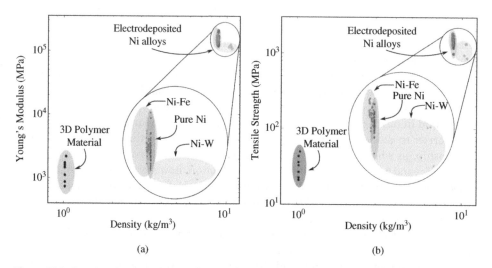

(a)   (b)

Figure 12.3 As-printed polymer 3D printer material properties juxtaposed with electrodeposited nickel and nickel-alloy materials on a log–log scale for (a) Young's modulus vs density and (b) tensile strength vs density.

provides an illustration of the consequences of such different material properties when designing a metal–polymer truss structure under a static load.

12.2.1 Case Study

Defining the expected types of failure in hybrid metal–polymer microtrusses requires derivation of individual analytical expressions for each expected failure mechanism. In

Figure 12.4 Cross-sections of struts in a tetrahedral microtruss beam measuring 1×0.5 m, optimally designed to support a central load of 700 N in three-point bending for the cases of: (a) 10-μm Ni on a polymer microtruss; (b) hollow 10-μm Ni microtruss; (c) 21-nm Ni on a polymer microtruss; (d) hollow 21-nm Ni microtruss.

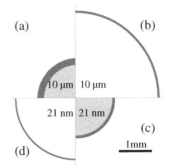

a previous study, we have done this for square pyramidal and tetrahedral microtrusses and developed failure mechanism maps and located trajectories of optimal design [20]. Using these models, an example is developed to illustrate the freedom in designing hybrid metal–polymer microtrusses: a beam in three-point bending is needed to span a gap measuring $1 \text{ m} \times 0.5 \text{ m}$ and hold a central load of 700 N. For a conventional polycrystalline nickel coating with a grain size of 10 μm on a polymer microtruss, the optimal architecture of the microtruss can be calculated and the resulting beam has a total mass of 3.2 kg.

Two different steps can be taken to improve the performance. First, the polymer core could be removed after electrodeposition (by either thermal decomposition or by chemical dissolution) or the grain size of the nickel could be decreased to the nanometer scale. The cross-sections shown in Figure 12.4b,c are for optimal designs in each of these cases. Both strut designs can carry the same load, and both have the same overall mass (2.0 kg). By combining grain-size reduction with polymer removal, one is able to achieve the scenario illustrated in Figure 12.4d, where the mass is reduced to 0.7 kg. This type of approach represents an idealization of the performance of the metal–polymer microtruss. In practice, material inhomogeneities act to reduce the expected performance. The circular shape and smooth coating assumed in the model are idealizations because the real structure can be rougher due to the step size of a layer-by-layer printing technique, generating some stress concentrations and hence reducing the maximum load-carrying capacity.

12.3 Lattice Dynamics

It is possible to take advantage of the manufacturing flexibility offered through the combination of 3D printed polymer and electrodeposited nanocrystalline metal in the tailoring of the wave-propagation properties of 3D lattices. The key is that, for normal material systems, the only method for changing the material properties—specifically the density and elastic constants—is to change the material. This means that there are a relatively small set of discrete material property choices. The use of nanometal-coated polymers *decouples* the effective bending stiffness of a coated polymer strut from the effective density. Changing the ratio of polymer radius to coating thickness can create a strut of equivalent effective density but differing bending stiffness. Not only does this generate a much broader range of effective "material" properties, but also makes the system amenable to optimization through efficient gradient-based methods.

In *Wave Propagation in Periodic Structures* [21], Brillouin describes plane-wave propagation in a variety of periodic media (such as electronic waves in a crystal lattice, and mechanical waves in macroscopic systems). Periodic structures often exhibit frequency band gaps; they block certain frequencies of waves, regardless of the direction of propagation [21]. This has led to the development of phononic band gap materials: materials engineered to reject certain undesired frequencies. These materials often consist of periodic inclusions placed in a host material and can be designed to have optimal band gap characteristics [22].

Periodic lattice structures can be designed as phononic band gap materials. This requires selecting an appropriate geometry—unit-cell topology and slenderness ratio—and an ideal combination of material properties—stiffness and density. Doing so allows for the use of these materials, not only as structural components, but also as acoustic filters or isolators. To design lattices with desirable wave-propagation properties requires a model of their behavior, which is described in detail in this section. The reference cells of octet lattices are modeled using a finite-element approach. The finite-element model treats each beam in the reference cells of these lattices as a Timoshenko beam with six nodal degrees of freedom; three in translation, and three in rotation. The Floquet–Bloch principles [21] are combined with this finite-element model to generate dispersion curves for the various lattice geometries. The general principles of lattice tailoring are then discussed.

12.3.1 Lattice Properties

Geometries of 3D Lattices

The geometry of a 3D periodic lattice is described similarly to that of a 2D lattice: by specifying the geometry of the unit cell and the periodicity vectors, which define how the unit cell is tessellated to obtain the lattice. The direct basis vectors, e_1, e_2, e_3, define the periodicity of the lattice. They are used as a basis to specify the position of cells in the lattice, relative to a reference cell. Each integer tuple (n_1, n_2, n_3) identifies a particular cell in the lattice, with position $n_1 e_1 + n_2 e_2 + n_3 e_3$ relative to the reference cell. Using this property, the position vector of any point $\mathbf{r}$ in the (n_1, n_2, n_3)-cell can be written as

$$\mathbf{r} = \mathbf{r}_r + n_1 e_1 + n_2 e_2 + n_3 e_3 \tag{12.1}$$

where $\mathbf{r}_r$ is the position vector of the corresponding point in the reference cell.

Plane-wave motion with wave vector κ, frequency ω, and amplitude A at a point $\mathbf{r}$ and at time t is governed by:

$$\mathbf{q}(\mathbf{r}) = A e^{i(\kappa \cdot \mathbf{r} - \omega t)} \tag{12.2}$$

The basis vectors that define the periodicity of the frequency are the reciprocal basis vectors, e_1^*, e_2^*, e_3^*. They are related to the direct basis vectors e_1, e_2, e_3 through:

$$e_i \cdot e_j^* = \delta_{ij} \tag{12.3}$$

where δ_{ij} is the Kronecker delta function. Thus, each reciprocal basis vector is normal to the plane defined by two of the direct basis vectors, and has a magnitude such that its projection on the third direct basis vector has unit length. Because lattices are periodic structures, the frequency of wave propagation ω is periodic with respect to the wave vector κ [21]. That is, for two wave vectors κ and κ' that are related as:

$$\kappa' = \kappa + n_1 e_1^* + n_2 e_2^* + n_3 e_3^* \tag{12.4}$$

for integral n_i, the frequency of wave propagation will be the same.

The reciprocal basis vectors define a reciprocal lattice. The fundamental unit cell of the reciprocal lattice is the first Brillouin zone [21]. The full frequency response of the lattice can be characterized by the frequency response to wave vectors in the first Brillouin zone. For any wave vector, a corresponding wave vector can be found in the first Brillouin zone with the same frequencies of propagation. Because of the periodicity of the frequency, *any* basic unit cell with the reciprocal vectors as the basis can be used as the first Brillouin zone. The simplest choice of the first Brillouin zone is the parallelpiped whose sides are the reciprocal basis vectors. (Strictly speaking, this choice of Brillouin zone excludes certain sections of the first Brillouin zone, as defined by Brillouin [21], but includes certain regions of the higher Brillouin zones that have the same frequency response.)

Figure 12.5 is a sketch of the reference cells, direct basis vectors, reciprocal basis vectors, and first Brillouin zone of the octet lattice. Because there are multiple ways of defining a specific lattice geometry, the choice of the particular reference cell and direct basis vectors of the sample lattice is arbitrary.

Effective Material Properties of Nanometal-coated Polymer Lattices

Because nanometal-coated lattices are not composed of a homogeneous material, it is necessary to calculate effective material properties for the composite configuration. The effective properties are calculated based upon equivalent bending and shear stiffness. The total bending stiffness of the lattice struts is a combination of the bending stiffness of the polymer core and the nanometal coating. There may be an additional metallic interlayer which can be accounted for in the same manner. For a composite cylinder with a solid core of radius r and coating of thickness t, the total bending stiffness EI_{tot} is given by:

$$EI_{tot} = E_p I_p + E_m I_m = \frac{\pi E_p r^4}{4} + \frac{\pi E_m ((r+t)^4 - r^4)}{4} \tag{12.5}$$

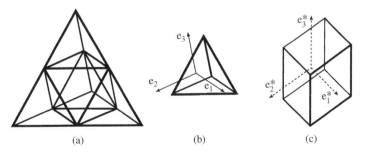

(a)   (b)   (c)

Figure 12.5 Octet lattice: (a) reference cells; (b) unit cell and direct basis vectors; (c) first Brillouin zone and reciprocal basis vectors.

with E_p and E_m the modulus of the polymer and metallic coating, respectively. The effective modulus is calculated by:

$$E_{eff} = \frac{4EI_{tot}}{\pi(r+t)^4} \tag{12.6}$$

Similarly, the effective shear modulus of the material is determined through:

$$G_{eff} = \frac{G_{tot}}{\pi(r+t)^2} = \frac{G_p r^2 + G_m((r+t)^2 - r^2)}{(r+t)^2} \tag{12.7}$$

where G_p and G_m are the shear moduli of the polymer and metal, respectively. This enables the calculation of an effective Poisson's ratio, v_{eff}. Employing Timoshenko's expression [23] for the shear correction factor κ for a solid circular beam,

$$\kappa = \frac{6(1 + v_{eff})}{7 + 6v_{eff}} \tag{12.8}$$

Finally, the density of the material must be calculated; the rule of mixtures is appropriate. With ρ_p and ρ_m the densities of the polymer and metal coating, the effective density is:

$$\rho_{eff} = \frac{\rho_p r^2 + \rho_m((r+t)^2 - r^2)}{(r+t)^2} \tag{12.9}$$

What is important to note is that, unlike for ordinary materials, where E and ρ are fixed, because r and t can be chosen independently, G_{eff} or E_{eff} and ρ_{eff} are decoupled. Henceforward, the simplified notations $G = G_{eff}$, $E = E_{eff}$ and $\rho = \rho_{eff}$ will be used.

Lamé's constants are related to the elastic moduli through:

$$\lambda = \frac{G(E - 2G)}{3G - E} \tag{12.10}$$

where E is the effective material modulus calculated above. Using Lamé's constants, Hooke's law becomes:

$$C_{ijkl}\epsilon_{kl} = \lambda\epsilon_{kk}\delta_{ij} + 2G\epsilon_{ij} \tag{12.11}$$

12.3.2 Finite-element Model

A finite-element model of the lattice unit cell, discretizing the lattice members as Timoshenko beam elements, is employed to develop the equations of motion for the reference cell. The nodal displacements are used to determine the kinetic and potential energies of the element. The total kinetic and potential energies in the reference cell can then be expressed in terms of the displacements of the nodes as the sum of the element energies. These nodal displacements form generalized coordinates, and allow for the Euler–Lagrange equation to be used to derive the equation of motion for the reference cell.

Displacement Field

In three dimensions, a node of a Timoshenko beam element has six degrees of freedom: three translations (u, v, w) and three rotations (ϕ, ψ, θ) (see Figure 12.6). The nodal

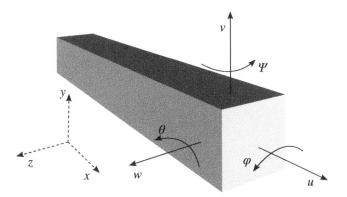

Figure 12.6 Notation for the nodal displacements of a Timoshenko beam element.

displacements for an element l with nodes A and B are $\mathbf{q}_l = [q_i] = [\mathbf{q}_A \;\; \mathbf{q}_B]^T$, where $\mathbf{q}_A$ and $\mathbf{q}_B$ are the nodal displacements at nodes A and B, respectively:

$$\mathbf{q}_A = [u_A \quad v_A \quad w_A \quad \phi_A \quad \psi_A \quad \theta_A]^T \tag{12.12}$$

$$\mathbf{q}_B = [u_B \quad v_B \quad w_B \quad \phi_B \quad \psi_B \quad \theta_B]^T \tag{12.13}$$

The displacements at any point x along the element at a time t are linear combinations of the nodal displacements:

$$u(x,t) = q_i(t)a_i(x)$$
$$v(x,t) = q_i(t)b_i(x)$$
$$w(x,t) = q_i(t)c_i(x) \tag{12.14}$$
$$\phi(x,t) = q_i(t)d_i(x)$$
$$\psi(x,t) = q_i(t)e_i(x)$$
$$\theta(x,t) = q_i(t)f_i(x)$$

The shape functions a_i through f_i are provided in Section 12.5.

The origin of the local coordinate system is on the centreline of the element. Hence a point $\mathbf{p}$ on the element in the undeformed state has position $[x \;\; y \;\; z]^T$. The corresponding point on the centreline is $\mathbf{c} = [x \;\; 0 \;\; 0]^T$. The displacement at $\mathbf{p}$ can be defined by specifying the displacement of $\mathbf{c}$ and the rotation of the line $\mathbf{p} - \mathbf{c} = [0 \;\; y \;\; z]^T$ about $\mathbf{c}$ [24]. Thus $\mathbf{p}' = \mathbf{c}' + \mathbf{R}(\mathbf{p} - \mathbf{c})$, where $()'$ represents vectors in the deformed state, and $\mathbf{R}$ is the rotation at $\mathbf{c}$. Using first-order approximations,

$$\mathbf{R} = \mathbf{R}_x \mathbf{R}_y \mathbf{R}_z = \begin{bmatrix} 1 & -\theta & \psi \\ \theta & 1 & -\phi \\ -\psi & \phi & 1 \end{bmatrix} \tag{12.15}$$

Because $\mathbf{c}' = \mathbf{c} + [u \;\; v \;\; w]^T$, the displacement field $\mathbf{u}$ is:

$$\mathbf{u} = \begin{bmatrix} u_1 \\ u_2 \\ u_3 \end{bmatrix} = \mathbf{p}' - \mathbf{p} = \mathbf{R}(\mathbf{p} - \mathbf{c}) + \mathbf{c}' - \mathbf{p} = \mathbf{R}(\mathbf{p} - \mathbf{c}) + \begin{bmatrix} u \\ v \\ w \end{bmatrix} + \mathbf{c} - \mathbf{p}$$

$$
= \begin{bmatrix} 1 & -\theta & \psi \\ \theta & 1 & -\phi \\ -\psi & \phi & 1 \end{bmatrix} \begin{bmatrix} 0 \\ y \\ z \end{bmatrix} + \begin{bmatrix} u \\ v \\ w \end{bmatrix} + \begin{bmatrix} 0 \\ -y \\ -z \end{bmatrix}
$$

$$
= \begin{bmatrix} u - y\theta + z\psi \\ v - z\phi \\ w + y\phi \end{bmatrix} \tag{12.16}
$$

Kinetic Energy

The kinetic energy of an element l is

$$
T_l = \frac{1}{2} \int_V |\mathbf{v}|^2 \, \rho \mathrm{d}V = \frac{1}{2} \int_M (\dot{u}_1^2 + \dot{u}_2^2 + \dot{u}_3^2) \, \rho \mathrm{d}V
$$

$$
= \frac{1}{2} \int_0^L \int_A ((\dot{u} - y\dot{\theta} + z\dot{\psi})^2 + (\dot{v} - z\dot{\phi})^2 + (\dot{w} + y\dot{\phi})^2) \rho \, \mathrm{d}A \, \mathrm{d}x
$$

$$
= \frac{1}{2} \int_0^L \int_A \left(\begin{matrix} \dot{u}^2 + y^2\dot{\theta}^2 + z^2\dot{\psi}^2 - 2y\dot{\theta}\dot{u} + 2z\dot{u}\dot{\psi} - 2yz\dot{\theta}\dot{\psi} + \\ \dot{v}^2 + z^2\dot{\phi}^2 - 2z\dot{v}\dot{\phi} + \dot{w}^2 + y^2\dot{\phi}^2 + 2yw\dot{\phi} \end{matrix} \right) \rho \, \mathrm{d}A \, \mathrm{d}x \tag{12.17}
$$

Assuming that the x-axis passes through the centroid of the cross-section of the element,

$$
\int_A y \, \mathrm{d}A = 0 \quad \int_A z \, \mathrm{d}A = 0 \quad \int_A yz \, \mathrm{d}A = 0
$$

$$
I_z = \int_A y^2 \, \mathrm{d}A \quad I_y = \int_A z^2 \, \mathrm{d}A \quad A = \int_A \mathrm{d}A \tag{12.18}
$$

producing an expression for the kinetic energy of the element:

$$
T_l = \frac{1}{2} \int_0^L A\rho(\dot{u}^2 + \dot{v}^2 + \dot{w}^2) + I_z\rho\dot{\theta}^2 + I_y\rho\dot{\psi}^2 + (I_z + I_y)\rho\dot{\phi}^2 \, \mathrm{d}x \tag{12.19}
$$

Substituting the displacements from Eq. (12.14)

$$
T_l = \frac{1}{2} \int_0^L \left[\begin{matrix} A\rho((\dot{q}_i a_i)^2 + (\dot{q}_i b_i)^2 + (\dot{q}_i c_i)^2) + \\ (I_z + I_y)\rho(\dot{q}_i d_i)^2 + I_y\rho(\dot{q}_i e_i)^2 + I_z\rho(\dot{q}_i f_i)^2 \end{matrix} \right] \mathrm{d}x
$$

$$
= \sum_{ij} \frac{1}{2}\dot{q}_i\dot{q}_j \int_0^L \left[\begin{matrix} A\rho(a_i a_j + b_i b_j + c_i c_j) + \\ (I_z + I_y)\rho d_i d_j + I_y\rho e_i e_j + I_z\rho f_i f_j \end{matrix} \right] \mathrm{d}x
$$

$$
= \frac{1}{2}\dot{q}_i\dot{q}_j m_{ij}, \tag{12.20}
$$

where

$$
m_{ij} = \int_0^L \left[\begin{matrix} A\rho(a_i a_j + b_i b_j + c_i c_j) + \\ (I_z + I_y)\rho d_i d_j + I_y\rho e_i e_j + I_z\rho f_i f_j \end{matrix} \right] \mathrm{d}x \tag{12.21}
$$

This expression is compactly expressed in matrix form:

$$
T_l = \frac{1}{2}\dot{\mathbf{q}}_l^T \mathbf{m}_l \dot{\mathbf{q}}_l \tag{12.22}
$$

Strain Potential Energy

The gradients of the displacements define the strains. Assuming uniform displacements across cross-sections,

$$\epsilon_{11} = \frac{\partial u_1}{\partial x} = u' - y\theta' + z\psi' \tag{12.23}$$

$$\epsilon_{12} = \epsilon_{21} = \frac{1}{2}\left(\frac{\partial u_1}{\partial y} + \frac{\partial u_2}{\partial x}\right) = \frac{1}{2}(v' - z\phi' - \theta) \tag{12.24}$$

$$\epsilon_{13} = \epsilon_{31} = \frac{1}{2}\left(\frac{\partial u_1}{\partial z} + \frac{\partial u_3}{\partial x}\right) = \frac{1}{2}(w' + y\phi' + \psi) \tag{12.25}$$

$$\epsilon_{22} = \epsilon_{33} = 0 \tag{12.26}$$

$$\epsilon_{23} = \epsilon_{32} = 0 \tag{12.27}$$

where $()'$ represents a partial derivative with respect to x.

The total strain potential energy U_l for the element l can be found by integrating the strain energy density $W = \frac{1}{2}C_{ijkl}\epsilon_{kl}\epsilon_{ij}$ over a given volume V.

$$U_l = \int \frac{1}{2}C_{ijkl}\epsilon_{kl}\epsilon_{ij}\,dV \tag{12.28}$$

As a consequence, the total strain energy in an element is:

$$\begin{aligned} U_l &= \frac{1}{2}\int \epsilon_{ij}(\lambda\epsilon_{kk}\delta_{ij} + 2G\epsilon_{ij})\,dV \\ &= \frac{1}{2}\int_v [\lambda\epsilon_{11}^2 + 2G(\epsilon_{11}^2 + 2\epsilon_{12}^2 + 2\epsilon_{13}^2)]\,dV \\ &= \frac{1}{2}\int_V \left[\begin{array}{c} (\lambda + 2G)(u' - y\theta' + z\psi')^2 + \\ G(v' - z\phi' - \theta)^2 + G(w' - y\phi' + \psi)^2 \end{array}\right]\,dV \end{aligned} \tag{12.29}$$

Incorporating λ into a shear correction factor κ,

$$U_l = \frac{1}{2}\int_V \left[\begin{array}{c} E(u'^2 - 2y\theta'u' + 2z\psi'u' - 2yz\theta'\psi' + z^2\psi'^2 + y^2\theta'^2) + \\ \kappa G(v'^2 - 2zv'\phi' - 2\theta v' + 2\theta z\phi' + \theta^2 + z^2\phi'^2) + \\ \kappa G(w'^2 - 2y\phi'w' + 2\psi w' + y^2\phi'^2 - 2y\phi'\psi + \psi^2) \end{array}\right]\,dV \tag{12.30}$$

These can be simplified through application of the expressions for area and moments of area in Eq. (12.18), generating:

$$U_l = \frac{1}{2}\int_0^L \left[\begin{array}{c} E(Au'^2 + I_y\psi'^2 + I_z\theta'^2) + \\ \kappa GA((v' - \theta)^2 + (w' + \psi)^2) + (I_z + I_y)\kappa G\phi'^2 \end{array}\right]\,dV \tag{12.31}$$

Again, the displacements from Eq. (12.14) can be substituted into this expression, producing:

$$U_l = \frac{1}{2}\int_0^L \left[\begin{array}{c} E(A(q_ia_i')^2 + I_y(q_ie_i')^2 + I_z(q_if_i')^2) + \\ \kappa GA((q_ib_i' - q_if_i)^2 + (q_ic_i' + q_ie_i)^2) + (I_z + I_y)\kappa G(q_id_i')^2 \end{array}\right]\,dx$$

$$= \sum_{ij} \frac{1}{2} q_i q_j \int_0^L \left[\begin{array}{l} E(Aa'_i a'_j + I_y e'_i e'_j + I_z f'_i f'_j) + (I_z + I_y)\kappa Gd'_i d'_j + \\ \kappa GA((b'_i - f_i)(b'_j - f_j) + (c'_i + e_i)(c'_j + e_j)) \end{array} \right] dx$$

$$= \frac{1}{2} q_i q_j k_{ij} \tag{12.32}$$

where the mass matrix m_{ij} is given by:

$$k_{ij} = \int_0^L \left[\begin{array}{l} E(Aa'_i a'_j + I_y e'_i e'_j + I_z f'_i f'_j) + (I_z + I_y)\kappa Gd'_i d'_j + \\ \kappa GA((b'_i - f_i)(b'_j - f_j) + (c'_i + e_i)(c'_j + e_j)) \end{array} \right] dx \tag{12.33}$$

In matrix form, this is expressed compactly as:

$$U_l = \frac{1}{2} \mathbf{q}_l^T \mathbf{k}_l \mathbf{q}_l \tag{12.34}$$

Collected Equation of Motion

The nodal displacements for each element are assembled into a global displacement vector $\mathbf{q}$ though standard finite-element procedures, with connectivity appropriate for the reference-cell geometry. The total kinetic, T, and potential, U, energies of the reference cell are found through the summation of the element energies. Similarly, the mass and stiffness matrices of each element are assembled into global mass and stiffness matrices $\mathbf{M}$ and $\mathbf{K}$. Both $\mathbf{M}$ and $\mathbf{K}$ are symmetric. thus:

$$U = \frac{1}{2} \mathbf{q}^T \mathbf{K} \mathbf{q} = \frac{1}{2} Q_i Q_j K_{ij}$$

$$T = \frac{1}{2} \dot{\mathbf{q}}^T \mathbf{M} \dot{\mathbf{q}} = \frac{1}{2} \dot{Q}_i \dot{Q}_j M_{ij} \tag{12.35}$$

The Lagrangian $\mathcal{L}$ of the reference cell is:

$$\mathcal{L} = T - U$$

$$= \frac{1}{2} \dot{Q}_i \dot{Q}_j M_{ij} - \frac{1}{2} Q_i Q_j K_{ij} \tag{12.36}$$

Differentiating the Lagrangian, we generate the derivatives:

$$\frac{\partial \mathcal{L}}{\partial \dot{Q}_i} = \dot{Q}_j M_{ij} \tag{12.37}$$

$$\frac{d}{dt} \left(\frac{\partial \mathcal{L}}{\partial \dot{Q}_i} \right) = \ddot{Q}_j M_{ij} \tag{12.38}$$

and

$$\frac{\partial \mathcal{L}}{\partial Q_i} = -Q_j K_{ij} \tag{12.39}$$

which are used in the Euler–Lagrange equations:

$$\frac{d}{dt} \left(\frac{\partial \mathcal{L}}{\partial \dot{Q}_i} \right) - \frac{\partial \mathcal{L}}{\partial Q_i} = F_i \tag{12.40}$$

to produce:

$$\ddot{Q}_j M_{ij} + Q_j K_{ij} = F_i \tag{12.41}$$

Returning to a matrix form:

$$M\ddot{q} + Kq = f \tag{12.42}$$

where f are the generalized forces corresponding to the generalized coordinates.

12.3.3 Floquet–Bloch Principles

Bloch's theorem states that the wave-propagation behavior in an arbitrary unit cell in a lattice is determined by the wave-propagation behavior in the reference cell. Thus the theorem can be used to impose boundary conditions that enforce a plane wave solution in an infinite lattice, and hence the behavior an infinite lattice can be modeled by examining only the reference cell. Bloch's theorem is a special case of the wave equation in a periodic medium; the classical equation describing the displacements q for plane-wave motion with wave vector κ, frequency ω, and amplitude A at a point r at time t is:

$$q(r) = Ae^{i(\kappa \cdot r - \omega t)} \tag{12.43}$$

This can be used to relate the displacement at a point r in the 3D lattice to the corresponding point in the reference cell r_j:

$$q(r) = q(r_j)e^{i\kappa \cdot (r - r_j)} \tag{12.44}$$

For a 3D periodic structure with basis vectors e_1, e_2, and e_3, the position vectors r and r_j, which refer to corresponding points in different unit cells, $r = r_j + n_1e_1 + n_2e_2 + n_3e_3$, where the n_i are integers. Hence:

$$i\kappa \cdot (r - r_j) = n_1 i\kappa \cdot e_1 + n_2 i\kappa \cdot e_2 + n_3 i\kappa \cdot e_3 = n_1\kappa_1 + n_2\kappa_2 + n_3\kappa_3 \tag{12.45}$$

and therefore

$$q(r) = q(r_j)e^{n_1\kappa_1 + n_2\kappa_2 + n_3\kappa_3} \tag{12.46}$$

This is a statement of Bloch's theorem [3]. The phase constants, determined by the wave vector and the basis vectors of the lattice, are the real parts of κ, while the imaginary parts are the attenuation.

For an infinite lattice, Bloch's theorem can be used to generate boundary conditions for the finite-element model, in the form of relationships between nodal displacements. In any reference cell, the finite-element nodes can be classified into one of the following three types [25]:

Internal nodes (with collected displacements q_i) are nodes that are internal to the reference cell, and are not shared with neighbouring cells.

Basis nodes (with collected displacements q_b) are nodes that are shared with neighbouring cells, but cannot be reached by traversing an integer number of lattice basis vectors from another basis node. In other words, for two basis nodes with distinct positions r_A and r_B, there exists no integer values n_1, n_2, or n_3 for which $r_A = r_B + n_1e_1 + n_2e_2 + n_3e_3$.

Boundary nodes (with collected displacements q_1, q_2, and so on) are nodes that are shared with neighbouring cells, and can be reached by traversing a certain combination of lattice basis vectors from a basis node. The subscripts on the collected displacements indicate the combination of basis vectors that must be traversed. For instance, the set of boundary nodes that can be reached by traversing e_1 from a basis node have collected displacements $\mathbf{q}_1$, the set of boundary nodes that can be reached by traversing $\mathbf{e}_1 + \mathbf{e}_2$ have collected displacements $\mathbf{q}_{12}$, and the set of boundary nodes reached by traversing $\mathbf{e}_1 + \mathbf{e}_2 + \mathbf{e}_3$ have collected displacements $\mathbf{q}_{123}$.

Based on the above classification and Bloch's theorem, the displacements at the *boundary nodes* can be expressed in terms of the displacements at the corresponding *basis nodes*. Consider a unit cell that has boundary nodes A, B, and C (with displacements $\mathbf{q}_A$, $\mathbf{q}_B$, and $\mathbf{q}_C$) that can be reached by traversing $\mathbf{e}_1 + \mathbf{e}_2$ from, respectively, basis nodes D, E, and F (with displacements $\mathbf{q}_D$, $\mathbf{q}_E$, and $\mathbf{q}_F$). Thus, there are three nodes, A, B, and C, which are categorized as 12-nodes with collected displacements $\mathbf{q}_{12}$:

$$\mathbf{q}_b = \begin{bmatrix} \mathbf{q}_D \\ \mathbf{q}_E \\ \mathbf{q}_F \end{bmatrix} = \begin{bmatrix} e^{\kappa_1+\kappa_2} & 0 & 0 \\ 0 & e^{\kappa_1+\kappa_2} & 0 \\ 0 & 0 & e^{\kappa_1+\kappa_2} \end{bmatrix} \begin{bmatrix} \mathbf{q}_A \\ \mathbf{q}_B \\ \mathbf{q}_C \end{bmatrix} = \mathbf{T}_{12}\mathbf{q}_{12} \tag{12.47}$$

Corresponding relations hold for all boundary nodes that are reached through any combination of basis vectors from a basis node, and hence, for the entire set of nodes in the reference cell, this relationship can be written:

$$\mathbf{q} = \mathbf{T}\tilde{\mathbf{q}} \tag{12.48}$$

where $\tilde{\mathbf{q}}$ represents the reduced set of generalized coordinates comprising the displacements at the internal and the basis nodes, $\mathbf{T}$ is a transformation matrix dependent on the wave vector κ, and $\mathbf{q}$ is the complete set of generalized coordinates, arranged as shown below:

$$\mathbf{q} = \begin{bmatrix} \mathbf{q}_i \\ \mathbf{q}_b \\ \mathbf{q}_1 \\ \mathbf{q}_2 \\ \mathbf{q}_3 \\ \mathbf{q}_{12} \\ \mathbf{q}_{13} \\ \mathbf{q}_{23} \\ \mathbf{q}_{123} \end{bmatrix}, \quad \mathbf{T} = \begin{bmatrix} \mathbf{I} & \mathbf{0} \\ \mathbf{0} & \mathbf{I} \\ \mathbf{0} & \mathbf{T}_1 \\ \mathbf{0} & \mathbf{T}_2 \\ \mathbf{0} & \mathbf{T}_3 \\ \mathbf{0} & \mathbf{T}_{12} \\ \mathbf{0} & \mathbf{T}_{13} \\ \mathbf{0} & \mathbf{T}_{23} \\ \mathbf{0} & \mathbf{T}_{123} \end{bmatrix}, \quad \tilde{\mathbf{q}} = \begin{bmatrix} \mathbf{q}_i \\ \mathbf{q}_b \end{bmatrix} \tag{12.49}$$

Using the above result in Eq. (12.42):

$$\tilde{\mathbf{M}}\ddot{\tilde{\mathbf{q}}} + \tilde{\mathbf{K}}\tilde{\mathbf{q}} = \mathbf{T}^H\mathbf{f} \tag{12.50}$$

where $()^H$ denotes the the conjugate transpose, and $\tilde{\mathbf{K}} = \mathbf{T}^H\mathbf{K}\mathbf{T}$ and $\tilde{\mathbf{M}} = \mathbf{T}^H\mathbf{M}\mathbf{T}$. This equation can be made homogeneous if $\mathbf{T}^H\mathbf{f}$ is shown to be $\mathbf{0}$.

Generalized Forces in Bloch Analysis

Consider an element AB in the reference cell of a lattice, bounded by nodes A and B. Suppose that traversing basis vector $\mathbf{e}$ leads to the corresponding element in the neighbouring cell CD, bounded by nodes C and D. For a wave in the lattice with wave vector κ, the nodal displacements can be related by Bloch's theorem:

$$\mathbf{q}_A e^{\kappa} = \mathbf{q}_C \quad \mathbf{q}_B e^{\kappa} = \mathbf{q}_D \tag{12.51}$$

where $\kappa = \mathbf{e} \cdot \boldsymbol{\kappa}$.

If the force at the ends of element AB depends only on the displacements at the nodes, then:

$$\mathbf{F}_{AB} = \mathbf{C}(\mathbf{q}_A - \mathbf{q}_B), \quad \mathbf{F}_{CD} = \mathbf{C}(\mathbf{q}_C - \mathbf{q}_D) \tag{12.52}$$

where $\mathbf{C}$ is a constant elastic matrix, and therefore:

$$
\begin{aligned}
\mathbf{F}_{CD} &= \mathbf{C}(\mathbf{q}_C - \mathbf{q}_D) \\
&= \mathbf{C}(\mathbf{q}_A e^\kappa - \mathbf{q}_B e^\kappa) \\
&= \mathbf{C}(\mathbf{q}_A - \mathbf{q}_B) e^\kappa \\
&= \mathbf{F}_{AB} e^\kappa
\end{aligned} \tag{12.53}
$$

Thus Bloch's theorem can be extended for use with forces, which enables the evaluation of $\mathbf{T}^H \mathbf{f}$.

Consider the generic 2D lattice in Figure 12.7. For compactness, this analysis is performed for a 2D system; it can be directly extended to three dimensions. Node A is taken to be the basis node. The collected vector of external forces $\mathbf{f}$ acting on the cell is:

$$
\mathbf{f} = \begin{bmatrix} \mathbf{F}_i \\ \mathbf{F}_b \\ \mathbf{F}_1 \\ \mathbf{F}_2 \\ \mathbf{F}_{12} \end{bmatrix} = \begin{bmatrix} 0 \\ \mathbf{F}_{EA} + \mathbf{F}_{FA} \\ \mathbf{F}_{AD} e^{\kappa_1} + \mathbf{F}_{FA} e^{\kappa_1} \\ \mathbf{F}_{AB} e^{\kappa_2} + \mathbf{F}_{EA} e^{\kappa_2} \\ \mathbf{F}_{AB} e^{\kappa_1 + \kappa_2} + \mathbf{F}_{AD} e^{\kappa_1 + \kappa_2} \end{bmatrix} \tag{12.54}
$$

with the corresponding transformation matrix:

$$
\mathbf{T} = \begin{bmatrix} \mathbf{I} & 0 \\ 0 & \mathbf{I} \\ 0 & \mathbf{T}_1 \\ 0 & \mathbf{T}_2 \\ 0 & \mathbf{T}_{12} \end{bmatrix} = \begin{bmatrix} 1 \\ e^{\kappa_1} \\ e^{\kappa_2} \\ e^{\kappa_1 + \kappa_2} \end{bmatrix} \tag{12.55}
$$

Figure 12.7 Forces applied to the reference cell of a generic 2D lattice.

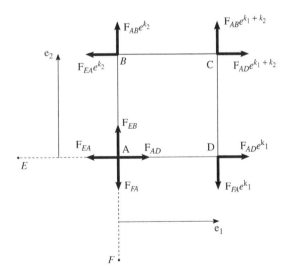

Evaluation of this expression generates:

$$
\mathbf{T}^H\mathbf{f} = \begin{bmatrix} 1 & e^{-\kappa_1} & e^{-\kappa_2} & e^{-\kappa_1-\kappa_2} \end{bmatrix} \begin{bmatrix} \mathbf{F}_{EA} + \mathbf{F}_{FA} \\ \mathbf{F}_{AD}e^{\kappa_1} + \mathbf{F}_{FA}e^{\kappa_1} \\ \mathbf{F}_{AB}e^{\kappa_2} + \mathbf{F}_{EA}e^{\kappa_2} \\ \mathbf{F}_{AB}e^{\kappa_1+\kappa_2} + \mathbf{F}_{AD}e^{\kappa_1+\kappa_2} \end{bmatrix}
$$

$$
= 2(\mathbf{F}_{EA} + \mathbf{F}_{FA} + \mathbf{F}_{AB} + \mathbf{F}_{AD})
$$

$$
= \mathbf{0}, \tag{12.56}
$$

by equilibrium at node A.

The same is true for other nodes and directions and in three dimensions. If $\mathbf{F}_b$ is the set of forces applied at the basis nodes and $\mathbf{F}_1$ is the set of forces applied at the boundary nodes reached by traversing $\mathbf{e}_1$, then

$$
\mathbf{F}_1 = \mathbf{T}_1\mathbf{F}_b \tag{12.57}
$$

$\mathbf{T}_1$ transforms a set of forces from a *basis node* to a *boundary node* [25]. For κ_1, κ_2, and κ_3 that are purely imaginary, the inverse transform can be performed by using the Hermitian operator (by virtue of the fact that $\mathbf{T}_1$ has only one nonzero element in each row and column). Thus $\mathbf{T}_1^H$ is the transform from a boundary node to a basis node, and $\mathbf{T}_1^H\mathbf{F}_1$ represents the set of forces applied at the basis node by elements in the $\mathbf{e}_1$ direction. Similar statements can be made about about $\mathbf{T}_2^H\mathbf{F}_2$, $\mathbf{T}_3^H\mathbf{F}_3$, $\cdots$ as well. From this,

$$
\mathbf{T}^H\mathbf{f} = \begin{bmatrix} \mathbf{I} & 0 & 0 & 0 & 0 & \cdots \\ 0 & \mathbf{I} & \mathbf{T}_1^H & \mathbf{T}_2^H & \mathbf{T}_3^H & \cdots \end{bmatrix} \begin{bmatrix} \mathbf{F}_i \\ \mathbf{F}_b \\ \mathbf{F}_1 \\ \mathbf{F}_2 \\ \mathbf{F}_3 \\ \vdots \end{bmatrix}
$$

$$
= \begin{bmatrix} \mathbf{F}_i \\ \mathbf{F}_b + \mathbf{T}_1^H\mathbf{F}_1 + \mathbf{T}_2^H\mathbf{F}_2 + \mathbf{T}_3^H\mathbf{F}_3 + \cdots \end{bmatrix} \tag{12.58}
$$

In the unforced case (with no external forces), the forces applied to the internal nodes are zero. Thus, $\mathbf{F}_i = 0$. $\mathbf{F}_b$ represents the external forces applied to the basis nodes, and $\mathbf{T}_1^H\mathbf{F}_1 + \mathbf{T}_2^H\mathbf{F}_2 + \mathbf{T}_3^H\mathbf{F}_3 + \cdots$ are the internal forces applied to the basis nodes. By equilibrium, their sum must be zero. Thus,

$$
\mathbf{T}^H\mathbf{f} = \mathbf{0} \tag{12.59}
$$

and Eq. (12.50) is shown to be homogeneous.

Reduced Equation of Motion

Substituting the above result in Eq. (12.50):

$$
\tilde{\mathbf{M}}\ddot{\tilde{\mathbf{q}}} + \tilde{\mathbf{K}}\tilde{\mathbf{q}} = \mathbf{0} \tag{12.60}
$$

For a plane wave propagating through the lattice, the displacements at the internal and basis nodes have the form

$$
\tilde{\mathbf{q}} = \mathbf{A}e^{i(\boldsymbol{\kappa}\cdot\mathbf{r}-\omega t)} \tag{12.61}
$$

Taking the real part of this expression

$$\tilde{q} = A\cos(\kappa \cdot r - \omega t) \tag{12.62}$$

$$\Rightarrow \ddot{\tilde{q}} = -\omega^2 A\cos(\kappa \cdot r - \omega t)$$

Thus

$$\tilde{K}A = \omega^2\tilde{M}A \tag{12.63}$$

Since $\tilde{K}$ and $\tilde{M}$ are Hermitian, Eq. (12.63) forms a generalized Hermitian eigenvalue problem, which can be be solved for the eigenvalue ω^2 and the eigenvector A for a particular wave vector (which determines T and hence $\tilde{M}$ and $\tilde{K}$). The eigenvectors and eigenvalues provide the mode shapes and frequencies of vibration of the lattice.

12.3.4 Dispersion Curves for the Octet Lattice

Equation (12.63) forms an eigenvalue problem for which the eigenvalue and eigenvector solutions are the frequency and wave form of modes of wave propagation through the lattice structures. Critically, both $\tilde{M}$ and $\tilde{K}$ are functions of the wave vector; since the frequencies of propagation will vary with the wave vector, the 3D lattice is dispersive. Because the frequency of wave propagation is periodic with respect to the wave vector, the complete dispersion characteristics can be determined from the examination of the wave vectors in the first Brillouin zone. In two dimensions, it can be shown that the band extrema occur along boundaries of the irreducible zone [26]. This section examines wave vectors from throughout the first Brillouin zone by discretizing the zone with a 3D mesh. The methods employed here are therefore applicable to lattice geometries for which band extrema do not occur on the edges of the irreducible Brillouin zone, and to lattice geometries with nonsymmetric reference cells.

Full dispersion curves for the nanometal–polymer lattices are four dimensional: three components of the wave vector and the set of natural frequencies associated with that wave vector. This would generate a set of surfaces in a 4D space. Instead, dispersion curves that plot the frequency of wave propagation against a single component of the wave vector are shown for all wave vectors in the first Brillouin zone. This is equivalent to examining the projection of the 4D frequency surface onto a single plane in the space spanned by k_1, k_2, k_3, and ω. Any frequency band gaps in such dispersion plots are immediately apparent.

Figures 12.8 and 12.9 plot the dispersion characteristics of the octet lattice, for radius-to-length ratios of 10 and 50, respectively. The material properties used were Young's modulus $E = 200$ GPa, density $\rho = 1$ Mg m^{-3}, and Poisson's ratio $\nu = 0.3$. To generate these figures, the 3D Brillouin zone was discretized into a network of points. The eigenproblem of Eq. (12.63) was solved for each one of those values of k, and the resulting eigenvalues ω were plotted alongside a component of k. The darkness of the points in the figures indicates how many dispersion surfaces in 4D space are mapped onto the single point in $k_1 - \omega$ space.

As is evident from the figures, the band structure of the octet lattice depends heavily on the radius-to-length ratio. Indeed, varying the elasticity and density of the lattice material has no effect on the band structure; it simply scales all frequencies by a constant factor.

As can be seen from Figure 12.8 for a radius-to-length ratio of 10, the octet lattice exhibits a thin but complete band gap centred at a normalized frequency of 6, which

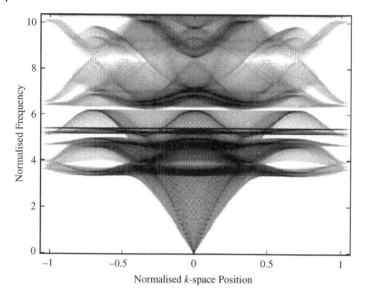

Figure 12.8 Dispersion curve for an octet lattice with a radius-to-length ratio of 10.

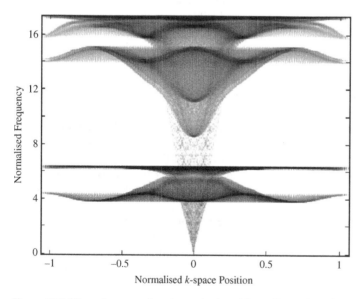

Figure 12.9 Dispersion curve for an octet lattice with a radius-to-length ratio of 50.

corresponds to 780 rad/s. The octet lattice does not have any complete frequency band gaps for a radius-to-length ratio of 50. However, several partial band gaps can be seen in Figure 12.9. Regions in the dispersion curve characterized by widely spaced dots or very light gray indicate that the frequency exhibits large changes for small changes in the wave vector.

12.3.5 Lattice Tuning

Bandgap Placement

Given the form of m_{ij} and k_{ij}, they can be written as

$$m_{ij} = \rho m'_{ij}$$

$$k_{ij} = E k'_{ij}$$

The same applies for the element mass and stiffness matrices.

$$\mathbf{m}_l = \rho \mathbf{m}'_l$$

$$\mathbf{k}_l = E \mathbf{k}'_l$$

If each element in the reference cell has the same density and Young's modulus, this can be said of the assembled mass and stiffness matrices of the reference cell, as well as the reduced mass and stiffness matrices. Thus

$$\mathbf{M} = \rho \mathbf{M}' \Rightarrow \tilde{\mathbf{M}} = \rho \tilde{\mathbf{M}}'$$

$$\mathbf{K} = E \mathbf{K}' \Rightarrow \tilde{\mathbf{K}} = E \tilde{\mathbf{K}}'$$

Using these in Eq. (12.63),

$$E \tilde{\mathbf{K}}' \mathbf{A} = \omega^2 \tilde{\mathbf{M}}' \mathbf{A}$$

$$\tilde{\mathbf{K}}' \mathbf{A} = \frac{\omega^2}{E/\rho} \tilde{\mathbf{M}}' \mathbf{A}$$

Since $\tilde{\mathbf{M}}'$ and $\tilde{\mathbf{K}}'$ are independent of density and Young's modulus, the quantity $\omega^2 \rho / E$ must be invariant as well. Thus, for a given lattice geometry with two different specific stiffnesses,

$$\frac{\omega_1^2}{E_1/\rho_1} = \frac{\omega_2^2}{E_2/\rho_2}$$

This relationship becomes useful when a lattice band gap is known to exist for a certain frequency, and it is desired to shift that band gap to some other target frequency. This can be achieved by simply changing the specific stiffness of the material, in accordance with the above relationship. Note that this relationship is valid for the 2D lattices in the paper by Phani et al. [3], as well.

Lattice Optimization

It is possible to use optimization techniques to adjust lattice properties, such as the cross-sections of the Timoshenko beams that make up the reference cell, in order to construct lattices that have strategically placed band gaps. If a known frequency must be filtered, the material and geometric properties can be adjusted to place a band gap at the desired frequency. Alternately, to block waves with the largest amplitudes, the band gaps should have the minimum possible frequency. The band gaps also should be as wide as possible, to block the widest range of frequencies.

The basic cost function for performing such an optimization is the relative size of the lowest band gap [22], defined as:

$$f = \frac{\Delta \omega_i}{|\bar{\omega}_i - x|}$$

$$\Delta\omega_i = \min\omega_{i+1} - \max\omega_i$$

$$\overline{\omega}_i = \frac{\min\omega_{i+1} + \max\omega_i}{2}$$

where ω_i refers to the set of frequencies in the ith band, $\Delta\omega_i$ is the width of the ith band gap, $\overline{\omega}_i$ is the location of the ith band gap, and x is the frequency that is desired to be blocked.

Unfortunately, this cost function requires a "dirty" optimization technique; the full band structure of a candidate lattice configuration must be calculated in order to evaluate the cost of the lattice, a procedure that is quite computationally intensive for complex, finely modeled 3D lattices. Consequently, optimization techniques with the minimum number of cost-evaluation steps, such as the method of moving asymptotes [27], are required.

12.4 Conclusions

Three-dimensional lattices can be made using a combination of well-understood manufacturing processes: rapid prototyping via stereolithography of polymers and electrodeposition of nanocrystalline metal. This enables the fabrication of complex 3D structures using ultra-high-performance material, which is typically impossible to obtain. Moreover, because of the geometric complexity and the fine control over material properties attained through this process, the resulting lattices can be tailored for a variety of structural and functional behaviors. In particular, weight can be minimized for a structure with a required stiffness or strength.

Here, we are primarily interested in tailoring the dynamic properties. We therefore develop a general method for analyzing wave propagation in 3D periodic lattices. This method models the reference cell of a lattice as a collection of Timoshenko beams and uses Floquet–Bloch principles to impose plane-wave boundary conditions. This procedure generates a generalized Hermitian eigenvalue problem, which needs to be solved for wave vectors inside the first Brillouin zone to obtain the possible frequencies of wave propagation in the lattice. There may exist certain frequencies for which there are no modes of propagation; these constitute frequency band gaps. The band gaps are the key characteristics of these dispersion curves. Because acoustic or vibration isolators are valuable structural elements, we would like the capacity to design band gaps in a particular frequency band.

Once the method for analysing the nanometal–polymer lattices is developed, there are several methods for tailoring the wave-propagation characteristics. The first is changing the topology of the unit cell, although this is not easily amenable to standard optimization techniques because it implies that the design variables are discrete. A second option, which is enabled by use of nanometal–polymer lattices, is to consider the radius of the polymer preform, the length of the struts, and the thickness of the coating as design variables. By doing so, a set of three continuous design variables is generated, and these enable the strut-slenderness ratio, the effective strut stiffness, and the effective strut density to be varied. This provides excellent design control while also being amenable to optimization through gradient-based methods.

12.5 Appendix: Shape Functions for a Timoshenko Beam with Six Nodal Degrees of Freedom

For a 1D Timoshenko beam with six degrees of freedom at each node (three translations u, v, and w and three rotations ϕ, ψ, and θ), the displacements at any point along the beam can be approximated by:

$$u(x, t) = q_i(t)a_i(x)$$
$$v(x, t) = q_i(t)b_i(x)$$
$$w(x, t) = q_i(t)c_i(x)$$
$$\phi(x, t) = q_i(t)d_i(x)$$
$$\psi(x, t) = q_i(t)e_i(x)$$
$$\theta(x, t) = q_i(t)f_i(x)$$

where

$$q_i = \begin{bmatrix} u_A & v_A & w_A & \phi_A & \psi_A & \theta_A & u_B & v_B & w_B & \phi_B & \psi_B & \theta_B \end{bmatrix}^T$$

$()_A$ and $()_B$ represents displacements at nodes A and B of the beam. The shape functions a_i–f_i are shown below [28]. Unspecified functions have a value of 0.

$$\xi = \frac{x}{L} \quad \eta_y = \frac{12EI_y}{\kappa GAL^2} \quad \mu_y = \frac{1}{1 + \eta_y} \quad \eta_z = \frac{12EI_z}{\kappa GAL^2} \quad \mu_z = \frac{1}{1 + \eta_z}$$

$$a_1 = 1 - \xi$$
$$a_7 = \xi$$

$$b_2 = \mu_z(1 - 3\xi^2 + 2\xi^3 + \eta_z(1 - \xi))$$
$$b_6 = L\mu_z\left[\xi - 2\xi^2 + \xi^3 + \frac{\eta_z}{2}(\xi - \xi^2)\right]$$
$$b_8 = \mu_z(3\xi^2 - 2\xi^3 + \eta_z\xi)$$
$$b_{12} = L\mu_z\left[-\xi^2 + \xi^3 + \frac{\eta_z}{2}(-\xi + \xi^2)\right]$$

$$c_3 = \mu_y(1 - 3\xi^2 + 2\xi^3 + \eta_y(1 - \xi))$$
$$c_5 = -L\mu_y\left[\xi - 2\xi^2 + \xi^3 + \frac{\eta_y}{2}(\xi - \xi^2))\right]$$
$$c_9 = \mu_y(3\xi^2 - 2\xi^3 + \eta_y\xi)$$
$$c_{11} = -L\mu_y\left[-\xi^2 + \xi^3 + \frac{\eta_y}{2}(-\xi + \xi^2)\right]$$

$$d_4 = 1 - \xi$$
$$d_{10} = \xi$$

$$e_3 = -\frac{6\mu_y}{L}(-\xi + \xi^2)$$

$$e_5 = \mu_y(1 - 4\xi + 3\xi^2 + \eta_y(1 - \xi))$$

$$e_9 = -\frac{6\mu_y}{L}(\xi - \xi^2)$$

$$e_{11} = \mu_y(-2\xi + 3\xi^2 + \eta_y\xi)$$

$$f_2 = \frac{6\mu_z}{L}(-\xi + \xi^2)$$

$$f_6 = \mu_z(1 - 4\xi + 3\xi^2 + \eta_z(1 - \xi))$$

$$f_8 = \frac{6\mu_z}{L}(\xi - \xi^2)$$

$$f_{12} = \mu_z(-2\xi + 3\xi^2 + \eta_z\xi)$$

References

1 D. J. Mead, "Wave propagation in continuous periodic structures: research contributions from Southampton, 1964-1995," *Journal of Sound and Vibration*, vol. 190, no. 3, pp. 495–524, 1996.

2 M. Ruzzene, F. Scarpa, and F. Soranna, "Wave beaming effects in two-dimensional cellular structures," *Smart Materials and Structures*, vol. 12, no. 3, pp. 363–372, 2003.

3 A. S. Phani, J. Woodhouse, and N. A. Fleck, "Wave propagation in two-dimensional periodic lattices," *Journal of the Acoustical Society of America*, vol. 119, no. 4, pp. 1995–2005, 2006.

4 V. Deshpande, N. A. Fleck, and M. F. Ashby, "Effective properties of the octet-truss lattice material," *Journal of the Mechanics and Physics of Solids*, vol. 49, pp. 1747–1769, 2001.

5 M. Arya and C. A. Steeves, "Bandgaps in octet truss lattices," in *Proceedings of the 23rd Canadian Congress on Applied Mechanics*, (Vancouver, Canada), 6–9 June 2011.

6 C. K. Chua, K. F. Leong, and C. S. Lim, *Rapid Prototyping: Principles and Applications.* World Scientific, 3rd ed., 2010.

7 S. Upcraft and R. Fletcher, "The rapid prototyping technologies," *Assembly Automation*, vol. 23, pp. 318–330, 2003.

8 S.-H. Ahn, M. Montero, D. Odell, S. Roundy, and P. K. Wright, "Anisotropic material properties of fused deposition modeling ABS," *Rapid Prototyping Journal*, vol. 8, pp. 248–257, 2002.

9 K. C. Ang, K. F. Leong, C. K. Chua, and M. Chandrasekaran, "Investigation of the mechanical properties and porosity relationships in fused deposition modelling-fabricated porous structures," *Rapid Prototyping Journal*, vol. 12, pp. 100–105, 2006.

10 T. A. Schaedler, A. J. Jacobsen, A. Torrents, A. E. Sorensen, J. Lian, J. R. Greer, L. Valdevit, and W. B. Carter, "Ultralight metallic microtrusses," *Science*, vol. 334, pp. 962–965, 2011.

11 L. M. Gordon, B. A. Bouwhuis, M. Suralvo, J. L. McCrea, G. Palumbo, and G. D. Hibbard, "Micro-truss nanocrystalline Ni hybrids," *Acta Materialia*, vol. 57, pp. 932–939, 2009.

12 M. Schlesinger and M. Paunovic, *Modern Electroplating, Fifth Edition*. John Wiley & Sons, 5th ed., 2010.

13 R. Weil, J. H. Lee, and K. Parker, "Comparison of some mechanical and corrosion properties of electroless and electroplated nickel phosphorus alloys," *Plating and Surface Finishing*, vol. 76, pp. 62–66, 1989.

14 H. Gleiter, "Nanocrystalline materials," *Progress in Materials Science*, vol. 33, pp. 223–315, 1989.

15 A. Robertson, U. Erb, and G. Palumbo, "Practical applications for electrodeposited nanocrystalline materials," *Nanostructured Materials*, vol. 12, pp. 1035–1040, 1999.

16 Y. Zhou, S. V. Petegem, D. Segers, U. Erb, K. T. Aust, and G. Palumbo, "On Young's modulus and the interfacial free volume in nanostructured Ni-P," *Materials Science and Engineering A*, vol. 512, pp. 39–44, 2009.

17 E. O. Hall, "The deformation and ageing of mild steel: III Discussion of results," *Proceedings of the Physical Society*, vol. 64, pp. 747–753, 1951.

18 N. J. Petch, "The cleavage strength of polycrystals," *Journal: Iron and Steel Institute*, vol. 174, pp. 25–28, 1953.

19 C. E. Carlton and P. J. Ferreira, "What is behind the inverse Hall–Petch effect in nanocrystalline materials?," *Acta Materialia*, vol. 55, pp. 3749–3756, 2007.

20 A. T. Lausic, C. A. Steeves, and G. D. Hibbard, "Effect of grain size on the optimal architecture of electrodeposited metal/polymer microtrusses," *To appear in Journal of Sandwich Structures and Materials*, 2014.

21 L. Brillouin, *Wave Propagation in Periodic Structures*. Dover Publications, 2nd ed., 1953.

22 O. Sigmund and J. S. Jensen, "Systematic design of phononic band-gap materials and structures by topology optimization," *Philosophical Transactions of the Royal Society of London*, vol. 361, pp. 1001–1019, 2003.

23 S. P. Timoshenko, *Strength of Materials*. Van Nostrand, 2nd ed., 1940.

24 Y. Luo, "An efficient 3D Timoshenko beam with consistent shape functions," *Advances in Theoretical and Applied Mechanics*, vol. 1, no. 3, pp. 95–106, 2008.

25 F. Farzbod and M. J. Leamy, "The treatment of forces in Bloch analysis," *Journal of Sound and Vibration*, vol. 325, pp. 545–551, 2009.

26 C. Kittel, *Elementary Solid State Physics: A Short Course*. Wiley, 1st ed., 1962.

27 K. Svanberg, "The method of moving asymptotes – a new method for structural optimization," *International Journal for Numerical Methods in Engineering*, vol. 24, no. 2, pp. 359–373, 1987.

28 A. Bazoune, Y. A. Khulief, and N. G. Stephen, "Shape functions of three-dimensional Timoshenko beam element," *Journal of Sound and Vibration*, vol. 259, pp. 473–480, 2003.

Index

Dynamics of Lattice Materials, First Edition. Edited by A. Srikantha Phani and Mahmoud I. Hussein.
© 2017 John Wiley & Sons Ltd. Published 2017 by John Wiley & Sons Ltd.

Printed and bound by CPI Group (UK) Ltd, Croydon, CR0 4YY

16/04/2025

14658475-0001